Advances in Delivery Science and Technology

Series Editor

Michael J. Rathbone

For further volumes:
http://www.springer.com/series/8875

Clive G. Wilson • Patrick J. Crowley
Editors

Controlled Release in Oral Drug Delivery

Springer

Editors
Clive G. Wilson
Univeristy of Strathclyde
Institute of Pharmacy and Biomedical Sciences
Glasgow, Scotland GR4 0RE, UK
c.g.wilson@strath.ac.uk

Patrick J. Crowley
Callum Consultancy LLC
Devon, PA 19333, USA
ktcrowley@msn.com

ISSN 2192-6204
ISBN 978-1-4614-1003-4 e-ISBN 978-1-4614-1004-1
DOI 10.1007/978-1-4614-1004-1
Springer New York Dordrecht Heidelberg London

Library of Congress Control Number: 2011934749

Printed on acid-free paper

Springer is part of Springer Science+Business Media (www.springer.com)

Preface

The ideal drug delivery system has been depicted as "getting the right amount of drug to the right place at the right time." It is now widely accepted that making drug available "immediately" following oral administration does not provide such idealized delivery in many cases. Consequently strategies, materials, and technologies have evolved to control delivery by delaying, slowing, pulsing, or delivering to a specific region of the gastrointestinal tract. The introductory chapter in this book traces the history and evolution of the concepts and achievements that has brought the discipline to where it is today.

Despite the advent of many useful release-modifying polymers and technologies, one of the greatest barriers to providing an appropriate release profile (and associated plasma presence) is the gastrointestinal tract. Knowing and understanding its structure, tissues, mechanics, and functions, and the limitations (and possibilities) that these present in attaining a target plasma profile is a prerequisite for successful dosage form design. Chapter 2 provides such perspectives.

Proving efficacy and safety for a novel molecule is currently so difficult, time consuming, and expensive that industrial R&D-based organizations are under pressure to "hurry" a drug to market when it is shown to be effective and safe. More nuanced effects may only become apparent after widespread use, prompting dosage form redesign. Hence, most current controlled release oral dosage forms are "second-generation" products. Whether shortcomings in "first-generation" (mostly "immediate release") products have contributed to "failures" in development cannot be stated but, conceivably, greater focus on a broader range of delivery options in phase 1 or phase 2 clinical trials, possibly allied with the use of relevant biomarkers, may offer hope that attrition can be reduced and performance optimized. Such possibilities are discussed from an industrial R&D perspective in Chapter 3.

It is clearly unrealistic (and expensive) to trial each and every delivery and formulation concept or system in human subjects. In vitro and animal studies are also valuable, both during exploratory and quality-monitoring phases. The range, predictive capabilities, and limitations of such models are presented and discussed in two chapters in this book.

The mechanisms by which release of drug from the dosage form is controlled can profoundly impact location, rate, and profile of release, and as a consequence, the plasma profile following absorption. Such behaviors can be influenced by release-modifying polymers or mixtures thereof, the presence of other materials and geometric (shape and size) factors. This is a wide subject area that is reflected in several chapters devoted to such topics.

Manufacturing technologies are immensely important for imparting reliability and consistency to dosage form performance. Quality of the release modifiers is also crucial. Many, being polymeric, may contain residues that could destabilize the drug (or other excipients). Hence, consistency of quality is an important consideration. It is heartening therefore that high-quality information on such phenomena is being generated and published by excipient providers to guide the formulation or manufacturing technologist on material performance. Chapters on polymers for matrices, capsules for controlled release, and multiparticulates in this book emanate from such sources and can provide useful guidance when considering "quality-by-design-based programs."

Fatty acids, fatty alcohols, and waxes that do not melt at body temperature are sometimes used to form insoluble matrices (possibly in combination with other materials) to slow release from dosage forms. Their capability for self-assembly in GI tract-like milieu is now evincing much interest, particularly with insoluble drugs, as controlled release platforms for the future. Hence, they merit a chapter on the topic, provided by probably the foremost group operating in this area (Monash University, Melbourne).

Regional delivery and control of drugs usually concern delivery to and absorption from the small intestine. On occasion, however, delivery in a controlled manner via the buccal cavity may be advantageous, in terms of onset of action or avoidance of hepatic metabolism. It is appropriate therefore that such "point of entry" (to the GI tract) delivery be allocated a chapter.

One of the "holy grails" for prolonging drug absorption (and consequent plasma presence) concerns retention of the dosage unit in the gastric region, drug being released gradually for absorption further along the GI tract. Gastroretentive strategies, devices, and performance are accordingly considered in a chapter.

Finally, drug delivery at "the other end" of the GI tract, both for local action and systemic absorption, must not be disregarded, particularly as lack of enzymatic activity in the colon makes it a tempting location for delivering peptides or other macromolecular entities. Hence, a chapter on colonic delivery is included.

These chapters have, as far as is possible, been formatted so that they can largely "stand alone." However, some repetition is inevitable as materials and mechanisms may be common to more than one strategy. Furthermore, the editors would like to stress that satisfactorily controlling drug release requires a "holistic" approach. Knowledge of the drug, the release-controlling agents, the mode and site for delivery, as well as the absorptive processes associated with oral delivery need to be factored into the dosage form design strategy for getting the correct amount of drug "to the right place at the right time." It is hoped therefore that insights contained in this volume provide the research scientist, formulation specialist, and manufacturing technologist with such broad-based information for dosage form design, manufacture, and control.

Scotland, UK — Clive G. Wilson
Devon, PA, USA — Patrick J. Crowley

Acknowledgements

The assembly, editing and general enhancement of many of the chapters in this book would not have been so thorough without the input of a number of people who provided wise counsel, added value and generally assured that the information and perspectives had validity. The individuals listed below generously gave their time, advice and expertise during chapter compilation and editing. We are most grateful for such input as it made our workload less daunting and added greatly to the quality of the information

Dr Kieran G. Mooney (Chapter 3)
Professor David J Brayden (Chapter 4)
Dr David P. Elder (Chapter 8)
Professor Howard N. Stevens (Chapter 9)
Professor Christopher J. H. Porter (Chapter 15)

The Editors

Contents

Contributors

Pieter Annaert Laboratory for Pharmacotechnology and Biopharmacy, Katholieke Universiteit Leuven, Leuven, Belgium

Patrick Augustijns Laboratory for Pharmacotechnology and Biopharmacy, Katholieke Universiteit Leuven, Leuven, Belgium

Daniel Bar-Shalom Faculty of Pharmaceutical Sciences, Department of Pharmaceutics and Analytical Chemistry, University of Copenhagen, Copenhagen, Denmark

Abdul W. Basit The School of Pharmacy, University of London, London, UK

Ben J. Boyd Drug Delivery, Disposition and Dynamics, Monash Institute of Pharmaceutical Sciences, Monash University, Parkville, VIC, Australia

Matthew D. Burke GlaxoSmithKline R&D, Harlow, Essex, UK

James M. Butler GlaxoSmithKline R&D, Harlow, Essex, UK

Christine Cahyadi Dipartimento Farmaceutico, Parma, Italy

Rick S. Chan LTS Lohmann Therapy Systems Corp., West Caldwell, NJ, USA

Li-lan Chen GlaxoSmithKline Consumer Healthcare, Parsippany, NJ, USA

Mark D. Coffin Platform Technology and Science, GlaxoSmithKline R&D, NC, USA

Gaia Colombo Dipartimento Farmaceutico, Parma, Italy

Paolo Colombo Dipartimento Farmaceutico, Parma, Italy

Patrick J. Crowley Callum Consultancy LLC, Devon, PA, USA

James DiNunzio Pharmaceutical and Analytical R&D Department, Hoffmann-La Roche, Inc, Nutley, NJ, USA

Shahin Fesharaki Research and Development, Watson Pharmaceuticals, Fort Lauderdale, FL, USA

Alexander T. Florence Professor Emeritus, School of Pharmacy, University of London, London, UK

James L. Ford School of Pharmacy and Biomolecular Sciences, Liverpool John Moores University, Liverpool, UK

Timothy Kutch K Kelly OCG at Johnson & Johnson, McNeil Consumer Healthcare, Fort Washington, PA, USA

Der-Yang Lee Johnson & Johnson, McNeil Consumer Healthcare, Fort Washington, PA, USA

Johan Adriaan Martens Center for Surface Chemistry and Catalysis, Katholieke Universiteit Leuven, Leuven, Belgium

Luigi G. Martini Institute of Pharmaceutical Sciences, King's College, London, UK

Emma L. McConnell MSD, Hoddesdon, Hertfordshire, UK

Jason T. McConville College of Pharmacy, The University of Texas at Austin, Austin, TX, USA

Randy Mellaerts Center for Surface Chemistry and Catalysis, Katholieke Universiteit Leuven, Leuven, Belgium

Anette Müllertz Department of Pharmaceutics and Analytical Chemistry, University of Copenhagen, Copenhagen, Denmark

Tri-Hung Nguyen Institute of Pharmaceutical Sciences, Monash University, Parkville, VIC, Australia

Hossein Omidian College of Pharmacy, Nova Southeastern University, Fort Lauderdale, FL, USA

Kinam Park Departments of Biomedical Engineering and Pharmaceutics, Purdue University, West Lafayette, IN, USA

Ali Rajabi-Siahboomi Colorcon Inc. Global Headquarters, Harleysville, PA, USA

Michael J. Rathbone School of Pharmacy, Griffith University, Brisbane, QLD, Australia

Michael A. Repka Department of Pharmaceutics, National Center for Natural Products Research, School of Pharmacy, University of Mississippi, University, MS, USA

Matthew Roberts School of Pharmacy and Biomolecular Sciences, Liverpool John Moores University, Liverpool, UK

Brigitte Skalsky Evonik Roehm GmbH, Darmstadt, Germany

Philip L. Smith PNPSmith Advisors, Haverford, PA, USA

Sven Stegemann Capsugel, Bornem, Belgium

Steven C. Sutton Department of Pharmaceutical Sciences, College of Pharmacy, University of New England, Portland, ME, USA

Sumalee Thitinan College of Pharmacy, The University of Texas at Austin, Austin, TX, USA

Sandip B. Tiwari Colorcon Inc. Global Headquarters, Harleysville, PA, USA

Michiel Van Speybroeck Laboratory for Pharmacotechnology and Biopharmacy, Katholieke Universiteit Leuven, Leuven, Belgium

Guy Van den Mooter Laboratory for Pharmacotechnology and Biopharmacy, Katholieke Universiteit Leuven, Leuven, Belgium

Clive G. Wilson Institute of Pharmacy and Biomedical Sciences, University of Strathclyde, Glasgow, Scotland, UK

Chapter 1
A Short History of Controlled Drug Release and an Introduction

Alexander T. Florence

Abstract The acquisition and the development of knowledge on how drugs exert their pharmacological effect, particularly information such as dose response, onset, and duration of action and pharmacodynamic and pharmacokinetic relationships has demanded attention to and stimulated interest in delivering drugs at rates (and to locations) that optimize their effects.

Controlling drug release from the medication has accordingly evolved as a multidisciplinary science, requiring expertise spanning such disciplines as polymer science, engineering technologies, and awareness of the complexities and vagaries of gastrointestinal conditions that affect transit of dosage forms. Much has been achieved but there is still much to learn. This chapter outlines the historical evolution of concepts, practices, and achievements in providing better delivery systems for drugs so that they are better medications.

1.1 Introduction

The history of the development of our understanding of the means to control the performance of drugs either through manipulation of the drug itself or through choice of modes of delivery can be illuminating for us today. The gradual understanding of the mechanisms of drug absorption following administration of medicines by a variety of routes, and the possible hazards from inappropriate doses and modes of delivery, provides the background for the development of today's varied array of controlled delivery systems. To ignore the history of any subject is to forget the evolution of thought and technique that has led to the inventions of the present, discussed in depth

A.T. Florence (✉)
Professor Emeritus, School of Pharmacy, University of London, London, UK
e-mail: ataylorflorence@aol.com

C.G. Wilson and P.J. Crowley (eds.), *Controlled Release in Oral Drug Delivery*, Advances in Delivery Science and Technology, DOI 10.1007/978-1-4614-1004-1_1,

in the chapters that follow. It would be wrong to imagine that the field of controlled drug release is a recent, even twentieth century affair. It may even be surprising to learn of the relatively slow evolution of the discipline. One needs only to consider the time that elapsed between the publication of the Noyes–Whitney equation in 1879 [1] and its utilization in pharmacy. The first studies on the effect of drug particle size on dissolution occurred in the 1950s. Even so, the introduction of dissolution tests into pharmacopeias occurred only in 1970s, after much discussion over the relative merits of disintegration and dissolution tests, which the author observed at first hand as a member at that period of the Chemistry, Pharmacy and Standards Subcommittee of the UK's Committee on Safety of Medicines (CSM). The example alone illustrates the slow advance from empiricism in pharmaceutical technology to the rigor introduced by, for example, Takeru Higuchi and his team some 60 years ago. Clearly, progress in most areas has been dependant on a growing and more precise appreciation of the underpinning technologies and biologies.

There could of course have been few truly oral controlled release delivery systems without the developments, from the nineteenth century onwards, in tabletting. Advances in materials science, especially polymer science has provided scientists with a large and growing range of materials with which to work, amongst which are carbon nanotubes and quantum dots, systems such as gels, hydrogels, films, polymer micelles, liposomes, dendrimers, and the like. Not all can be discussed in this short introduction. Another influence has been the increasing precision of imaging technology that has aided progress towards defining and refining the behavior of delivery systems in vivo. Nevertheless, there was considerable at least implied knowledge of the fate of therapeutic agents and dose forms in the nineteenth century and early part of the twentieth century despite the lack of facile analytical techniques and the virtual absence of the ability to monitor medicines in vivo, apart from X-ray photography.

From the early days of the twentieth century to the mid-1950s it might be claimed that pharmaceutics was concerned primarily with the science and practice of the manufacture of dosage forms at small and large scale and with the preparation of galenicals. However, it would be wrong to conclude that it had no regard for the fate or influence of the dosage form in vivo. It is interesting to note that in the early days of the twentieth century most reports on medicinal product performance emanated from work in human patients or volunteers, before the widespread use of experimental animals. The literature therefore emphasized clinical outcomes or physiological measures. Analytical science rarely allowed the measurement of plasma drug levels, which today remain substitutes for performance.

The 1924 edition of Martindale and Wescott's *The Extra Pharmacopeia* [2] discusses enteric coating of tablets as a means of minimizing the adverse effects of drugs on the intestinal mucosa, and enhancing action:

> Various substances have been proposed for the coating of pills, tablets and capsules to render them insoluble in the stomach but soluble in the intestines, i.e. on reaching the duodenum. Drugs, for example, which irritate the mucous membrane and the administration of which is liable to induce vomiting, and substances intended to act solely on the intestines and the anthelmintic drugs, have been so given. Keratin, as usually employed, seldom brings about the desired effect.

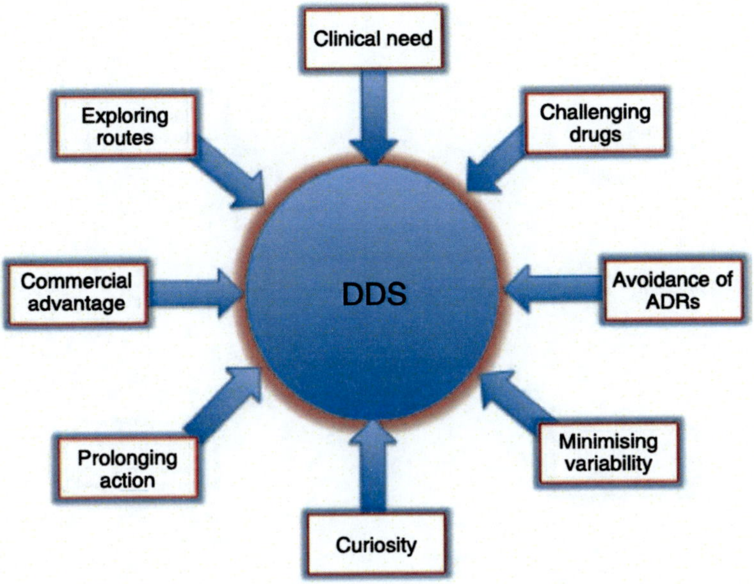

Fig. 1.1 The driving forces for the development of optimized or controlled release systems. Clinical need is key, enabled by new technologies. Curiosity-driven research will provide the seeds of novel ideas, materials, and approaches

Here there is clear concern for the efficacy of formulations and the therapeutic consequences of their properties. Clinical need and clinical benefit are of course the main driving forces for developments in controlled release. But there are other drivers also, not least scientific curiosity, as outlined in Figs. 1.1 and 1.2. The discipline must allow and indeed encourage research which has no obvious and immediate applicability in delivery. The word "novel" is often misused in the titles of papers in the modern literature. Truly novel systems and approaches are nonetheless required to achieve the goal of individualized or personalized medicine, for as Nobel Laureate and physicist Pierre-Gilles de Gennes reminded us: *It was not by perfecting the candle that electricity was invented* [3].

One of the notable trends in the field has been the move from a certain empiricism to a more theoretical, mathematical, and materials science approach to the design and understanding of drug delivery and drug delivery systems. We have yet to reach the stage where the behavior of controlled release systems, let alone conventional systems, can be predicted by any given equation or mathematical formulation. There is as yet no theory of everything biopharmaceutical, and there probably never will be. The ultimate test of performance will always be determined in individual human subjects.

In observing the literature on the development of the discipline of controlled release there are three critical elements: the drug, its formulation, and its route of administration. These are of course intertwined (Fig. 1.3) and can be deconstructed. The target is rightly emphasized now, and at the organ and even cellular level there

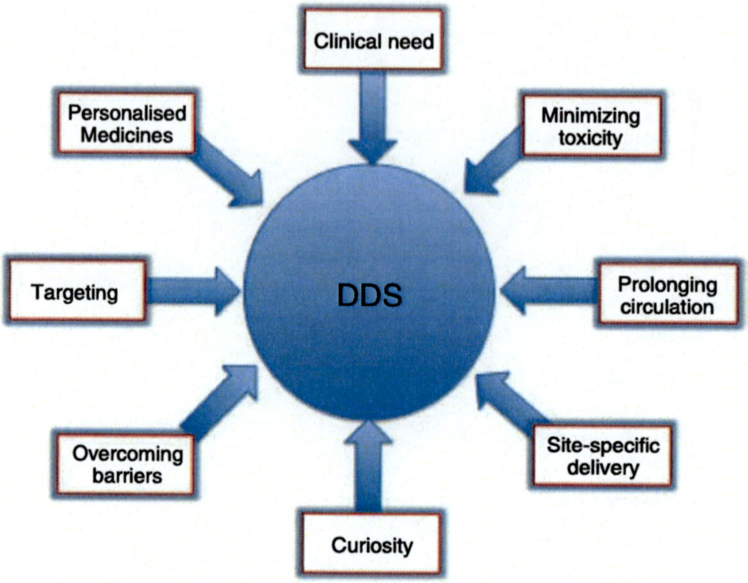

Fig. 1.2 A version of Fig. 1.1 listing other ambitions for controlled drug delivery in terms of advanced systems targeting to specific sites in the body, minimizing variation, overcoming barriers, or responding to the need for more personalized medicines/dose forms

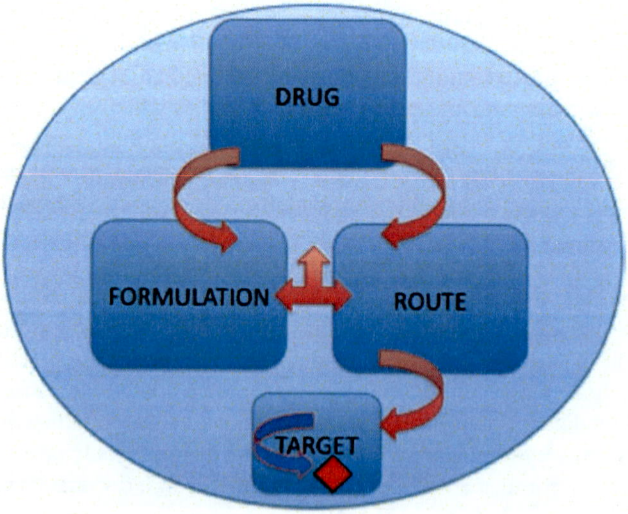

Fig. 1.3 The four key elements of drug delivery: the drug, the formulation, the route of administration, the target organ and cellular or intracellular target

are targets within targets. The two elements that can be manipulated technologically are the drug and the formulation, although one can temporarily enhance the permeability of barrier membranes, it is sometimes a risky strategy. More attention certainly needs to be paid to the matching of the active to particular delivery systems and devices. Again the early observation of clinical outcomes has led to the awareness of the formulation and its ability to enhance or indeed sometimes jeopardize outcomes, a subject that has not perhaps had the emphasis it should have had in the pharmaceutical sciences, but which is of course of vital importance to each patient.

In assembling the material for this chapter it has not always been possible to determine the exact date of any invention or development: sometimes the date recorded is the date of the granting of a patent or the publication of a key paper. Sadly, the scientists who are the true inventors and discoverers are not always revealed, especially in clinical papers describing the testing of the delivery system on patients.

Although this book is primarily focused on oral delivery, many technologies cross route boundaries. What is clear is that much of the relevant history of controlled release depends on the early attempts to administer drugs by various routes, and from the clinical study of outcomes. For example, early views that subcutaneous (sc) and intramuscular (i.m) administration of drugs resulted only in local action were dispelled by the Scots physician Benjamin Bell [4], who in the *Edinburgh Medical Journal* of 1858 opined that "absorption from the enfeebled stomach may not be counted on; we possess in subcutaneous injections a more direct rapid and trustworthy mode of conveying our remedy in the desired quantity to the circulatory blood."

Preliminary studies have frequently led, as in the case of insulin and penicillin to a realization that early formulations were inadequate and that other approaches were needed to overcome either toxicity or short duration of action or both. Hence the history of controlled drug delivery includes those earlier "uncontrolled" or "conventional" systems. While the subject has come far, it has in no way reached its apogee. The plasma level versus time diagram used for explaining the pharmacokinetic advantages of controlled release systems, such as sustained release delivery forms usually purveys an ideal that has not been always been reached in spite of much ingenuity and invention. Figure 1.4 shows a more complex picture for four subjects. The reality is still that inter- and intrapatient variability (in minimum effective dose, maximum tolerated dose, and physiology) is still an issue with some drugs and some delivery systems. The lower diagrams in Fig. 1.4 shows such a situation for oral controlled release nicotine. Data have to be scrutinized with care. Figure 1.5 shows pharmacokinetic data for a transdermal patch studied in a number of human subjects, and the different impression one obtains from such individual data sets, compared to mathematically manipulated data, often plotted logarithmically. When they first became available transdermal patches were implied to offer much tighter control of levels of absorption.

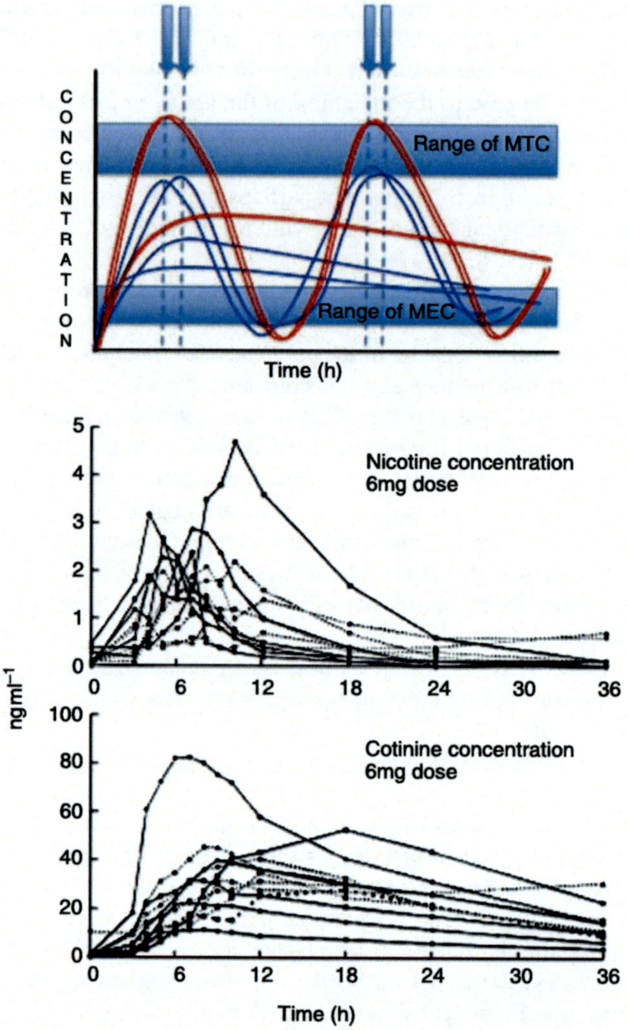

Fig. 1.4 *Upper plot*: An idealized representation of the plasma concentration–time profiles for a conventional release oral dosage form and a sustained release form, attempting to show four individual responses and the range of minimum effective concentrations (MEC) and maximum tolerated concentration (MTC), both of which vary from patient to patient. The two identical plasma curves could represent two individuals whose MECs and and MTCs differ. The *arrows* point to different t_{max} values. *Lower plot*: Results from an oral form of nicotine (5 mg dose) designed for release in the colon. Plasma levels of both nicotine and its metabolite cotinine are shown. From Green JT, Evans BK et al. (1999) An oral formulation of nicotine for release and absorption in the colon: its development and pharmacokinetics. Br J Clin Pharmacol 48:485–493 with permission

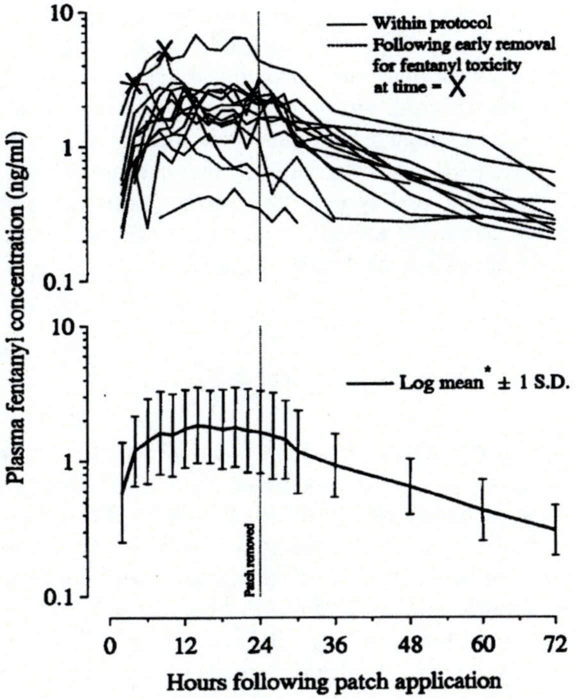

Fig. 1.5 Plasma concentrations of fentanyl delivered from a Cygnus™ patch showing individual data on the *upper diagram* and the mean data below. The results shown here are not to critique the particular product but to demonstrate strikingly how controlled release products do not always behave for one reason or another as designed to do. The data above are from [5]. The variability has been confirmed in other studies including evidence of toxicity

1.1.1 Terminology of Controlled Release

The vocabulary used in the controlled release field to describe delivery systems is paradoxically uncontrolled: it is diverse, flexible, and overlapping. An attempt at a lexicon follows:

- Controlled release: suggests true control of drug release rates
- Sustained release: suggests prolonged release and prolonged plasma levels
- Prolonged release: as above
- Modified release: suggests release rates which are different from fast release, but it is not a precise descriptor
- Pulsatile release: means the release of more than one dose of drug from a given system
- Timed release: suggests release of drug after a specified period of time
- Triggered release: applies to systems from which drug release is stimulated by an external or endogenous signal

In the following sections, the history of various approaches to controlled release, in its broadest sense, is discussed. Judah Folkman who made significant contributions to the field was, however, wrong to state in the abstract to his reminiscences in 1990 [6] that "the first controlled release system was developed in 1962." No doubt he meant that the first silicone implant system for the release of drugs and proteins was. Allan Hoffman's excellent review of the origins and evolution of controlled release drug delivery systems [7] also takes this as a starting point, but here we take the story back at least 130 years to emphasize the step-wise growth of the subject. Hoffman's account, well illustrated and detailed, highlights key players in the field as it developed thereafter. The majority of us are the laborers, contributing a few bricks here and there, usually as a result of the work of research students.

Given that the oral route is not only the most widely used route but that this book is devoted to oral systems, this account begins with oral delivery systems. But, as stated earlier, the history of other routes cannot be ignored for there are some general principles to be derived from them.

1.1.2 Oral Delivery Systems

Writing in 1961, Lazarus and Cooper [8] stated that the possibility of favorably modifying the therapeutic response of drugs administered by mouth "provide a powerful stimulus to research workers in pharmacy and medicine. Here was an open sesame of scientific opportunity further triggered by vistas of economic rewards." But they added trenchantly that the speed with which pharmaceutical laboratories jumped on the bandwagon resulted in "a plethora of ideas and products, many of which should have remained unborn." Some, simply put had no real clinical benefit. Approximately, 180 different prolonged release products were available in the USA in 1961. There was and is indeed a problem with the prescribing and dispensing of (generic) prolonged release products of the same active [9], as while the drug release rate from conventional release tablets can be defined within a narrow range, modified release products, such as with theophylline *legitimately* provide a spectrum of rates of delivery; because of their different physical compositions there is a risk of quite different pharmacokinetics due to potential differences in GI transit or the effect of food [10, 11]. Of course the variety of release rates provides overall a palette from which clinicians can choose, but the appropriate pharmacokinetic information is not always available to prescriber or pharmacist to allow reasoned choices. The recent trend for the pharmaceutical industry to introduce controlled release products at or near the end of the patent life for their original invention sullies the field, by the introduction of products whose purpose is to protect a brand name and to outwit generic manufacturers. Why, one asks, was there no clinical need for such a product in the previous decade of product life, when suddenly it becomes an advance of great benefit to all concerned?

Considerable ingenuity has been demonstrated in the development of oral controlled release systems over the years with much overall benefit to patients.

Early work to alter the rate of release and the duration of action of products depended very much on coating pills, capsules, and latterly tablets when these emerged from the compression process invented by Brockedon [12]. His invention not initially intended for pharmaceutical use, but was soon recognized by Brockedon himself, who contacted the *Pharmaceutical Journal* about potential products from it [13]. The advent of tablets seemed to have raised some opposition, especially in France according to Pariente [14] where they were given the epithet of "*barbarismes pharmaceutiques*" – engendered in part by the failure of some products to disintegrate, the sarcastically named *comprimé perpetual*. Soon the technical problems in compressing powders became evident and Dunton obtained a patent in 1875 [15], in which he suggested the processing of materials by drying before compression and the use of lubricants to reduce cohesion between powder and die. Dunton states that the forces of adhesion are often greater than the forces of cohesion such that the mass breaks up readily without lubricant. Another patent by Sauter employed starch as a disintegrant, and as early as 1878 we have a patent on coating tablet cores by compression, so "that the powder covering holds by cohesion, thus producing a seamless coated medicament [16]." The coat was designed to protect the active, to disguise unpleasant tastes without changing the solubility of the drug. It is a small step to the incorporation of a second drug in the outer compressed coat to be released at a faster rate. The stage was set for more developments.

First we survey of the use of coatings to change the performance of formulations for oral use. Early alternative inventions included Ellzey's drug-containing capsule placed inside a larger capsule with an intervening alkaline material [17], perhaps a forerunner of capsules containing mini-tablets? There were some diversions into gelatin tablets [18, 19] which may in retrospect have been a useful advance, perhaps as a future platform for pediatric and geriatric formulations.

1.1.3 Coating as a Means to Control Drug Release

Figure 1.6 summarises the many ways in which control of drug release has been achieved by "enveloping" drugs as opposed to modifying the form of the drug as discussed later.

1.1.3.1 Early Enteric Coating Materials

The use of keratin-coated pills is reported in a *Lancet* paper in 1893 [20]. Although the process is credited to the German dermatologist, Dr Paul Unna, who first marketed such products [21], he himself did not claim to have originated the idea of enteric coating. The early alternative to keratin was salol (phenyl salicylate) and the two were compared in a study in 1917; salol was found to be superior in the delivery of emetine bismuth iodide [22]. Cowen [23] states that what made Unna's work "truly revolutionary" was the concept that "the form of the medicament could be used to influence, if not determine, its substantive effects." This is of course the kernel of controlled release formulation with its ability to influence therapeutic outcomes. While Unna himself demonstrated the effect of his coating system in vitro, there were frequently expressed concerns about the quality of such systems, particularly on product aging. An early noninvasive technique – X-ray examination – was used to observe the disintegration in vivo of keratin coated systems in 1938 [24]. This technique has of course been used many times since, showing delayed transit of dosage forms in the esophagus [25], the lack of disintegration of enteric coated tablets of potassium chloride [26], the esophageal retention of barium sulfate tablets [27], and later Channer's work on esophageal transit of formulations [28].

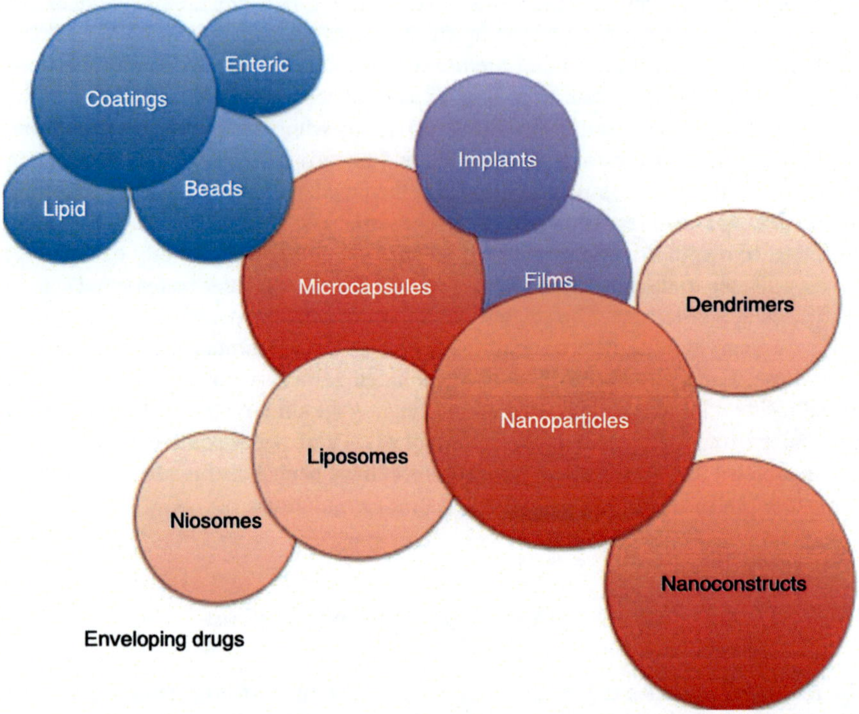

Fig. 1.6 Schematic of the approach to controlled release through enveloping the drug in a carrier either by coating or by preparation of microcapsules, nanoparticles, dendrimers, and nanoconstructs of increasing diversity, implants and films

Ingenuity was shown by those who estimated whether enteric coated pills or tablets actually disintegrated in vivo by determining the presence of drugs or markers such as iodides, and methylene blue in urine. Thomson and Lee [29] suggest, perhaps tongue in cheek, that a foolproof method of determining the lack of disintegration is the recovery of the dosage form intact in the feces [30], but they themselves provide an account of sound methods of ensuring in vitro that enteric coated products were reliable and standardized, with tests of disintegration in acid pepsin, and both acid and alkaline pancreatin [31].

1.1.3.2 Wax Coatings

A number of formulations appeared comprising granules or beads coated with waxes of varying structures. Glycerides have a greater effect on retarding release in vitro as they become less polar, glyceryl monostearate being less effective than the distearate and the tristearate most effective [32]. The SKF Spansule™ product was one of the early systems to reach the market (circa 1945). The product contained a range of granules or beads coated with different thicknesses of a wax coating, releasing drug over a period of 10–12 h. By the 1950s there was a growing range of such products, not all of which again were beneficial clinically. Nitroglyn™ a preparation of coated granules of nitrogyclerin (but elsewhere described as a porous plastic matrix tablet [33]) gave disappointing results given the "logic behind the product" in alleviating exercise-induced cardiac stress, while a long-acting formulation of pentaerythritol (Peritrate™) provided marked responses for 5–6 h after administration, but only after a latent period of 60–90 min [34]. However, Peritrate lost *all* of its activity when given with food, early indication of the differences between disintegrating and non-disintegrating controlled release forms and their transit behavior in the gastrointestinal tract.

1.1.3.3 Polymer Coatings and Matrices

Polymer coats are now used widely in modified release coated tablets, materials such as cellulose acetate phthalate (CAP) from 1938 [35], hydroxypropyl cellulose phthalate (HP50 and HP55), and mixtures of ethyl cellulose and hydroxypropyl methylcellulose (HPMC) [36] whose physicochemical characteristics such as solubility parameters, pH dependency of solubility, solubility in a range of organic solvents, and rheology were reviewed by Rowe [37]. By varying the proportion of ethylcellulose and HPMC, disintegration times varying from 10 min (ratio: 25:75) to 5.75 h (ratio: 75:25) could be obtained. At Smith, Kline and French, Swintosky recalls his contributions after he joined the company in 1953 to the use of hydrogenated castor oil–ethyl cellulose systems for coating pellets and tablets.

The Eudragit™ polymers are copolymers of acrylic acid or methacrylic acids and their esters. They were first marketed by Rohm and Hass in 1953. The properties of these polymers can be adjusted by variations in their chemistry. Eudragit type L and S polymers are copolymers of methacrylic acid and methacrylic esters in different ratios. The ratio of the free carboxylic groups to the ester groups is approximately 1:1 in Eudragit type L and about 1:2 in Eudragit type S. Therefore, Eudragit type L polymer is more acidic than Eudragit type S polymer. Eudragit L 12.5 was available from 1954 and new materials were introduced at intervals, one of the latest being Eudragit FS 30 D in 1999. The Eudragit polymers can be used as tablet coating or as matrix materials.

A wide range of matrix systems have been developed, from hydrophobic inert polymers such as polyethylene, PVC, ethyl cellulose, and acrylates and their copolymers to various lipid matrix systems, and of course hydrophilic matrices and biodegradable matrices.

As will be seen from other chapters in this book, oral controlled release products can be divided into those whose drug release properties are controlled by diffusion, dissolution, erosion of the matrix, osmotic pressure, ion-exchange reactions, or sometimes a combination of processes. The backgrounds to some of these are discussed below.

1.1.3.4 Osmotic Pumps

The application of sound physical chemical principles led to the invention of the ingenious series of capsule and tablet systems whose driving force for drug release was osmotic pressure; semipermeable membranes controlled ingress of water, and a laser-drilled hole a mode of escape of the drug [38]. While Rose and Nelson [39] produced a prototype for veterinary use in 1955, whose driving force was the osmotic pressure difference between a saturated solution of Congo Red against water, it was the work of T. Higuchi and colleagues at Alza who drove the work forward [40, 41] to provide both prototypes and, in 1982 the first product based on an "elementary osmotic pump" for the delivery of indomethacin (Osmosin). This drug seemed a likely candidate for a system which accurately controlled the release of the active, given its propensity to cause serious effects in the small intestine, like many other nonsteroidal anti-inflammatory drugs. An occasional patient suffered from intestinal perforations as a result of long-term intake of NSAIDs [42]. Between December 1982 and the end of June 1983, 400,000 prescriptions had been written for Osmosin in the UK, but there was a high rate of reporting of adverse effects to the UK's Committee on Safety of Medicines (CSM), some 200 by August of that year, amongst them reports of intestinal bleeding and perforation [43]. Two cases of perforation distal to the duodenum (unusual with other NSAIDs) suggested that the Osmosin formulation exposed new areas of the bowel to the drug. The CSM noted that the formulation contained 158 mg of potassium bicarbonate a known irritant. Other reports accumulated in the literature [44, 45]. Of one report it was said [46] that as the two patients concerned were 70 years old, "it is unfortunate that a new drug (sic) should fall into disrepute as a result of inadequate attention to dosage, unwarranted polypharmacy, or lack of concern for the patient's age." The product was withdrawn from the UK market in 1984. This short life is highlighted here not to diminish the invention, but to emphasize that as Laidler and colleagues [47] pointed out at the time, "new formulations of drugs may have unexpected side effects," a lesson that history tells us and that current experience reinforces. This episode is also a warning that whatever the merits of any dosage form, care has to be taken with its use. Few are panaceas. The effects produced by novel devices and formulations are not always as we anticipate, and sometimes difficult to fathom in hindsight, but in the case of Osmosin the effect of drug and excipient was no doubt exacerbated by the single exit point of the active in some orientations in the GI tract. This is very true of the delivery of endogenous molecules which we seek to insert into cavities that these molecules would never normally see, or accumulate in organs and tissues that the free drug would not. The 30 years of development of osmotic systems and a detailed analysis of the many successful products relying on this and related technology are comprehensively reviewed by Malaterre et al. [48]. Already by 1995 over 240 US patents had been granted [49] for variants and improvements on this initial concept that has spawned a new generation of controlled release devices.

1.1.3.5 Ion-Exchange Resin Complexes

Oral depot therapy with two long-acting dexamphetamine-resin salt formulation (Dexten™ and Barbidex™) was described by Abrahams and Linnell in 1957 [50].

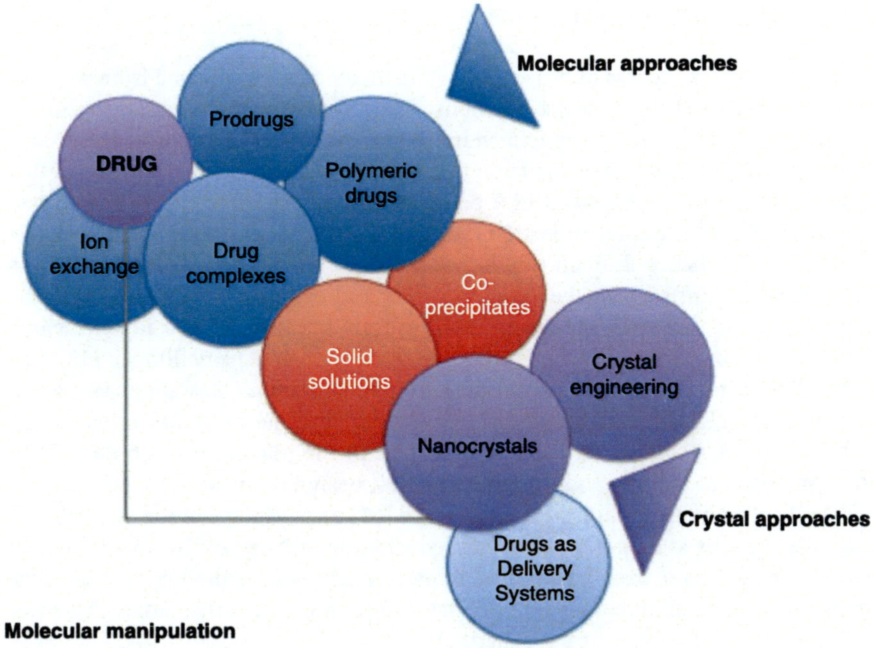

Fig. 1.7 Molecular manipulation of the active to achieve control of release rate from dosage forms, in some cases in increase solubility and thus rate of solution, e.g., by forming nanocrystals

Their paper showed in vitro release of the drug from the resin over 12 h using the techniques adopted by Chaudhry and Saunders [51]. With a rather overblown introduction to their paper, Abrahams and Linnell state that the many advantages of sustained release oral therapy "demand no emphasis in recommendation; they include certainty of continuity, no relapse due to forgotten dose, and convenience." While the last is certainly true, one could debate the other points. A forgotten dose is a forgotten dose and more sophisticated means of reminding patients to take their medication at the right time is one of the necessary developments of recent years. Reviews on the pharmaceutical applications of ion-exchange resins are useful in exploring the background and uses of these complexes [52, 53].

1.1.4 Intramuscular and Subcutaneous Injections

The history of the i.m. and s.c. routes of drug administration is illuminating in that they spurred developments of salts and esters of drug substances and of viscous vehicles in attempts to reduce the rate of drug release from depots and to prolong action. Figure 1.7 represents various approaches that have been used to modify the properties of the active to prolong action.

1.1.4.1 Drugs and Drug Salts

Much of the early success of prolonged acting medicines was derived from work on intramuscular and subcutaneous injections. The problems with the administration of the then new penicillins spurred much invention. In the early days of penicillin, ice bags were used at the site of injection to prolong release [54] but it was found that calcium penicillin G dispersed in a beeswax–peanut oil vehicle [55] gave blood levels for 24 h after a single injection [56]. Beeswax is not absorbed and the vehicle is also viscous, hence difficult to inject without warming. Others sought insoluble salts of penicillin, such as the bismuth, silver and even mercury salts [57] to be injected as an aqueous suspension. An advance came from Lilly Research Laboratories in 1948 with the development of procaine penicillin G [58, 59]. Crystalline penicillin G was also suspended in an aluminum stearate peanut oil or sesame oil gel, and it was found that suspending procaine penicillin in these gels produced measurable quantities of antibiotic in the plasma for 4–6 days [60]. A comparison of the influence of three vehicles, peanut oil, water, and peanut (arachis) oil gelled by aluminum stearate revealed [61] an advantage of the oily vehicles over the aqueous at longer time periods, concluding that the arachis oil–aluminum stearate preparation had the optimal clinical outcomes, removing "the greatest drawback in penicillin therapy – i.e. the necessity for frequent injections." This was especially important in the treatment of children [62].

The particle size of the active in these formulations was found to be of importance: the smaller the size of the particles the greater the prolongation of action, the optimal in the arachis oil–stearate gels being less than 5 μm in diameter [63]. This may have seemed counterintuitive, but has been confirmed and explained with model nonaqueous systems much later by Crommelin and de Blaey [64].

Benzathine penicillin G, the *N,N′*-dibenzylethylene diamine salt of penicillin G, with a lower aqueous solubility provided an alternative to procaine penicillin [65, 66] but administered orally in children produced peak levels at 1 h with a rapid decline over 3–6 h.

1.1.4.2 Drug Esters

Minto et al. [67] remind us that since the late 1950s administration of androgens such as testosterone and 19-nortestosterone in esterified form has been by i.m. injection [68]. These provided sustained release. Not all oil-soluble agents give prolonged release on injection in oily solvents [69], clearly because one key parameter is the partition coefficient between oil and tissue fluid [70]. There are many other factors. Minto and colleagues explored the difference in human subjects of nandrolone as its decanoate and phenylpropionate. Injected into the gluteal muscle 4 ml of the phenylpropionate ester was superior to the decanoate. Such studies have only been made possible by advances in analytical technology. Similarly comparisons of fluphenazine hydrochlroide, its enanthate, and decanoate were carried out in detail some 15 years after the introduction of these long chain esters [71]; the decanoate produced the longest half-times and the difference in the kinetics was ascribed to the differences in the kinetics of release from the oily depot. This topic has been addressed recently in a review [72]. One factor can be the amount of oil; testosterone propionate had a greater intensity of action (in rats) and a longer duration of action when given in 0.8 ml sesame oil, compared to 0.2 ml, whereas testosterone itself showed the converse result [73].

1.1.5 Subcutaneous Implants

In 1937, Deansley and Parkes studied the effect of implanted hormones as compressed tablets of the testosterone, testosterone propionate, progesterone, and oestrone and oestradiol [74, 75]; Noble [76] subcutaneously implanted crystals of synthetic oestrogens, Bishop a 14-mg tablet of oestrone [77] acting over 4 weeks. In experimenting with more water soluble drugs, a search for suitable matrix materials began; 90% cholesterol was used in some experiments [78]. Folkman then of the National Naval Medical Center, Bethesda, MD reported on his work with silicone (Silastic™) implants in 1964 [79]. The idea came for Silastic-based prolonged drug release systems when during "in vitro studies on artificial heart valves [when] it was noted that silicone rubber possessed the property of absorbing certain dyes from solution and subsequently giving off these dyes." Following studies on the diffusion of steroids through polysiloxane tubes [80] this led to Upjohn's silicone elastomer vaginal contraceptive rings (1968) containing medroxyprogesterone acetate [81].

ICI Pharmaceuticals in the UK pioneered PLGA implants in their Zoladex™ (goserelin) product [82, 83], overcoming considerable formulation difficulties in relation to peptide solubility in the polymer matrices, and problems in controlling the morphology of the drug–polymer mixtures. By 1992, similar formulations had emerged for nafarelin, leuprolide, buserelin, and triptorelin.

Deansley and Parkes found no local tissue reactions but did find that a coat of connective tissue grew "tightly round the tablet in the course of time," a recurrent problem today. Testosterone was found to be absorbed faster from a compressed tablet than its propionate; Parkes [84] found that indeed that the duration of effect of an endocrine preparation was inversely proportional to its solubility in the tissues. The absorption from oestrone tablets represented about 15% per month. With the appropriate sized tablets they speculated that release could last for up to 2 years. A commentary [85] at the time refers to the problem of encapsulation of implanted tablets in some cases preventing further drug release after 3 months, the use of esters to further prolong the activity of such implanted tablets and the need to avoid their breakup in vivo.

More advanced in material terms, the medicated wafers that Brem pioneered, namely the Gliadel™ wafer, is a polyanhydride implant containing carmustine (BCNU) for the treatment of malignant melanoma. The drug is released by a combination of diffusion and erosion of the matrix, a 20:80 molar ratio of 1,3-bis (p-carboxyphenoxy)propane (CPP) and sebacic acid (SA) [86]. Langer discusses implantable controlled release systems in his 1983 review [87] and 21 years later describes with Brem, Cima, and colleagues in vivo release from a microelectromechanical system (MEMS) [88].

Drug eluting stents might be considered to be an extension of implanted systems. Stents were introduced in the late 1970s by Andreas Grüntzig [89] for percutaneous transluminal coronary angioplasty (PTCA), or balloon angioplasty, in which a catheter was introduced through a peripheral artery and a balloon expanded to dilate the narrowed segment of artery. However because of the problems of recurrent stenosis, drugs such as sirolimus, paclitaxel, and actinomycin D, amongst others were incorporated into polymer layers on the scaffold of the stents. Their delivery at the site of action at controlled rates reduces the intimal thickening. These products have had considerable success [90].

1.1.6 Aerosols

While aerosols are not at the forefront of sustained release, they display clear examples of modifying the activity of active molecules through technology. The complexity of the aerosol device and the formulations used, not least the particle size of the emitted droplets, have a large impact on bioavailability. Release of drug is controlled by many factors. These complexities were recognized over 60 years ago, an editorial in *The Lancet* [91] stating "it is hardly surprising that the status of inhalation therapy has still to be defined." The aerosol "must be of a particle size suitable for ingress to the required depth, but the weight of drug carried by these small particles must be as great as possible" [92]. Presciently, the editorial goes on to point out that "the amount of solution leaving the nebulizing phial is by no means the same amount actually passing into the trachea." But the origins of aerosol therapy go back further. Beigel [93] recounts the French origin of atomizing water in hydrotherapy and Sales-Giron's invention of an atomizer which caused a "sensation" at the Academie de Médicine in Paris. The Academy some years later declared that not only the vapor but the chemicals it contained reached not only the trachea but the "cells of the lungs." Later Beigel recognized the importance of the depth of respiration on the deposition and the rapidity of absorption of drugs from the lung, hence his expression of the need for caution in dosing [94], still an issue today.

It was probably not until the advent of penicillin that inhalation became an important route of delivery. Nebulizers powered by compressed air or oxygen produced particles from the visible to the submicron in diameter. Mutch's exposition on the inhalation of chemotherapeutic substances discusses the fate of particles in some detail [95] and refers to early in vitro experiments by Heubner [96] on aerosol particle deposition. Presaging pharmacopeial artificial throats and other valuable apparatus in use today, the complex glass apparatus of Heubner with "bends, forks and narrowings and expansions" to resemble the complexity of the respiratory tract, demonstrated that particles of 5 μm reached its furthermost ends. Indeed in 1925, 5 μm diameter particles were presumed to reach the bronchus [97]. The rubber bag of the Collison inhaler, the use of which Mutch describes, probably served as the spacers used today. Nearly 60 years later, a *Lancet* editorial spoke of the nebulizer epidemic in the UK, expressing concern about appropriate doses of beta-agonists by nebulizer and by metered dose inhaler [98].

While the particle size distribution for maximal aerosol penetration had been determined using solid particles, it was a significant finding that large porous particles greater than 5 μm in diameter could be inspired into the far reaches of the lung and avoid the natural expulsion mechanisms [99]. This was attributed to the smaller surface to volume ratio, the lower tendency of large particles to aggregate and the fact that larger particles tend to exit dry powder inhalers as single entities.

1.1.7 Topical Application of Drugs

The advent of the transdermal patch had to wait until the transdermal absorption was fully understood, suitable polymers and adhesives were available and the ingenuity of the inventors brought them to fruition. In the 1920s it was becoming fashionable to give drugs by the skin, acknowledging that "the difficulty of absorption is often great," [100] which for some agents could not be overcome by choice of vehicle. Rothman wrote a masterly review in 1943 of the principles of percutaneous absorption [101]. In relation to debates on what properties a drug must have to penetrate the skin, he quotes the fact that quantitatively a degree of solubility in water and a degree of solubility in lipid is required. There was much activity in research into insulin absorption through the skin [102, 103], but perhaps tongue in cheek, Rothman concludes from that work that "under carefully controlled experimental conditions negative results are prevalent, any positive results being ascribed to damage to the skin." Can we criticize that work when with better knowledge of, for example, the molecular properties of insulin and its fate in vivo considerable efforts are put into studies of oral absorption? It perhaps sends another message to researchers in the present day that just because one invests effort in research it does not mean that success will follow.

The advent of the sulfonamides provided a further impetus for such work [104], much of it covered in a historical perspective by Hadgraft and Lane [105]. Rothman surmises that the role of the vehicle is secondary to the properties of the drug. This has been the subject of considerable research since. The unimportance of the vehicle is not sustainable [106]. The importance of both the vehicle and the particle size of suspended corticosteroids was shown by Sarkany and colleagues at the Royal Free Hospital in London [107]. Even in the 1940s it was proposed to use what Herman and colleagues [108] named "penetrasols" (propylene glycol and a synthetic surfactant inter alia). Katz and Poulsen did much to determine the effect of propylene glycol on penetration, their experimental work following on the formal approaches of Higuchi (1960) and colleagues in applying physical–chemical principles to the issue, determine the importance of the thermodynamic activity of the drug. The importance of using appropriate methodology to obtain accurate and reproducible results has been emphasized [109]. A report on the failure to detect absorption of hydrocortisone applied to human skin [110] may have resulted from such problems, as it was demonstrated by others in the same year [111].

1.1.7.1 Iontophoresis

While the importance of tailoring vehicles to drugs in percutaneous delivery was recognized, so too was the possibility of enhancing absorption by physical means such as electrophoresis or iontophoresis, an effect demonstrated for strychnine in 1888 and for dyes in 1890, but as recalled by Rothmann dramatically demonstrated by Leduc at the start of the twentieth century [112]: rabbits painted with a solution of strychnine nitrate survived while those to which an electrophoretic current was applied died. There was considerable interest in the use of iontophoretic techniques in ophthalmology [113]. Fleming refers to it as iontotherapy [114] when used for the medication of the cornea, the bulbar conjunctiva, or the everted eyelids. A "device for inserting medicaments into the body by iontophoresis" was the subject of a US Patent in 1934 [115]. Later work confirmed this effect [116] not only with dyes but with sulfonamides [117]. The more recent interest in iontophoresis has its origins in this work. Iontophoretic enhancement of the transport of peptides was investigated in the 1990s [118] and lately the relative contributions of electrorepulsion and electroosmosis to the movement of drugs across the skin has been revisited [119].

1.1.7.2 Transdermal Patches

While transdermal patches seem to have little direct relevance to oral delivery, they employ the same polymer technology as many oral dose forms. The crucial elements are the rate controlling membrane and the nature of the drug in the reservoir. There are some direct applications of course as with the patent for an ingestible film [120]. The first design and application of a transdermal patch derived from Alza. Alejandro Zaffaroni's "bandage for administering drugs" was for "use in the continuous administration of systemically active drugs by absorption through the skin or oral mucosa" [121]. In one variant, the drug is contained in permeable microcapsules in the adhesive layer of the patch [122].

1.1.7.3 Microneedles

Microneedles were proposed first by Gerstel and Place [123] but progress had to wait until these could be produced in the 1990s and then tested [124]. Much of the work on these needles has emanated from Prausnitz's laboratories [125].

1.1.7.4 Hydrogels and Other Systems of Delivery to the Eye

One hundred years of pilocarpine's use in ophthalmology was celebrated in 1976. There are many problems with the installation of conventional eye drops by either carer or patient and new ways of administering drugs to the eye were sought. The possibility of using gel contact lenses was proposed in 1965 [126]. Hydrophilic gels were first synthesized and proposed for biological use by Wichterle and Lim 5 years earlier [127]. In 1974, Alza's Ocusert device gained FDA approval. A patent for it was granted in 1971 [128], and studies conducted on systems with different release rates [129]. Intimations of such systems might be seen in the 1883 US Patent of Wadleigh for concave medicated gelatin eye disks to obviate the two problems with eye drops, the difficulty in administration and the lack of a reasonable duration of action [130].

1.1.8 Microencapsulation

Microcapsule technology can be applied by a number of routes, oral and parenteral, so it is considered separately here. It was on July 5, 1955 that the National Cash Register Company (NCR) gained a patent on the process of microencapsulation a key component of their carbonless copy paper. This was to have consequences for an area of controlled release technology. Originally developed for ink containing microcapsules, it soon became a means of enveloping drugs for slow release and as components of artificial cells through the pioneering work of T.M.S. Chang [131]. They are still presenting challenges today as delivery vehicles of biologicals derived from cells [132]. In 1961, NCR were granted a patent for dual-walled oil containing microcapsules [133]. There followed a large number of patents, including one which [134] claimed a method of "localizing a therapeutic agent at a preferred treatment site within an organism by injecting said agent in association with a magnetically responsive substance and concentrating the agent and substance at the treatment site by application of magnetic fields," the carrier being microcapsules of up to 5 μm in diameter. The Fuji Photo Film Company was in the field too, with patents in 1972 [135] and one in 1976 [136] for producing aspirin-containing microcapsules with walls of ethylcellulose, polyacrylic acid, polymethylacrylic acid inter alia.

Kramer's paper in 1974 was one of the first to describe microspheres for achieving specificity in drug delivery [137]. Ekman and Sjöholm published several papers on the topic of immobilization of macromolecules in microcapsules, the first in 1975 [138]. 225 μm ethylcellulose microcapsules of mitomycin C, following studies in dogs, were progressed to treat a renal cell carcinoma patient by infusion into the ileac and femoral arteries [139, 140]. Another key paper appeared in *Science* [141] demonstrating the use of albumin beads to provide sustained release over 20 days of progesterone. Takeda Chemical Industries' interest in the topic is evidenced by one of their patents [142] which has a Japanese application priority date of 1983. Okada later published data on the leuprolide PLGA microspheres which after a single s.c. or i.m. injection produced biological effects for more than 1 month in rats [143]. The development of a 3-month preparation was described later [144]. Other proteins have been used as the basis of microspheres, including hemoglobin, transferrin, albumin/polyaspartic acid, and casein were reported at the time [145]. At least one method of preparation of microspheres, that of coacervation preceded the NCR patents and is based on the work of Bungeberg de Jong [146] on simple and complex coacervation in the late 1940s. Gelatin microcapsules prepared by these techniques are discussed in Burgess and Carless's paper [147]. With all the interest in microsystems, it was timely that Dapper and Thies worked on statistical models for estimating the kinetics of drug release [148]. The progress of the microcapsule field was changed by the interest in nanosystems, which are outside the scope of this chapter, but whose history has been considered by Kreuter [149] who worked with Peter Speiser [150] in whose ETH laboratories nanocapsules and "nanoparts" were first suggested for pharmaceutical application. Biodegradable polycyanoacrylate nanocapsules were reported by Couvreur et al. [151] and attention largely switched to these systems and similar systems. New advances in microsphere-based systems were reviewed in 1997 by Hanes, Cleland, and Langer [152].

1.1.9 Conclusions

This exploration of the historical background to today's less empirical science of controlled drug delivery has clearly not delved into all aspects of the subject. History does inform today's research in that we can recognize that a variety of factors about the performance of controlled release formulations have been known for a long time, yet in spite of this we are still grappling with attempts to make our products more predictable in outcome, especially by the oral route where gastrointestinal transit and patient to patient variability in physiology, diet, ethnicity compound matters. We can see that early attempts, for example, to administer insulin by the skin, as well as orally with saponins, cause us pause to consider whether all our endeavors with oral or inhaled insulin are realistic. Nature sometimes cannot be beaten and we need sometimes to make a deal with it. This is not a defeatist stance, but one of challenge. If we alter the nature of absorbing membranes, even momentarily, if we administer endogenous molecules into body spaces that do not normally admit such substances, if we accumulate drug in certain organs, there are bound to be some unexpected adverse effects, as with the Osmosin product, an unfortunate combination of drug and osmotic agent in a rather beautiful technology. There are numerous successful oral osmotic pumps now available in the clinic. On the other hand the tentative steps with transdermal medications given the work put into understanding the nature of the skin barrier, there is now a raft of devices for systemic therapy.

The great challenge is to provide medications for individuals or subgroups of patients much more flexible in design than our monolithic systems have been in the past. This will require a mix of technological approaches, many of which are discussed in the chapters in this book.

Omitted from this survey are the pioneering theoretical developments of scientists such as T. Higuchi [153], W.I. Higuchi [154], Simonelli [155], Peppas [156], and others, who have allowed a more rational approach to, and explanation of, the design and behavior of often complex controlled release systems.

Each route of delivery and each system is deserving a history of its own. The chapters that make up this book will all allude to the pioneering developments that have presaged today's intriguing systems.

Note Late on in the writing of this chapter the historical review by P.I. Lee and J-X. Li, Evolution of oral controlled release tablets, in H. Wen and K. Park (eds) Oral Controlled Release: Formulation Design and Drug Delivery, Wiley: New York, 2010 came to my attention. Although many of the sources will be the same I trust that the accounts are sufficiently different to be complementary.

Acknowledgments I thank Mrs. Michelle Wake and Mr. Peter Field of the Library of The School of Pharmacy, University of London for retrieving copies of some of the older literature. I am also indebted to all those who have written historical reviews on many of the topics covered here.

There have been two Schools of Pharmacy with which I have been associated, at the University of Strathclyde, Glasgow, and in the University of London. In the former I began my research in 1962, so as both a student and researcher have seen some of the history discussed above unfold. My interest in pharmaceutics was piqued by the work that had been done in Glasgow, in wartime on transfusion and pyrogens, on transdermal absorption, atracurium, not to mention the work of my mentor Peter Elworthy on surfactant systems. In London its pedigree included the work of Prof Leonard Saunders in the 1960s with proto-liposomes, diffusion and ion-exchange drug resonates, formed a little part of the history, not all of it recounted here.

References

1. Noyes AA, Whitney WR (1897) The rate of solution of solid substances in their own solutions. J Am Chem Soc 19:930–934
2. Martindale WH, Wescott WW (1924) The extra pharmacopoeia, vol 1, 18th edn. H.K. Lewis, London, pp 682–683
3. de Gennes P-G (2004) La Memoire du College de France. "*Ce n'est pas en perfectionnant la bougie que l'on invente l'électricité*"
4. Bell B (1858) The therapeutic relations of opium and belladonna to each other. Edinb Med J 4:1–7
5. Fiset P, Cohane C, Browne S, Brand SC, Shafer SL (1995) Biopharmaceutics of a new transdermal fentanyl patch. Anesthesiology 83:459–469
6. Folkman J (1990) How the field of controlled-release technology began, and its central role in the development of angiogenesis research. Biomaterials 11:615–618
7. Hoffman AS (2008) The origins and evolution of "controlled" drug delivery systems. J Control Rel 132:153–163
8. Lazarus J, Cooper J (1961) Absorption, testing and clinical evaluation of oral prolonged-action drugs. J Pharm Sci 50:715
9. Florence AT (1992) Generic medicines: a question of quality. In: Wells FO (ed) Medicines: responsible prescribing. Queen's University Press, Belfast, pp 63–83
10. Florence AT (1987) Sustained release theophylline preparations: pitfalls in generic prescribing. In: Fairfax AJ (ed) Therapeutics in respiratory medicine. Royal Society of Medicine Symposium Series 117, pp 23–32
11. Jonkmann JHG (1989) Food interactions with sustained release theophylline preparations. Clin Pharmacokinet 16:162–179
12. Brockedon T (1843) Shaping pills, lozenges and black lead in dies. GB Patent 9977
13. Brockedon's Patent. Process Pharm J 3:1843–1844
14. Pariente L (1996) Comprimés Chap. 6 in *Naissance et Evolution de Quinze Formes Pharmaceutiques*. Editions Louis Pariente, Paris, pp 94–118
15. Dunton J (1875) Improvement in the manufacture of pills. US Patent 168,240
16. Carter C (1878) Improvement in coated compressed medicaments. US Patent 207,013
17. Ellzey EF (National Aniline Company) (1931) Device for administering medicaments. US Patent 1,815,902

18. Kraus F (1870) Improvements in putting up medicines. US Patent 108,034
19. Bird JC (American Cyanamid) (1948) Casting gelatin tablets. US Patent 2,552,027
20. Ritchie J (1893) Brief notes of several cases of acute diarrhoea treated with keratin-coated carbolic acid pills. Lancet 142:1305–1306
21. Dumez AG (1921) A contribution to the history of the development of enteric capsules. J Am Pharm Assoc 10:372–376. cited by B.E. Jones (2004), The history of the medicinal capsule, in Podczeck F, Jones BE (eds) Pharmaceutical capsules, 2nd edn. Pharmaceutical Press, London
22. Lille DG, Shepheard S (1917) A report on the treatment of *Entamoeba histolytica* "carriers" with emetine bismuth iodide giving a comparison between keratin coated and salol-coated pills. Lancet 190:418–419
23. Cowen DL (2002) Swallowing the pill. *Apothecary's Cabinet* 4, p 6
24. Gorley JT, Lee CO (1938) A study of enteric coatings. J Am Pharm Assoc 27:379–384
25. Praetorius E, Faber JH (1950) Om tabletters henfald og passage geenem esophagus af ventrikel. Ugeskrift for Laeger 112:628–631
26. Freestone DS (1969) Formulation and therapeutic efficacy of drugs used in clinical trials. Lancet 294:98–99
27. Evans KT, Roberts GM (1976) Where do all the tablets go? Lancet 308:1237–1239
28. Channer KS (1990) Adverse reactions to the transit of dosage forms. In: Florence AT, Salole EG (eds) Formulation factors in adverse reactions. Wright, London, pp 75–95 (and references therein)
29. Thompson HO, Lee CO (1944) History, literature and theory of enteric coatings. J Am Pharm Assoc Sci Ed 34:135–138
30. Grönberg J (1919) Acta Med Scand 52:227–268
31. Thompson HO, Lee CO (1944) A study of enteric coatings. J Am Pharm Asssoc Sci Ed 27:138–142
32. Jack D (1962) Prolongation of drug action. Pharm J 188:581–585
33. Riseman JF, Altman G, Koretsky S (1958) Nitroglycerin and other nitrates in the treatment of angina pectoris. Circulation 17:22–39
34. Russek HI, Zohman BL, Drumm AE et al (1955) Long-acting coronary vasodilating drugs: Metamine, Paveril, Nitroglyn and Peritrate. Circulation 12:169–175
35. Genung LB (1938) (Eastman Kodak) US Patent 2,123,460
36. Greminger GK et al (1959) (Dow Chemical Co) US Patent 2,887,440
37. Rowe RC (1984) Material used in the film coating of oral dosage forms. In: Florence AT (ed) Critical reports on applied chemistry, Materials used in pharmaceutical formulation. Blackwells, Oxford
38. Theeuwes F (1975) Elementary osmotic pump. J Pharm Sci 64:1987–1991
39. Rose S, Nelson JF (1955) A continuous long term injector. Aust J Exp Med 33:415–429
40. Theeuwes F, Higuchi T, Zaffaroni A, Michaels AS (1973) Belgian Patent 800,485
41. Theeuwes F, Higuchi T(1974) (Alza Corp) US Patent 3,845,770
42. Bjarnason I, Macpherson AJS (1994) Intestinal toxicity of non-steroidal anti-inflammatory drugs. Pharmacol Ther 62:145–157
43. The Committee on Safety of Medicines (Aug 1983) Osmosin (Controlled release indomethacin) Current Problems, No 11
44. Cree IA, Walker MA et al (1985) Osmosin and ileal ulceration: a case report. Scott Med J 30:40–41
45. Day TK (1983) Intestinal perforation associated with osmotic slow release capsules. Br Med J 287:1671–1672
46. Celin A (1984) Intestinal perforation associated with osmotic slow release capsules. Br Med J 288:240–241
47. Laidler P, Maslin SC, Gilhome RW (1985) What's new in Osmosin and intestinal perforation? Pathol Res Pract 180:74–76
48. Malaterre V, Ogorka J, Loggia N, Gurny R (2009) Oral osmotically driven systems: 30 years of development and clinical use. Eur J Pharm Biopharm 73:311–323

49. Santus G, Baker RW (1995) Osmotic drug delivery: a review of the patent literature. J Control Rel 35:1–21
50. Abrahams A, Linnell WH (1957) Oral depot therapy with a new long-acting dexamphetamine salt. Lancet 27:1317–1318
51. Chaudhry NC, Saunders L (1956) Sustained release drugs from ion-exchange resins. J Pharm Pharmacol 8:975–986
52. Elder DP (2005) Pharmaceutical applications of ion-exchange resins. J Chem Ed 82:575–587
53. Mahore JG, Wadher KJ et al (2010) Ion exchange resins: pharmaceutical application and recent advancement. Int J Pharm Sci Rev Res 1:8–13
54. Anon (1949) Prolonged penicillin action. Lancet 253:872–873
55. Romansky MJ, Rottman GE (1944) A method of prolonging the action of penicillin. Science 100:196–198
56. Romansky MJ, Rottman GE (1945) Penicillin blood levels for twenty four hours following a single intramuscular injection of calcium penicillin in beeswax and peanut oil. N Engl J Med 233:577–582
57. Monash S (1947) The use of bismuth, silver and mercury salts of penicillin for the prolongation of penicillin blood levels. J Invest Dermatol 9:157
58. Sullivan NP, Symmes AT, Millar HC, Rhodehamel HW Jr (1948) A new penicillin for prolonged blood levels. Science 107:169–170
59. Salivar CJ, Hedger FH, Brown EV (1948) Crystalline procaine penicillin. J Am Chem Soc 70:1287
60. Boger WP, Flippin HF (1949) Penicillin plasma concentrations. J Am Med Assoc 139:1131
61. Young MY, Andrews GWS, Montgomery DM (1949) Procaine penicillin. Lancet 253:863–865
62. Emery JL, Rose IM, Stewart SM, Wayne EJ (1949) The use in children of procaine penicillin with aluminium monostearate. Br Med J 1:1110–1112
63. Thomas EW, Lyons RH et al (1949) Newer repository penicillin products. J Am Med Assoc 137:1517–1519
64. Crommelin DJA, de Blaey CJ (1980) *In vitro* release studies on drugs suspended in non-polar media. I. Release of sodium chloride from suspensions in liquid paraffin. Int J Pharm 5:305–316 (see also idem. II. The release of paracetamol and chloramphenicol from suspensions in liquid paraffin. Int J Pharm 6:29–42
65. Szabo JL, Edwards CD, Bruce WF (1951) N.N' dibenzylethylene diamine penicillin: preparation and properties. Antibiot Chemother 1:499
66. Seifter J, Glassman JM, Begany AJ, Blumenthal A (1951) Antibiot Chemother 1:504–508
67. Minto CF, Howe C et al (1997) Pharmacokinetics and pharmacodynamics of nandrolone esters in oil vehicle: effects of ester, injection site and injection volume. J Pharmacol Exp Ther 281:93–102
68. Junkman K (1957) Long acting steroids in reproduction. Recent Progr Horm Res 13:380–419
69. Kelly M (1947) Failure of oil-soluble anaesthetics to give prolonged analgesia. Lancet 249:710–711
70. Larsen SW, Thomsen AE, Rinvar E, Friis GJ, Larsen C (2001) Effect of drug lipophilicity on in vitro release from oil vehicles using nicotinic acid esters as model prodrug derivatives. Int J Pharm 216:83–93
71. Curry SH, Whelpton R, de Schipper PJ, Vranckx S, Schiff AA (1979) Kinetics of fluphenazine after fluphenazine dihydrochloride, enanthate and decanoate administration to man. Br J Clin Pharmacol 7:325–331
72. Larsen SW, Larsen C (2009) Critical factors influencing the *in vivo* performance of long-acting lipophilic solutions- impact on *in vitro* release method design. AAPS J 11(4): 762–770
73. Honrath WL, Wolff A, Meli A (1963) The influence of the amount of solvent (sesame oil) on the degree and duration of action of subcutaneously administered testosterone and its propionate. Steroids 2:425–428

74. Deanesly R, Parkes AS (1937) Factors affecting the effectiveness of administered hormones. Proc R Soc B Med 124:279–298
75. Deanesly R, Parkes AS (1943) Further experiments on the administration of hormones by the subcutaneous implantation of tablets. Lancet 242:500–502
76. Noble RL (1938) Effect of synthetic oestrogenic substances on the body growth and endocrine organs of the rat. Lancet 232:192–195
77. Bishop PMF (1938) A clinical experiment in oestrin therapy. Br Med J 1:939–941
78. Parkes AS (1942) Administration of non-steroid substances by the implantation technique. J Endocinol 3:220–233
79. Folkman J, Long DM (1964) The use of silicone rubber as a carrier for prolonged drug therapy. J Surg Res 4:139–142
80. Dzuik PJ, Cook B (1966) Passage of steroids through silicone rubber. Endocrinology 78:208
81. Mischell DR (1975) Intravaginal rings for contraceptive use. An editorial comment. Contraception 12:249–251
82. Hutchinson FG, Furr BJA (1990) Biodegradable polymer systems for the sustained release of polypeptides. J Control Rel 13:279–294
83. Furr BJA, Hutchinson FG (1992) A biodegradable delivery system for peptides; preclinical experience with the gonadotrophin-releasing hormone agonist Zoladex™. J Control Rel 21:117–128
84. Parkes AS (1938) Effective absorption of hormones. Br Med J 1:371–373
85. Anon. (1941) Absorption of implanted hormone tablets. Lancet 238:318
86. Dang W, Daviau T, Ying P et al (1996) Effects of Gliadel wafer initial molecular weight on the erosion of wafer and release of BCNU. J Control Rel 42:83–92
87. Langer R (1983) Implantable controlled release systems. Pharm Ther 21:35–51
88. Li Y, Shawgo RS, Tyler B et al (2004) In vivo release from a drug delivery MEMS device. J Control Rel 100:211–219
89. Grüntzig A, Senning A, Siegenthaler WE (1979) Nonoperative dilatation of coronary-artery stenosis: percutaneous transluminal coronary angioplasty. New Engl J Med 301:61–68
90. Muller DWM, Topol EJ (1991) Implantable devices in the coronary artery. Trends Cardiovasc Med 1:225–232
91. Anon. (1949) Aerosols by inhalation. Lancet 254:660
92. Rooth G (1949) Inhalation of liquid aerosols. Acta Med Scand 133(Suppl. 228)
93. Beigel H (1865) On the inhalation of atomized fluids. Lancet 86:35–36
94. Beigel H (1865) On the inhalation of atomized fluids. Lancet 86:114–116
95. Mutch N (1944) Inhalation of chemotherapeutic substances. Lancet 244:775–780
96. Heubner W (1920) Z ges Exp Med 10:35
97. Hückel R (1925) Z ges Phys Ther 30:317
98. Anon. (1984) The Nebuliser epidemic. Lancet 324:789–790
99. Edwards DA, Hanes J, Caponetti G et al (1997) Large porous particles for pulmonary delivery. Science 276:1868–1872
100. Anon. (1929) Atopham administered by inunction. Lancet 2:413–414
101. Rothman S (1943) The principles of percutaneous absorption. J Lab Clin Med 28:1305–1321
102. For example: Telfer SV (1923) The administration of insulin by inunction. Br Med J 1:715
103. Major RH, Delp M (1936) Cutaneous absorption of insulin. Proc Soc Exp Biol Med 35:429
104. Zondek B, Bromberg YM, Shapiro B (1942) Percutaneous application of sulfonamides in animals and men. Proc Soc Exp Biol Med 50:116
105. Hadgraft J, Lane M (2005) Skin permeation: the years of enlightenment. Int J Pharm 305:2–12
106. Shelmire JB (1958) Observations on the role of vehicles in percutaneous absorption. AMA Arch Derm 78:191–195
107. Sarkany I, Hadgraft JW, Caron GA, Barret CW (1965) The role of vehicles in the percutaneous absorption of corticosteroids. Br J Dermatol 77:569–575

108. Herrmann F, Salzberger MB, Baer RL (1942) New penetrating vehicles and solvents. Science 96:451–452
109. Gemmell DHO, Morrison JC (1957) The release of medicinal substances from topical applications and their passage through the skin. J Pharm Pharmacol 9:641–656
110. Smith CC (1955) Failure to demonstrate absorption of hydrocortisone when topically applied to the human skin. Ann NY Acad Sci 61:517–519
111. Malkinson FD, Ferguson EH (1955) Percutaneous absorption of hydrocortisone-4-C^{14} in two human subjects. J Invest Dermatol 25:281–283
112. Leduc S (1900) Introduction des substances medicamenteuses dans la profondeur des tissus par le courant electrique. Ann d'Electrobiol 3:545–560
113. Erlanger G (1929) Iontophoretic medication in ophthalmology. Arch Phys Ther 20:16
114. Fleming N (1943) Iontotherapy (Ionic medication, iontophoresis, ionization) as an aid in ophthalmic therapeutics. Br J Ophthalmol 27:354–360
115. Deutsch D (1934) Device for inserting medicaments into the body by iontophoresis. US Patent 1,967,927
116. Abramson HA, Engel MG (1942) Skin reactions. XII. Patterns produced in the skin by electrophoresis of dyes. Arch Dermatol Syph 44:190
117. Clark WG, Strakosch EA, Nordlum C (1942) Penetration of sulfonamides through intact skin by iontophoresis and other means of local application. Proc Soc Exp Biol Med 50:43
118. Hoogstraate AJ, Srinivasan V, Simms SM, Higuchi WI (1994) Iontophoretic enhancement of peptides: behaviour of leuprolide versus model compounds. J Control Rel 31:41–47
119. Guy RH, Kalia YN, Delgado-Charro MB et al (2000) Iontophoresis: electrorepulsion and electroosmosis. J Control Rel 64:129–132
120. Yang RK et al (2008) (Monosol Rx) Process for making an ingestible film. US Patent 7,357,891
121. Zaffaroni A (1971) (Alza Corporation). Bandage for administering drugs. US Patent 3,598,122
122. Zaffaroni A (1971) (Alza Corporation) Bandage for administering drugs. US Patent 3,598, 123
123. Gerstel MS, Place VA (1976) Drug delivery device, US Patent 3,964,482
124. Henry S, McAllister D, Allen MG, Prausnitz MR (1998) Microfabricated microneedles: a novel method to increase transdermal drug delivery. J Pharm Sci 87:922–925
125. Prausnitz MR (2004) Microneedles for transdermal drug delivery. Adv Drug Deliv Res 56:581–587
126. Sedlácek J (1965) Possibility of the application of ophthalmic drugs with the use of gel contact lenses. Ceskoslovenska oftalmologie 21:509–512
127. Wichterle O, Lim D (1960) Hydrophilic gels in biological use. Nature 185:117–118
128. Ness RA (Alza Corporation) Ocular insert, US Patent 3,618, 604
129. Armaly MF, Rao KR (1973) The effect of pilocarpine Ocusert with different release rates on ocular pressure. Invest Ophthalmol 12:491–496
130. Wadleigh JC (1883) Gelatine eye-disk. US Patent 273,410
131. Chang TMS (1964) Semipermeable microcapsules. Science 146:524–525
132. Orive G, Hernández RM et al (2004) History, challenges and perspectives of cell microencapsulation. Trends Biotechnol 22:87–91
133. Brynko C, Scarpelli JA (1961) (National Cash register Company) Process of making dual-walled oil-containing capsules. US Patent 2,969,331
134. Figge FHJ et al (1969) (AMP Inc) Method of administering therapeutic agents. USP 3,474,777
135. Kitajama M et al (1972) (Fuji Photo Film Co. and Toyo Jozo Co.) Encapsulation method. US Patent 3,691,090
136. Kitajima M, Kondo A, Arai F (1976) (Fuji Photo Film Co.) Process of producing aspirin-containing capsules. US Patent 3,951,851
137. Kramer PA (1974) Albumin microspheres as vehicles for achieving specificity in drug delivery. J Pharm Sci 63:1646–1647
138. Eckman B, Sjöholm I (1975) Use of macromolecules in microparticles. Nature 257:825–826

139. Kato T, Nemoto R, Mori H, Kumagai I (1978) Microencapsulated mitomycin C therapy in renal cell carcinoma. Lancet 314:479–480
140. see also Kato T, Nemoto R (1978) Preparation and properties of microencapsulated mitomycin C. IRCS Med Sci 6:311
141. Lee TK, Sokolski TD, Royer GP (1981) Serum albumin beads: an injectable biodegradable system for sustained release of drugs. Science 213:233–235
142. Okada H et al (1987) Prolonged release microcapsule production, US Patent 4,652,441
143. Okada H, Heya T, Ogawa Y, Shimamoto T (1988) One-month release injectable microcapsules of a luteinizing hormone-releasing hormone agonist (leuprolide acetate) for treating experimental endometriosis in rats. J Pharmacol Exp Ther 244:744–750
144. Okada H, Doken Y, Ogawa Y, Toguchi H (1994) Preparation of three-month depot injectable microspheres of leuprorelin acetate using biodegradable polymers. Pharm Res 11:1143–1147
145. Chen Y, Willmott N, Anderson J, Florence AT (1988) Haemoglobin, transferrin and albumin/polyaspartic acid microspheres as carriers for the cytotoxic drug adriamycin. I. Ultrastructural appearance and drug content. J Control Rel 8:93–101
146. Bungenberg de Jong BH (1949) Crystallisation-coacervation-flocculation. In: Kruyt HR (ed) Complex colloid systems. Colloid Science, Vol 2. Elsevier, Amsterdam, pp 232–255, 335–429
147. Burgess DJ, Carless JE (1985) Manufacture of gelatin/gelatin coacervate microcapsules. Int J Pharm 27:61–70
148. Dapper T, Thies C (1978) Statistical models for controlled release microcapsules: rationale and theory. J Membr Sci 4:99–113
149. Kreuter J (2007) Nanoparticles: a historical perspective. Int J Pharm 331:1–10
150. Speiser P (1976) Controlled release of drugs from microcapsules and nanocapsules. Acta Pharm Suec 13(Suppl):35
151. Couvreur P, Kante B, Roland M et al (1979) Polycycanoacrylate nanocapules as potential lysosomotropic carriers: preparation, morphological and sorptive properties. J Pharm Pharmacol 31:331–332
152. Hanes J, Cleland JL, Langer R (1997) New advances in microsphere-based single-dose vaccines. Adv Drug Deliv Rev 28:97–119
153. Higuchi T (1963) Mechanism of sustained-action medication. Theoretical analysis of rate of release of solid drugs dispersed in solid matrices. J Pharm Sci 52:1145–1149
154. Higuchi WI (1967) Diffusional models useful in biopharmaceutics. Drug release rate processes. J Pharm Sci 56:315–324
155. Singh P, Desai SJ, Simonelli AP, Higuchi WI (1967) Release rates of solid drug mixtures dispersed in inert matrices: I. Non-interacting drug mixtures. J Pharm Sci 56:1542–1547
156. Peppas N (1984) Mathematical modelling of diffusion processes in drug delivery polymeric systems. In: Smolen VF, Ball L (eds) Controlled drug bioavailability, vol 1. Wiley, New York, pp 203–237

Chapter 2
The Organization of the Gut and the Oral Absorption of Drugs: Anatomical, Biological and Physiological Considerations in Oral Formulation Development

Clive G. Wilson

Abstract Oral drug delivery remains the mainstay of patient treatment although the candidate drugs of the new millennium are becoming increasingly difficult to formulate for good systemic absorption. The area of oral delivery therefore represents an important area of innovation for pharmaceutical formulation including modulating solubility, exploiting windows of absorption and increasing bioavailability in a robust manner to attempt a more predictable outcome.

In order to deliver an active pharmaceutical ingredient to facilitate systemic exposure, the drug must be presented in a dosage unit that contains an accurate dose of a specified active pharmaceutical ingredient which remains intact to the point of administration. On dosing, the pharmaceutical phase must be undone appropriately: the drug must be liberated at the correct rate, escaping degradation and metabolism and reach sufficient concentrations in the target tissue. The exposition of the pharmacist's art is then completed in the lumen of the gut and therefore an understanding the organization of the organ system, at a macroscopic level, is of great relevance. In this chapter, the general integration of anatomy and motility with regard to the interaction of the dosage form will be considered. The biochemical and biophysical elements of absorption of the drug substance will not dealt with in detail in this chapter but by other books in this series.

2.1 Background

The gut is primarily designed for the absorption of nutrients which are presented in a complex and varied matrix comprising protein, carbohydrate, fat, minerals and vitamins in different proportions. The components must be extracted by batch

C.G. Wilson (✉)
Institute of Pharmacy & Biomedical Sciences, University of Strathclyde,
Glasgow, Scotland, UK
e-mail: c.g.wilson@strath.ac.uk

C.G. Wilson and P.J. Crowley (eds.), *Controlled Release in Oral Drug Delivery*,
Advances in Delivery Science and Technology, DOI 10.1007/978-1-4614-1004-1_2,

processing, which involves fluid secretion of liquids providing an optimum milieu for the enzymes to work in a controlled sequence. If a foodstuff is energy rich but difficult to process, motility must be slowed to allow presentation at an appropriate rate with mixing patterns predominating over propulsive activity during digestion. This has to be achieved in the proximal regions of the gut, particularly the jejunum and ileum. At the end of the ileum, secretion is lower and assimilation is the main physiological activity. Finally in the colon water, salts and remaining nutrients must be extracted to conserve the ionic balance of cellular fluids.

The gut of mammals evolved into specialist herbivores, fairly inefficient carnivores and balanced omnivores who were able to take advantage of high calorific densities in flesh and nuts by processing in the fore-gut and to extract significant nutrients from pulverized and enzyme treated vegetables using bacterial populations of the hind-gut. This diversity required a range of enzymes to be available and control of exposure to allow efficient processing. The early diet contained seeds from berries which were poisonous, and nature has preserved protective functions throughout evolution of the mammals to man. Thus we recognize poisonous alkaloids as bitter by taste and have several protective mechanisms to avoid toxin exposure including, in the last resort, vomiting.

Our earliest medicines were derived from plant stuff, and of varied potency. The poor analytical techniques hampered quality control and thus the medicines were dangerous to use. The replacement of plant extracts by chemically synthesized drugs, which were obtained at high purity, and were single entities, made the materials easier to use as pharmacons. Doses of the chemically derived drugs could be relatively large (those which were more potent and hard to detect were still commonly referred to by the public as poisons) and although knowledge of the importance of hepatic metabolism and renal excretion was well established 60 years ago, we knew little of more subtle defense mechanisms. As pharmacological knowledge was refined and medicines became more potent, scientists became aware of protection at the prehepatic, intestinal level including efflux and drug metabolizing systems, which attempt to avoid exposure to xenobiotic materials.

These comments emphasize a couple of important principles which must be always considered. First, the gut is designed to process food and some component of the drug's absorption profile is likely to be affected by the sequence of meals. Second, if the drug concentration is sufficiently low, it may be processed by the protective guardians that reduce exposure.

The basic design of the gut is a long muscular tube with specialized areas for digestion and storage. The plan of the gut is illustrated in Fig. 2.1. As shown, the gut is a long tube supplied by arteries and drained by veins and a lymphatic trunk, all of which are supported in a mesentery, which are folds of the peritoneum attached to the abdominal wall. The small intestine is the major site of nutrient and anutrient absorption. Although uptake occurs in stomach tissue, the contribution of direct gastric absorption to bioavailability is small, and slow delivery into the upper gastrointestinal tract is far more important.

In adults the length of the gut is approximately 7 m and the large intestine 1.5 m in length. Differences in length are apparent at death, when inherent tone is lost.

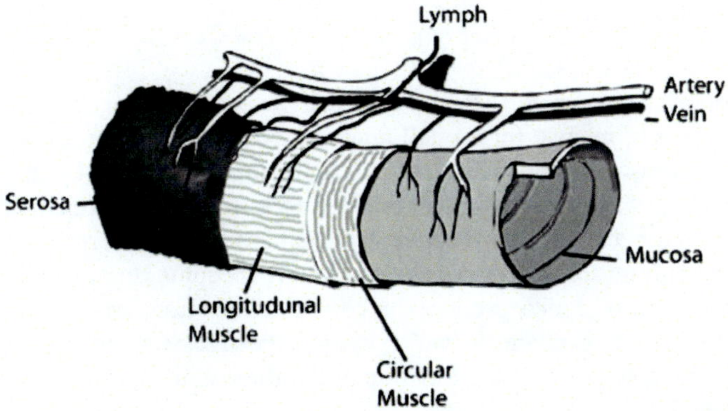

Fig. 2.1 Illustration of the plan of the gastrointestinal tract showing arrangement of mucosa and muscles

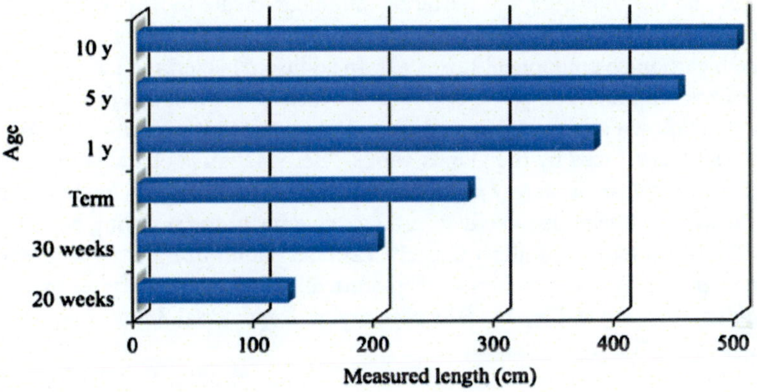

Fig. 2.2 The growth of the human intestine. Measurements made at necropsy. From data of [1]

A study of 1,010 small intestines at autopsy by Weaver, Austin and Cole was used to construct the data shown in Fig. 2.2 describing the growth of the small intestine to adulthood [1].

Functionally, the gut is divided into a preparative and primary storage region (mouth and stomach), a secretory and absorptive region (the midgut), a water reclamation system (ascending colon) and finally a waste-product storage system (the descending and sigmoid colon). The whole structure loosely fills the abdomen, with the esophageo-gastric junction just below the diaphragm. The pyloric sphincter area and the cardia provide points of attachment and help fix the ends of the stomach; however, when posture changes or the stomach is filled with food, organs such as the stomach can change shape and therefore their position in the abdomen. This generates potential differences in emptying patterns in supine, prone and upright positions.

2.2 Buccal Delivery

The first port of call to consider in oral drug delivery is the buccal cavity, and buccal delivery remains of interest for a small range of drugs used for cardiovascular control, smoking cessation and pain control. The primary function of the mouth is guarding of the gut by moistening the food to a soft, shaped bolus: the mucosa must therefore be tough and act as protective layer rather than an absorptive membrane. In areas of maximum abrasive stress, the mucosa will become keratinized. Prolonged exposure to tobacco smoke produces excess keratinization, as does poor dental hygiene. The water inlet channels, which hydrate the digesta, must have high capacity and react instantly: this is the function of the three main sets of glands assisted by minor glands.

Saliva is a viscous, watery fluid which is hypo-osmotic compared to plasma. One to two liters are discharged every day into the mouth and the composition and pH varies with the rate of secretion as illustrated in Fig. 2.3. The pH as shown varies from 7.4 and 6.2; however, the bacterial action can create local pockets where the pH falls below 5 and the tooth enamel starts to demineralize. Saliva acts as a diluent and the bicarbonate component raises pH. In addition, bacteria in the dental plaque metabolize components in saliva and raise the local pH: when this protection is lost a condition known as *xerostomia*, a diffuse and severe caries results.

The saliva produced by the glands varies. "Serous" saliva contains more protein particularly amylase, is watery and subserves the sense of taste by beginning digestion; the saliva stream also needs to produce mucins to resist drying at rest and to lubricate the structures to allow speech. Taste sensation in the tongue, palate and upper esophagus provide an input to the brain allowing involuntary responses such as gagging, retching and excess salivation to remove material. In the dog, the mouth is also used for thermo-regulation.

The classical routes of buccal delivery are summarized in Fig. 2.4 and specific examples are given in Chap. 16. The access to saliva, the variation in patterns of keratinization and squamous cell thickness, and the abrasive forces associated with speech and chewing are important factors in variation in performance.

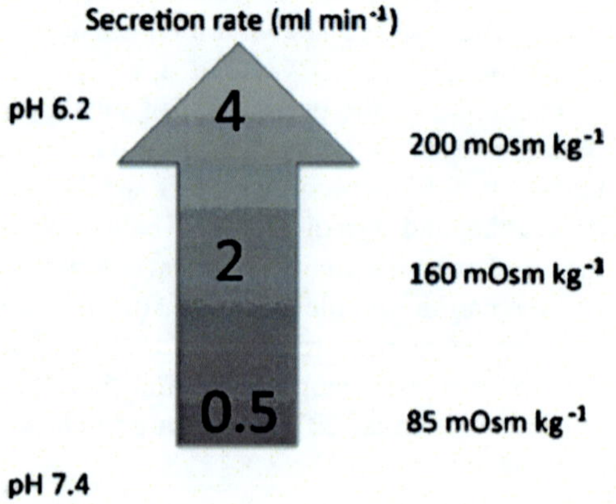

Fig. 2.3 The change in saliva pH and osmolality with increasing flow

Fig. 2.4 Buccal routes of delivery

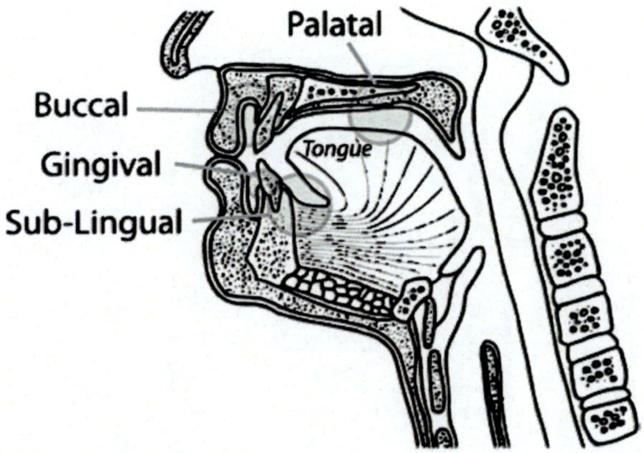

The mucosa or inner lining of the mouth is divided into four zones. The first part of the gut has a lining of the squamous epithelium which extends from the mouth to the stomach. The many layers of cells are analogous to dermal tissue and drugs will only penetrate if residence is prolonged. The exception is the tissue under the tongue, as used in sublingual delivery, where the epithelium is thin. The vessels of the face drain directly to the heart and thus avoid the hepatic portal system, which provides a number of obvious advantages.

An important property is mouth feel and taste, since the released drug will be in intimate contact with the tongue. Variability in performance may be associated with changes in saliva flow and movements of the mouth when talking. The marked variation in the thickness and keratinization of the epithelial lining is also exaggerated in rodents, and pig and dog are more suitable models for human buccal tissue. The characteristics of buccal delivery are summarized in Fig. 2.5.

2.3 The Esophagus

The esophagus is approximately 40 cm long in the adult, passing through the diaphragm at approximately 38 cm. The surface of the esophagus is a squamous epithelium with a protective function as in the mouth and has few if any glands. The morphology changes sharply at the junction with the stomach into secretory epithelium.

After the dosage form leaves the buccal cavity, movement through the esophagus is normally complete within 10 s. The voluntary maneuver is handed over to a complex autonomic sequence in the cricopharynx, followed after swallowing by short secondary peristaltic waves, which serves to attempt to clear the esophagus. The efficiency of clearance may be influenced by several factors, including the outside surface of the dosage form, the age of the subject and pre-existing disease. Conditions such as type 1 diabetes reduce the amplitude of peristaltic waves and further exacerbate the problems of esophageal clearance, particularly for solid swallows [2]. The elderly often report problems in attempting to swallow

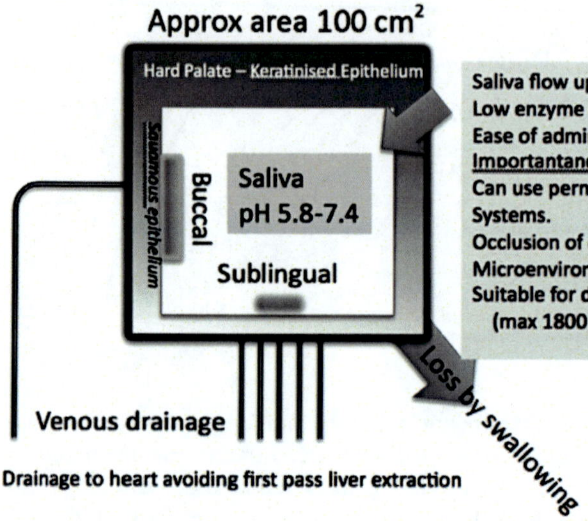

Fig. 2.5 General plan of buccal physiology. Note that the tissues at the top of the mouth are much less permeable than the sublingual area. Buccal systems are used along the gum margin and cheeks and are generally sustained delivery systems, whereas sublingual systems are fast release, as they cannot be anchored

large objects, in part influenced by previous unsuccessful attempts but influenced by the increased stiffness and lower muscle compliance. The elderly have little "swallowing reserve" but experience fewer problems in clearing a liquid bolus compared to a solid mass. It is a common practice in nursing homes to crush medications for dysphagic patients, despite the fact that controlled release formulations are specifically designed not to be damaged prior to ingestion. Although, large tablets are commonly identified as problematic, small flat and buoyant dosage forms are particularly likely to cause problems in the elderly because of the inability to complete the swallowing maneuver.

The coating of tablets to identify the product, to protect the integrity of the dose or to mask bitterness or appearance is a principal activity in the manufacture of oral formulations. The film coat can be functional as for enteric release products or esthetically pleasing and the mouth feel emphasizes the "swallowability" of the product. Channer and Virjee (1985) showed that the clearance of plain, sugar-coated, enteric-coated and film coated tablets in 34 patients was strongly influenced by coating and by posture [3]. The authors reported 100% clearance of film coated tablets in 13 s; for the plain uncoated formulation full clearance was observed in only 60% of subjects at this time. The findings also confirmed their earlier report that oval coated tablets showed the fastest esophageal transit in the erect position, even when swallowed with low volumes of water [4]. A recent interesting article nicely illustrates the importance of shape factors and organoleptic issues on the swallowing of large dosage forms [5].

2.4 The Stomach

The gut contains two reservoirs, in which the tube structure of the gut is modified to accommodate gut contents for longer periods of time. The first, the stomach, allows a regulated supply of calories to the small intestine by control of rate of emptying according to food type. The arrangement of the human gut is illustrated in Fig. 2.6, with the stomach sitting below the diaphragm, nestled by lobes of the liver (removed from the illustration) and the greater curvature of the stomach placed just above the transverse colon. The position of the cardia and the pyloric sphincter are usually fixed but as the stomach is filled, the fundus changes shape by receptive relaxation and on lying down, the proximal stomach falls into the abdomen cavity remaining lower than the distal stomach.

The stomach is lined by a secretory epithelium which is covered by a thick, relatively impermeable layer of gastric mucus. This is the second type of mucosal structure with a longitudinal cell structure; however, the tightness of the intercellular junctions restricts significant passive diffusion, even for small well-absorbed molecules such as ethanol. At the epithelial surface, cells secrete bicarbonate such that a pH gradient is created across the strongly adherent mucus, produced by goblet cells.

2.4.1 Gastric pH

In compendial terms, the pH of the stomach contents is mimicked as a hydrochloric acid solution of 1.0, 1.2 or 1.8. At rest, the stomach pH varies: in a large study of

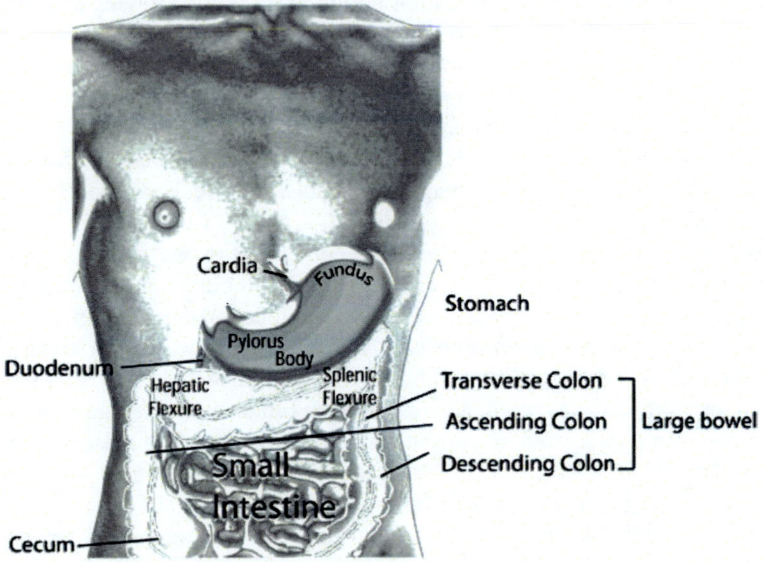

Fig. 2.6 Diagram of the features of the gastrointestinal tract showing location in the abdomen

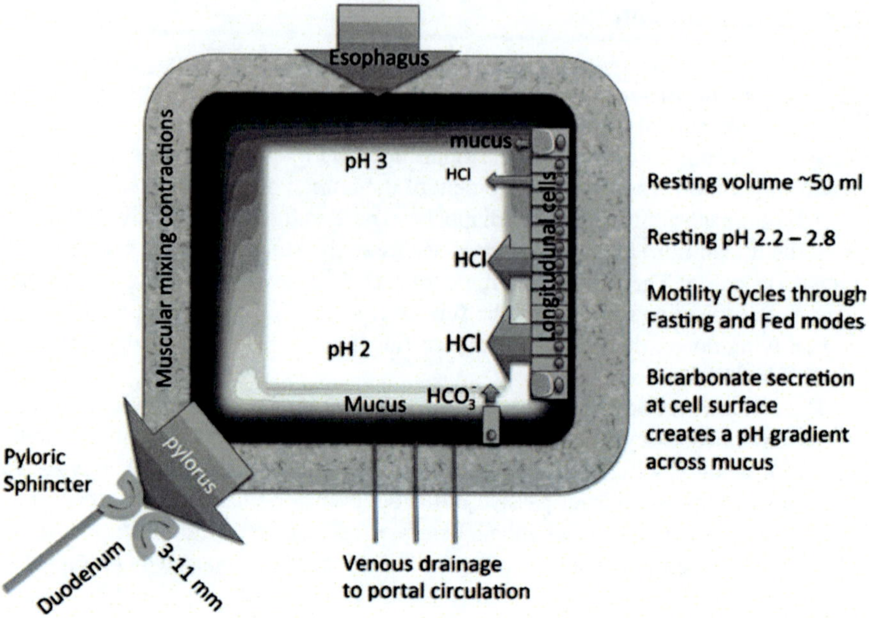

Fig. 2.7 Key gastric features

685 volunteers, Feldman and Barnett reported that the median basal pH for females was 2.79±0.18 and that for males was 2.18±0.18 [6]. In adults, the population of parietal cells is decreasing which will lead to elevated pH in the elderly. As the drugs encountered for oral medication are often weak electrolytes and many are salts of bases, the pH change between the stomach and the small intestine will exert important effects on systemic exposure. Moreover, differences will emerge when considering absorption from the fasted compared to the fed state. If a medication is taken with water, the pH will be elevated temporarily by dilution, returning to baseline at around 20 min post-imbibing. Intake of the meal will cause acid secretion but meal components dictate the magnitude of the response. Food processing acts as a sustained delivery mechanism regulating the supply of materials by controlling gastric emptying. Backwash of duodenal contents into the stomach will cause decrease in surface tension, which can further aid solubilization, which may either subsequently increase absorption or the rate of compound degradation.

The change in pH environment of the upper gastrointestinal tract is very important in oral drug delivery, although the import of regional variations both within the stomach as an organ and between the lumen and unstirred water layer is sometimes not appreciated. A current research direction is the preservation of the super-saturated state to avoid precipitation, particularly of bases, on change of media from the gastric milieu to intestinal fluid.

The pH gradients within an organ and between the lumen and the unstirred water layer next to epithelium can vary by at least a pH unit (Fig. 2.7). In the stomach, such differences are very large as the parietal cell mass decreases in the fundus raising the pH. In the stomach, such differences are very large as the parietal cell mass decreases in the fundus raising the pH.

The volume of the stomach swells by relaxation of the fundus to accommodate a meal and food layers without significant mixing if the viscosity is high enough. The resting volume is very low – around 50–100 ml but intake of food causes it to relax to accommodate between 1 and 1.5 L. The maximum volumes recorded are around 4 L in man.

2.4.2 *Regulating Gastric Emptying*

The duodenum regulates the supply of material from the stomach to the small intestine. Fat, high salinity and highly acid solutions cause the duodenal wall pressure to increase and slow down the exit of the gastric contents. Because the gastroduodenal system regulates the exit of the slurried contents from the stomach, the transit time from duodenum to the caecum is relatively constant. The diameter of the pyloric opening varies according to the nature of the gastric contents. When taken with water in a fasted individual, the time of tablet emptying will be highly variable. Tablets will be emptied at various times after ingestion according to posture, volume of fluid taken and the calorific value of food taken before or with the dosing. If they disintegrate and dissolve, pulses of material will appear regularly in the small intestine at a rate determined by the meal, with the rise to intestinal pH (Fig. 2.8).

Pellets and disintegrated dosage forms empty from the stomach as either a series of pulses when fasted or distributed in the meal when fed [7]. The emptying of pellets is much more predictable in the fasted state as illustrated in Fig. 2.9. Tablets that remain intact will empty at very variable times when fasted but eating a light meal reduces the variability in emptying as illustrated.

Large tablets will stay in the stomach for prolonged periods of time especially in the more elderly subject where the laxity in stomach becomes predominant.

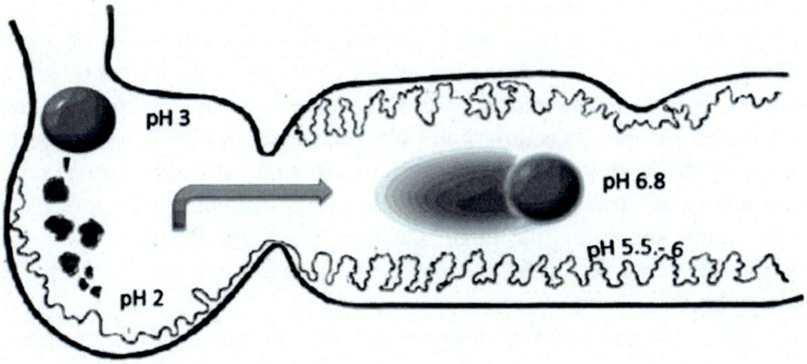

Fig. 2.8 pH and motility in the upper gastrointestinal tract

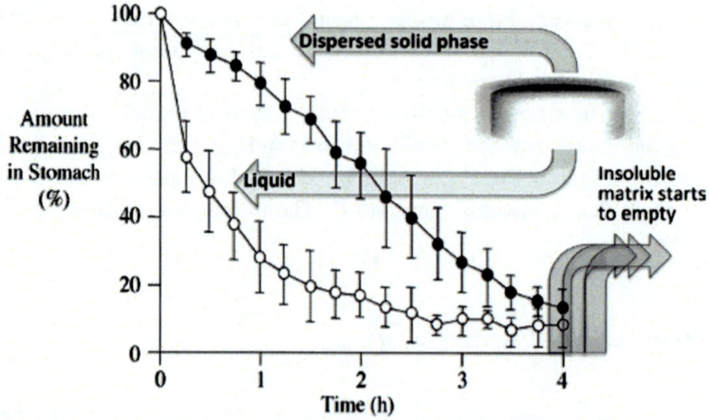

Fig. 2.9 Emptying of tablet components with a meal. Dissolving drug will follow the liquid emptying curve, disintegrated API the disperse phase. Any large fragments or intact tablets will exit with the housekeeper sequence

When food is taken, this discrimination is more extreme as the effective diameter of the pylorus decreases and large tablets are retropulsed back into the body of the stomach at the end of a gastric contraction. Smaller particulates are emptied in the mass of the food and the presentation of the dose in a dispersed system is a function of calorific load and mass of the gastric contents [8]. If food is eaten throughout the day after a heavy breakfast and subsequent meals, then in some individuals a conventional enteric-coated ibuprofen will remain intact up to the end of the day having neither disintegrated nor emptied [9].

Drug, ejected with chyme from stomach will be absorbed in the first highly permeable part of the intestine; however, transit through this region occurs rapidly. Thus, whilst high drug absorption can be demonstrated under in vitro conditions, it is most probable that duodenal absorption occurs when the dose remains in the body of the stomach. When recumbent, the fundus or first part of the stomach is positioned lower in the abdomen than the pyloro-duodenal sphincter. As a consequence, drug released in the upper stomach may not appear in the systemic circulation until a postural movement allows flow through to the distal stomach and out into the intestine.

Once the intake of food stops and blood sugar and free fatty acids decrease, the "housekeeper sequence" (migrating myoelectric complex) is initiated which serves to remove debris. This powerful peristaltic wave causes powerful contractions against an open pylorus. In scintigraphy studies, in young people who have fasted, this is evident about 2 h postdosing (i.e. around 10 a.m.). As tablets travel down the gut, the movement slows and periods of stasis are common just before the tablet leaves the ileum and enters the large intestine. Eating food later on will cause a gastrocolic reflex (see Sect. 2.6.2), enabling the contents to move from small intestine to large bowel.

This mechanism, colloquially known as the housekeeper sequence or more properly as the migrating motor complex (MMC) can be recorded externally with

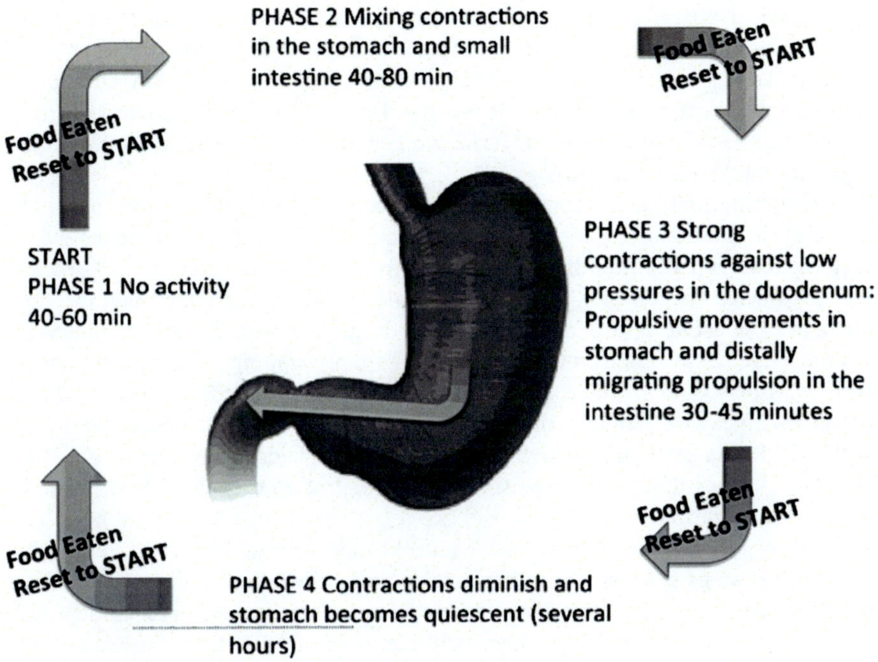

Fig. 2.10 The migrating myoelectric complex or housekeeper sequence

electrodes on the abdomen (Fig. 2.10). This was first described in the literature following the first experiments described by Code and Martlett and by Bull and colleagues [10, 11]. The mixing activity gives way to strong propulsive waves, which migrate through the small intestine. The strong contractile activity during phase III of the MMC is an important factor limiting the retention of dosage forms, but the cycle may be interrupted and reset by the intake of food.

2.4.3 Utilization of Upper Windows of Absorption

Exploitation of areas of the gastrointestinal tract where maximum absorption occurs by deliberate formulation efforts has been attempted using several mechanisms including ligand association, bioadhesion and physical properties (gastroretention). It was thought that prolonged positioning of a dose form within a specific area of the intestine would allow exploitation of transporter populations expressed differentially along the length of the gut. Although binding to the small intestine in vitro is readily achieved using a variety of ligands, the motility and pattern of lumenal flow may restrict access, except in the distal gut during periods of stasis. The relative importance of pH effects versus differential transporter expression has often been a subject of debate. Woodley has commented that using everted sacs, his group has noted marked differences in the pattern of absorption of xenobiotic compounds along the gastrointestinal tract but he suspected that so-called "windows of absorption" are largely a phenomenon related to solubility and pH [12]. Any attempt to modulate the point or time of release is, however, still a useful endeavor as it may result in increased patient benefits. It relies on construction of a dosage form utilizing a controlled release technology.

A few important drugs show an apparent window of absorption, with best permeability in the duodenal segment. Since transit through this region of the gut is very rapid – typically less than 5 min – the formulator must attempt to keep the delivery device for a prolonged period of time in the stomach such that the first segment of intestine is continually bathed in drug. A simple test was proposed to test the usefulness of gastroretentive devices – simply sipping a formulation over a prolonged period to examine a change in the pharmacokinetic parameters [13]. Lewis describes an example of this maneuver and compared the exposure following an oral IR formulation of acyclovir with a solution of the drug sipped over 4 h [14]. The $AUC_{0\text{-inf}}$ was doubled for the sipping administration compared to the simple tablet administration.

Floating systems are therefore most successful if the patient is fed and upright. A general strategy is to float or to expand due to the liberation of gas into a gelling structure such as alginic acid. There are problems if the subject is recumbent and turns onto the left as the floating layer will empty out of the stomach ahead of the rest of the gastric contents [15].

2.5 The Intestine

The third type of mucosa is the secretory/absorptive mucosa of the intestine, designed for the digestion of food and assimilation of smaller building blocks of fats, proteins and carbohydrates. Materials such as glucose, vitamins and essential amino acids must be actively scavenged from the intestine by active transport processes. Some drug absorption routes can utilize these pathways, such as valine-based prodrugs, but most drug absorption occurs at least in significant part by passive or facilitated diffusion (Fig. 2.11).

The small bowel is divided into three parts, the first 20–30 cm is termed the duodenum, the second 2.5 m the jejunum and the final 3.5 m the ileum. The mucosa of the small intestine has a surface area which is greatly increased by the folds of Kerckring, villi and microvilli (brush border) and is about 200 m^2 in an adult. The surface of the mucous membrane of the small intestine possesses about 5 million villi, each about 0.5–1 mm long. Although the villi are often described as "finger-like," their shape changes along the gut and duodenal villi are shorter and broader than those found in the jejunum. Further down the gut the villus height decreases. Diet and environment markedly affect mucosal morphology.

The epithelium, which covers the intestinal villi, is composed of absorptive cells, goblet cells, a few endocrine cells and tuft or calveolated cells. The absorptive cells or enterocytes are tall, columnar cells, with their nuclei located close to their base.

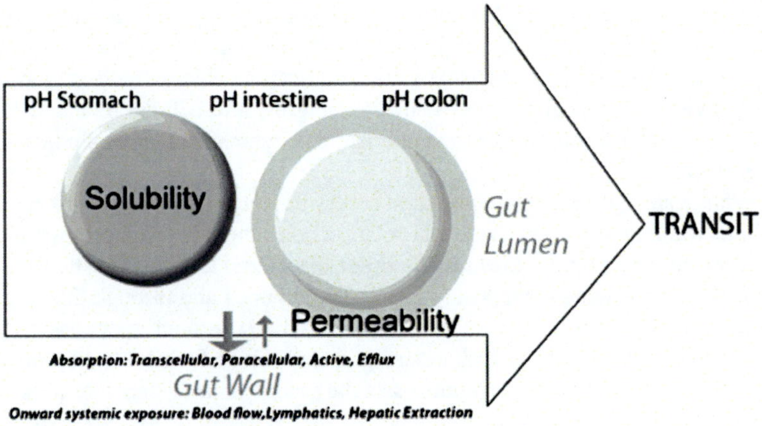

Fig. 2.11 Summary of the transit, solubility and permeability interactions

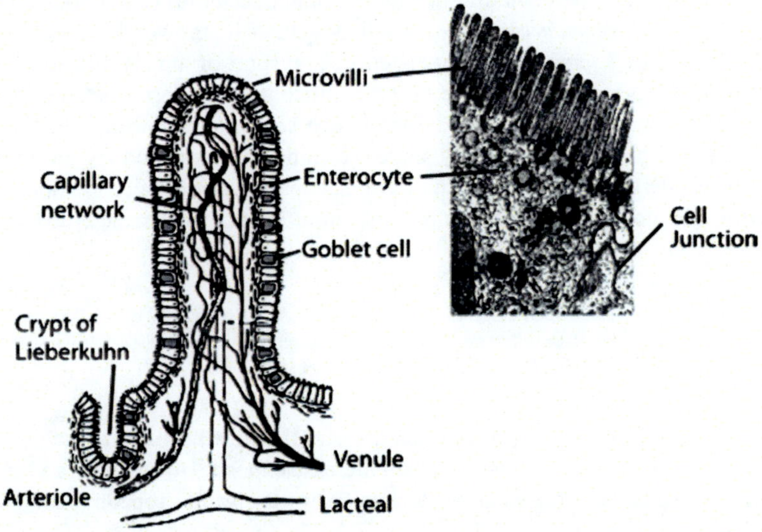

Fig. 2.12 Structure of intestinal villus, showing microstructure of villus surface and enterocyte junctions

The principal permeability barrier is represented by the luminal surface of the brush border, or microvilli as shown in Fig. 2.12. Most drugs are absorbed by passive diffusion in their unionized state. The pH of the small intestine determines the degree of ionization and hence controls the efficiency of absorption; this is the basis of the pH-partition theory of drug absorption. Protein binding at the serosal side of the epithelium helps maintain a concentration gradient by binding the absorbed drug, which is then removed by blood flow from the absorption site.

Between cells epithelial brush borders come into close contact and under the electron microscope it appears as if the membrane is fused. However, functionally the tight junctions are not sealed but are permeable to water, electrolytes and other charged or uncharged molecules up to a certain size. The size of the "pore" varies along the length of the gastrointestinal tract and can be calculated from recoveries of polyethylene glycols of various molecular weights. Intercellular transport may be important for oligosaccharides and small peptides, which is an area of considerable current interest.

There is a special mode of permeation across the intestinal wall in which the cell membranes are not involved. Intestinal cells are continuously produced in the crypts of Lieberkühn and migrate towards the tip of the villus. During digestion the cells are sloughed off leaving a temporary gap at the cell apex and through this gap large particles can slip into the circulation. This has been termed "persorption." The observation that large objects such as starch grains can be found in the blood after a meal of potatoes or corn is often quoted as the *prima facie* evidence of persorption or phagocytosis.

Although the absorption of most drugs can be explained by passive diffusion, some compounds have specific transport mechanisms. An example is the absorption in the intestine of some penicillin derivatives, e.g. cyclacillin (1 aminocyclohexyl-penicillin). This process is saturable, proceeds against an unfavorable concentration gradient and shows temperature dependence. Transport of amoxycillin is also carrier mediated but it is not an active process. Since these materials are xenobiotics, the transport mechanism is probably one which serves some other function in the body. The two penicillins probably share the same carrier since they are mutually competitive. Digitalis and other cardioselective glycosides also demonstrate a behavior not compatible with simple partition theory which suggests carrier-mediated transport.

2.5.1 Movement of the Dosage form Along the Gut

Muscular contractions in the wall of the small intestine have to achieve two objectives: first, stirring of the contents to increase exposure to enzymes and second, to bring the lumenally digested products close to the wall, propelling indigestible material towards the distal gut. To accomplish this, movements of the gut consist of a mixture of annular constricting activity (segmentation) together with peristaltic movements, which are of both long and short propagation types. Measurements indicate that there are only small perturbations caused by meal components such as fat. Early emptying of partially digested lipid, initiated by gastric lipase and perhaps backwash of proximal intestinal contents into the stomach, initiates the ileal brake, which is discussed later in the book by Boyd and colleagues.

The pattern of movement through the small intestine was first nicely illustrated by the work of Lydia Kaus and the Manchester team of Fell, Taylor and colleagues [16]. Progress in this area has been facilitated by techniques including scintigraphy but more significantly by magnetic moment monitoring. Small intestinal transit of magnetized units in the small intestine is characterized as a series of rapid movements in the proximal intestine, becoming more quiescent in the distal intestine. Scintigraphy shows a plug flow of pulses through the small intestine, stopping occasionally and bunching of material at the ileocaecal junction with a reduction in the dispersed volume.

The extent to which shape controls gastrointestinal transit is important as illustrated for pellet and single unit emptying of the stomach; however, in the small intestine formulations appear to travel at approximately the same rate. Measurements indicate that there are only small perturbations caused by meal components such as fat. Early emptying of partially digested lipid, initiated by gastric lipase and perhaps backwash of proximal intestinal contents into the stomach, initiates the ileal brake. Following administration of a light meal, movement through the proximal gut is rapid and longer periods of stasis become evident as the formulation enters in the terminal ileum. Bunching of the formulation label is noticeable at the ileocaecal junction, immediately before entry into the caecum. Eating initiates propulsive activity and approximately 15 min after a meal, pulses of activity can be recorded in the sigmoid colon. Essentially, material is swept forward from the small intestine to clear a path for gastric effluent. The ability to small intestinal transit time remains annoyingly elusive and therefore attempts at extending the therapeutic time window after a single dose focus on gastric residence and retention in the ascending colon. Both factors appear to be influenced by bowel habit and in very young children, there is evidence that establishing a normal microflora has an action on gastric emptying and colic of the new born [17]. In nonulcer dyspepsia, constipation is a common observation [18].

2.5.2 *Intestine: Can Transit Be Modulated?*

The primary area of the gut for drug absorption is the small intestine; with an absorptive flux around 10–20 times that of the large bowel [19]. The slow absorption associated with those compounds with poor solubility or intrinsic slow dissolution in Class 2 of the BCS classification still prompts scientists in biopharmaceutics to look for methods of increasing small intestinal transit time. Although binding to the small intestine in vitro is readily achieved using a variety of ligands, the motility and pattern of lumenal flow may restrict access, except in the distal gut during periods of stasis. Small studies in dogs have suggested that carbomer 934 may achieve bioadhesion in the intestine but there is an important methodological problem in that a prolongation of gastric emptying time will result in later (clock) arrival at the end of the small intestine [20]. The material presented at the end of the intestine will be retained until the next large migrating movement of the bowel.

Using imaging techniques including gamma scintigraphy and magnetic moment imaging show that movement gradually slows as the dosage form moves from duodenum to ileum, with periods of stasis and sluggish movement [21]. The movement is an inherent feature of the gastrocolic reflex and appears to be difficult to modulate. Sluggish movement in the gastrointestinal tract is associated with blockages-bezoars, due to ingestion of fruit (especially unripe persimmons, phytobezoars), hair (trichobezoars) and mixtures of tablets with an anticholinergic effect (pharmacobezoars).

The opposite extreme – fast transit – is also evident. In volunteer studies with the drug gefitinib, the pharmacokinetics was shown to be highly variable [22]. Subsequently, it was appreciated that there was a subgroup of about a fifth of the whole panel who displayed a pharmacokinetic profile following single oral dose that was significantly different to their peers. The shape of the elimination profile in this subset was also different, showing a monophasic elimination pattern rather than the biphasic pattern observed in the majority of subjects. A study was conducted using radiolabeled tablets to examine the relationship of gastrointestinal transit and drug absorption in the subgroup and matched normal volunteers [23]. The rapid clearance cohorts were shown to have a faster mean gastric emptying T90% (37 min vs. 74 min) and shorter small intestinal transit time (156 min vs. 204 min). Mean plasma C_{max} was lower (99.2 ng/ml vs. 116 ng/ml) and AUC almost half in the rapid clearance group (2,162 ± 81 ngh/ml vs. 4,996 ± 64 ngh/ml).

Overall, using a wide range of markers, meals, tablets and pellets, the small intestinal transit time in normal, healthy volunteers is between 3 and 4 h. As might be expected, the presence of nutrients in the gut alters motility – drinking glucose solutions or Intralipid® increases contraction of the gut significantly. Both increase contractions to the same extent, with the duration of the increase dependent on caloric activity [24]. Modulation of transit by food produces relatively modest effects in xenobiotic absorption and it is unlikely that this is a fruitful area for consideration.

2.6 The Colon and Drug Delivery

The importance of the colon varies in mammals according to the nature of their diet. Thus true carnivores have a short colon with a small caecum, whereas large ruminants have a high capacity rumen for fermentation. The appendix in humans is vestigial and apparently unimportant in the human nutritional process. On opening the abdomen, the large colon is usually easily visible because the transverse loop has a very antral position in the abdominal cavity and may contain gas. Figure 2.13 illustrates the main physiological features of the colon. The bacterial fermentation of ingested soluble carbohydrates yields carbon dioxide, and in some individuals if the redox potential is low enough, hydrogen and methane.

Compared to the small intestine it is shorter – 1.5 m rather than 5 m – and the lumen is wider, without the extra surface area provided by the folds of Kekring and the villi. The absorptive capacity for drugs is therefore markedly reduced but this can be balanced by the long periods of residence in the ascending colon. The major regions of the colon are the right or ascending colon; the transverse colon which is folded in front of the ascending and descending arms by the hepatic and splenic flexures; the descending colon which stores feces and finally the rectum and anus. Overall the length of the human colon is approximately 150 cm, but only the last

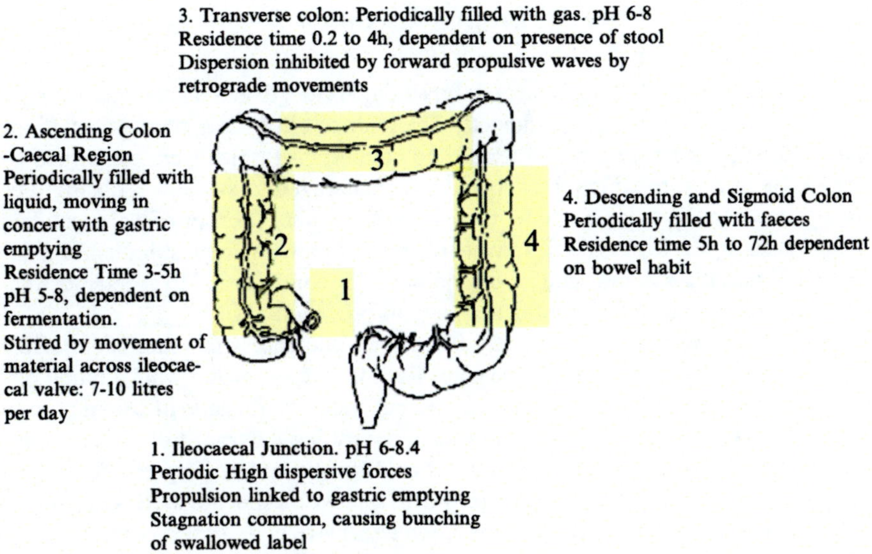

Fig. 2.13 Schematic of colon transit. From [25]

30 cm is accessible from the anus, since the folding of the splenic flexure resists material entering the transverse colon if rectal delivery of large volume enemas is attempted. Targeting the first half of the colon is therefore difficult from a physiological perspective; however, the bacterial population provides a step change in luminal environment with a different set of metabolic enzymes to aid selective release. As an incentive, drug delivery to the colon has often been an attractive goal for peptide delivery as it is supposed that the lack of digestive enzymes would facilitate absorption. A drawback is the lack of fluid for dissolution and the environment is moist rather than full of fluid, with normal maximal water content of 30 ml recoverable postmeal from the caecum [26]. When empty, the colon is collapsed with little motility but the transverse section may extend with gas following fermentation of the carbohydrate. The terminal segments may be occupied by stool and little drug absorption can occur from the distal regions under these conditions.

The wall of the ascending colon when scraped with a pH electrode gives an alkaline reading as high as pH 8, caused by secretion of bicarbonate by a sodium-dependent bicarbonate secretion which is non-chloride ion dependent [27]. This secretion of the bicarbonate would be expected to render the colon alkaline, but this is balanced by the bacterial fermentation of carbohydrate to short chain fatty acids, particularly in the caecum and right colon. Studies with reliable pH electrodes implanted on the colon wall during colonoscopy in areas free of debris indicate that patients with a normal bowel have a more acidic right colon (pH 7.05 ± 0.32), followed by a more alkaline transverse colon (pH 7.42 ± 0.51), becoming more acid moving towards the rectum (pH 7.15 ± 0.44). The lumen pH mirrors the changes of the wall, but remains consistently more acidic [28]. Press and colleagues (1998) report values illustrated in Fig. 2.12 [28, 29].

2.6.1 Regional Transit Through the Colon

The environment of the large bowel differs along its length and it is only in the right colon where conditions are sufficiently favorable to allow drug absorption. In the clinic, the first measurements of ascending colon transit times were performed by long tube studies, in which the subject was encouraged to swallow a dosing tube orally, down to the caecum. Under these conditions, very short proximal ascending loop transit times, 87.6+27.0 min were observed following the instillation of a liquid bolus into the caecum. Intubation and the CCK administered to accelerate transit were probably significant influences on this procedure [30].

The transverse colon is frequently full of gas, and access to water is extremely limited. In the descending colon, the consolidation of fecal matter would inhibit dissolution and absorption of drug through the gut wall. The division of colonic transit into regional areas is therefore important in describing the transit of dosage forms through the colon and the possible impact on drug absorption. Targeted delivery of drugs to the terminal gut has been employed to achieve a variety of therapeutic objectives including to delay delivery to the colon to achieve high local concentrations in the treatment of diseases of the distal gut; to delay delivery to treat acute phases of disease at the appropriate time of day (chronotherapy or chronopharmaceutics); to utilize sustained delivery to reduce dosing frequency and historically, in the hope that compliance would improve.

2.6.2 The Gastrocolic Reflex

It has been noted in scintigraphic studies that ingestion of food whilst a tablet was in the ascending colon tended to move the unit into the transverse colon, or if the tablet was in the transverse colon, it moved it further along [31]. This provided a good illustration of the propulsive ileocolic reflex, sometimes mistakenly termed the gastrocolic reflex. Misiewicz, in a classical paper on colonic motility, referred to his earlier study [32, 33] and pointed out that this phenomenon occurs in patients who had undergone a total gastrectomy and therefore gastrin is unlikely to be involved.

2.6.3 Problems of Low Motility and a Lack of Water

In the colon, only the first parts of the colon (ascending and transverse loops) contribute to drug absorption when the drug is delivered from the orad end and fluid levels are very restricted. Both Reppas and Weitschies estimate, by different methods, about 20–30 ml of liquid is available for dissolution (Weitschies W. and Reppas C. Personal communications). Disease states such as diarrhea associated with hypersecretion of fluid by the small intestine can be simulated by administering 20 g of lactulose for 3 days. When a capsule containing tablets and beads is then given to young volunteers, there is a marked increase in the dispersion and dissolution in the transverse colon [34], suggesting that lack of water usually restricts the surface area following release in the distal gut.

Pulsincap® represented an interesting concept in the delivery in which a swellable plug hydrated and ejected from the delivery device exposing the contents with remarkable accuracy in vitro. It seemed to be the ideal solution to targeting the proximal and mid colon, a region which is not accessible from the anus as the splenic flexure prevents ingress of enemas. It proved a good tool for illustrating the problems of colonic drug delivery. Studies using the Pulsincap system [35] were carried out with the objective of targeting the distal colon with a pulsed delivery of a transcellular probe (quinine) and [^{51}Cr]-EDTA, a paracellular probe. In these studies, subjects were dosed at 10 p.m. to ensure delivery to the descending colon by lunchtime the following day. The site of release was identified by incorporating [^{111}In]-labeled resin into the unit and imaging the subjects by scintigraphy. Fifteen hours after nocturnal administration, the majority of the delivery systems were situated in the proximal colon at their predicted release time and had not advanced further than a similar set of systems viewed only 6 h after dosing. This relative stagnation appeared to reflect the lack of propulsive stimuli caused by the intake of food, and the effect of sleep in reducing colonic electrical and contractile activity. A further problem was the poor ingress of water into the system.

Eventually this system was abandoned and the only device of this type in development was the Egalet® system. This had two end plugs at end and the earliest iteration had similar problems to the Pulsincap®. By changing the dimensions of the device, the problem of water access was overcome [36].

2.6.4 The Bacteria of the Colon

At birth, the colon is sterile but gradually becomes populated with microbes from maternal supply and the environment, dominated by species acquired from the mother's vaginal microbiota including *Lactobaccilus*, *Prevotella* or *Sneathia* spp, whereas C section babies show those of the skin surface, including the *Staphylococcus* genus [37]. The thick mucus lining of the colon provides a structural and metabolic support for the bacteria, partitioning the microbiota from the underlying epithelium. In diseases of the colorectal mucosa, the normal biochemistry of this human-bacterial balance is lost and is difficult to re-establish [38].

The colonic microflora secretes a number of enzymes which are capable of hydrolytic cleavage of glycosidic bonds. These include β-D-glucosidase, β-D-galactosidase, amylase, pectinase, xylanase, α-D-xylosidase, and dextranases. In addition, there are scission reactions catalyzed azo-reductases secreted by the anaerobes but these are more generally used in a prodrug approach for example 5-ASA adducts such as balsalazide and olsalazine. The biodegradable polysaccharides can be employed (1) in the formulation matrix, or (2) as a coat, alone and in combination. Many of these polysaccharides have limited release control properties due to high water solubility. Hence, they are employed in formulations in two ways (1) combination with synthetic nonbiodegradable polymers, especially acrylates or (2) synthetic modification such that solubility is decreased [23].

2.7 Disease and Gut Transit

An issue for all systems relying on consistent transit times is the issue of diseases affecting gastrointestinal motility. The change produced by increased hydrodynamic action – conditions such a diarrhea – will impact on sophisticated zero-order release formulations such as osmotic pumps. For the pumps inadequate retention may occur in some patients, leading to less optimal clinical outcomes. For example, the median GI transit time for both oxprenolol and metoprolol Oros drug delivery systems has been reported as 27.4 h, with individual times ranging from 5.1 to 58.3 h [39]. The possibility of inadequate GI retention of the nifedipine GITS is perhaps more likely in patients who have pre-existing GI motility disorders or who are taking other medications that enhance GI motility. In many patients who are hospitalized, slow transit is often a problem, usually associated with gastric stasis or intestinal trauma. Clearly, any strategy based on control within the formulation will be susceptible to intrinsic factors based on the characteristics of the patients receiving treatment.

References

1. Weaver LT, Austin S, Cole TJ (1991) Small intestinal length: a factor essential for gut adaptation. Gut 32(11):1321–1323
2. Holloway RH, Tippett MD, Horowitz M et al (1999) Relationship between esophageal motility and transit in patients with type 1 diabetes mellitus. Am J Gastroenterol 94:3150–3157
3. Channer KS, Virjee JP (1985) The effect of surface coating of tablets on oesophageal transit. Br J Pharm Pract 1985:9–14
4. Channer KS, Virjee JP (1984) The effect of formulation on oesophageal transit. J Pharm Pharmacol 37:126–129
5. Bar-Shalom D (2010) Seeds of change: thoughts on making solid dosage forms easier to swallow in "Tablets and Capsules" in Tablets & Capsules 8:29–31
6. Feldman M, Barnett C (1991) Fasting gastric pH and its relationship to true hypochlorhydria in humans. Dig Dis Sci 36:866–869
7. Davis SS, Hardy JG, Taylor MJ, Whalley DR, Wilson CG (1984) A comparative study of gastrointestinal transit of a pellet and tablet formulation. Int J Pharm 21:167–173
8. Reilly S, Wilson CG, Hardy JG (1987) The influence of food on gastric emptying of multiparticulate dosage forms. Int J Pharm 34:213–216
9. Wilson CG, Washington N, Greaves JL, Kamali F, Rees JA, Sempik AK, Lampard JF (1989) Bimodal release of drug in a sustained release Ibuprofen formulation: A scintigraphic and pharmacokinetic open study in healthy volunteers under different conditions of food intake. Int J Pharm 50:155–161

10. Code CF, Marlett JA (1975) The interdigestive myo-electric complex of the stomach and small bowel of dogs. J Physiol 246(2):289–30
11. Bull JS, Grundy D, Scatcherd T (1987) Disruption of the jejunal migrating motor complex by gastric distension and feeding in the dog. J Physiol 394:381–392
12. Lacombe O, Woodley J, Solleux C, Delbos JM, Boursier-Neyret C, Houin G (2004) Localisation of drug permeability along the rat small intestine, using markers of the paracellular, transcellular and some transporter routes. Eur J Pharm Sci 23:385–391
13. Burke MD, Wilson CG (2006) Clinical protocol design: gastroretentive dosage forms. Drug Deliv 6(8):26–31
14. Lewis LD, Fowle AS, Bittiner SB, Bye A, Isaacs PE (1986) Human gastrointestinal absorption of acyclovir from tablet duodenal infusion and sipped solution. Br J Clin Pharmacol 21(4):459–462
15. Jenkins JRF, Hardy JG, Wilson CG (1983) Monitoring antacid preparations in the stomach using gamma scintigraphy. Int J Pharm 14:143–148
16. Kaus LC, Fell JT, Sharma H, Taylor DC (1984) The intestinal transit of a single non-disintegrating unit. Int J Pharm 20:315–323
17. Indrio F, Riezzo G, Raimondi F, Francavilla R, Montagna O, Valenzano ML, Cavallo L, Boehm G (2009) Prebiotics improve gastric motility and gastric electrical activity in preterm newborns. J Ped Gastroenterol Nutr 49:258–261
18. Hammer J, Talley NJ (2006) Disturbed bowel habits in patients with non-ulcer dyspepsia. Aliment Pharm Ther 24:405–410
19. Wilson CG, Washington N, Greaves JL, Washington C, Wilding IR, Hoadley T, Sims EE (1991) Predictive modelling of the behaviour of a controlled release Buflomedil HCl formulation using scintigraphic and pharmacokinetic data. Int J Pharm 72:79–86
20. Fu J, Sun X, Zhang ZR (2002) Study on of bioadhesive property of carbomer 934 by a gamma camera in vivo. World J Gastroenterol 8(1):176–179
21. Wilson CG, Weitschies W (2009) Modern drug delivery: physiological considerations for orally administered medications. In: van de Waterbeemd (ed) Drug bioavailability, estimation of solubility, permeability, absorption and bioavailability. Wiley-VCH, Weinheim, Germany pp 571–595
22. Swaisland HC, Smith RP, Laight A, Kerr DJ, Ranson M, Wilder-Smith CH, Duvauchelle T (2005) Single-dose clinical pharmacokinetic studies of gefitinib. Clin Pharmacokinet 44:1165–1177
23. Wilson CG, Mukherji G, Shah HK (2008) Biopolymers and colonic delivery. In: Rathbone M, Hadgraft J, Roberts M (eds) Modified release drug delivery technology, 2nd edn. Swarbrick J (ed) Drugs and the pharmaceutical sciences series. Marcel Dekker, New York, pp 315–329
24. Von Schonfeld J, Evans DF, Renzing K, Castillo FD, Wingate DL (1998) Human small bowel motor activity in response to liquid meals of different caloric value and different chemical composition. Dig Dis Sci 43:265–269
25. Wilson CG (2002) Colon drug delivery. In: Rathbone M, Hadgraft J, Roberts M (eds) Modified release drug delivery technology. Swarbrick J (ed) Drugs and the pharmaceutical sciences series. Marcel Dekker, New York, pp 217–222
26. Diakidou A, Vertzoni M, Goumas K, Abrahamsson B, Söderlind E, Dressman JB, Reppas C (2009) Characterisation of the contents of the ascending colon to which drugs are exposed after oral administration to healthy adults. Pharm Res 26:2141–2151
27. Sullivan SK, Smith PL (1986) Bicarbonate secretion by the rabbit proximal colon. Am J Physiol 251:G436–G445
28. McDougall CL, Wong R, Scudera P, Lesser M, DeCosse JJ (1993) Colonic mucosal pH in humans. Dig Dis Sci 38:542–545
29. Press AG, Hauptmann IA, Hauptmann L, Fuchs B, Fuchs M, Ewe K, Ramdori G (1998) Gastrointestinal pH profiles in patients with inflammatory bowel disease. Aliment Pharmacol Ther 12:673–678
30. Barrow L, Spiller RC, Wilson CG (1981) Pathological influences on colonic motility: implications for drug delivery. Adv Drug Deliv Rev 7:201–218

31. Hodges LA, Connolly SA, Band J, O'Mahony B, Ugurlu T, Turkoglu M, Wilson CG, Stevens HNE (2009) Scintigraphic evaluation of colon targeting pectin–HPMC tablets in healthy volunteers. Int J Pharm 370:144–150
32. Misiewicz JJ (1975) Colonic motility. Gut 16:311–314
33. Holdstock DJ, Misiewicz JJ (1970) Factors controlling colonic motility: colonic pressures and transit after meals in patients with total gastrectomy, pernicious anaemia or duodenal ulcer. Gut 11:100–110
34. Hebden JM, Gilchrist PJ, Perkins AC, Wilson CG, Spiller RS (1999) Stool water content and colonic drug absorption: contrasting effects of lactulose and codeine. Pharm Res 16:1254–1259
35. Stevens HNE, Wilson CG, Welling PG, Bakhsaheee M, Binns JS, Perkins AC, Frier M, Blackshaw EP, Frame MW, Nichols DJ, Humphrey MJ, Wicks SR (2002) Evaluation of Pulsincap to provide regional delivery of dofetilide to the human GI tract. Int J Pharm 236:27–34
36. Bar-Shalom D, Wilson CG, Washington N (2009) Chronotherapy using Egalet® technology. In: Bi-Botti C Youan (ed) Chronopharmaceutics: science and technology for biological rhythm-guided therapy and prevention of diseases. Wiley, Hoboken, NJ, pp 165–173
37. Dominguez MG, Costello E, Contreras M, Magris M, Hidalgo G, Fierer N, Knight R (2010) Delivery mode shapes the acquisition of the initial microbiota across multiple body habitatas in the newborn. Proc Natl Acad Sci USA. doi:10.1073/pnas.1002601107
38. Pearson JP, Brownlee IA (2010) The interaction of large bowel microflora with the colonic mucus barrier. Int J Inflamm 2010:321426
39. Sathyan G, Hwang S, Gupta SK (2000) Effect of dosing time on the total intestinal transit time of non-disintegrating systems. Int J Pharm 204:47–51

Chapter 3
Controlling Drug Release in Oral Product Development Programs: An Industrial Perspective

Luigi G. Martini and Patrick J. Crowley

Abstract Most new drugs for oral administration are released rapidly from the dosage form. This may give "peak-trough" plasma profiles not suited to the drug's mode of action or side effect profile. Traditionally, there were few options for better design to optimize rate or time of release. However, advances in many areas of drug evaluation are now identifying opportunities to modify drug release during earlier phases of development to optimize therapeutic efficacy and reduce undesirable effects. It should also result in better success rates in drug development programs as well as better medications for patient treatment.

This chapter explores possibilities for modifying release during Preclinical, Phase 1, Phase 2, and Phase 3 programs. Opportunities to improve performance or find new indications for mature drugs, by controlling release continue to emerge. Such possibilities are also considered. The chapter is written, not only for the product design (formulation) specialist but also for scientists involved in all aspects of drug evaluation and development. Consequently, its focus is on the breadth rather than depth of the topic. Other chapters provide more comprehensive accounts of specific approaches, technologies, and modes of evaluation.

3.1 Background

It is rare for a novel drug to be developed as a modified release product by first intent. With the exception of acid unstable drugs like the Proton Pump Inhibitors, where delaying release by enteric coating or time-controlled release is mandated by poor stability at low pH, most novel drugs have been first formulated as "immediate

P.J. Crowley (✉)
Callum Consultancy LLC, Devon, PA, USA
e-mail: ktcrowley@msn.com

C.G. Wilson and P.J. Crowley (eds.), *Controlled Release in Oral Drug Delivery*, Advances in Delivery Science and Technology, DOI 10.1007/978-1-4614-1004-1_3,

Table 3.1 Effectiveness of medications

Therapeutic area	Success rate (%)
Alzheimers	30
Analgesics (Cox-2)	80
Asthma	60
Depression (SSRI)	62
Diabetes	57
HIV	47
Hypertension	40
Oncology	25
Rheumatoid arthritis	50
Schizophrenia	60

Reproduced with permission from [1]

release" products. Dosage forms where drug release is modified have usually been second-generation products that took account of knowledge accumulated during widespread use.

This situation was understandable. Research and Development programs to find and evaluate novel molecular constructs as medicinal agents invariably focus on novel modes of action. Many R&D-based organizations have programs that focus on common targets, indicated by current knowledge of the clinical condition and its alleviation. Such commonality of approach makes "first to market" an imperative. Follow-on compounds are considered to be "me too" and unlikely to gain premium pricing where national health bodies negotiate cost.

These and other drivers have meant that clinical programs, being invariably rate limiting in drug development are often designed for speed of completion. There is pressure to quickly follow successful volunteer (Phase 1) studies with trials in patients to ascertain safety and efficacy. A positive outcome in patients in Phase 2 trials then brings pressure to hurry to and through Phase 3 to registration and commercialization. Lingering in Phase 1 or Phase 2 to perform sequential or iterative evaluations to better characterize behavior and effects in humans can delay such progression. Such "speed to market" (or to failure) strategies have probably have caused many drug candidates to perform inadequately, because of inappropriate "drug delivery." They are discarded, as a consequence, being replaced by a followup compound that is then subjected to the same evaluation cycle. Even candidates that ultimately attain product status may be less-than-ideal as medications (Table 3.1 and [1]).

It is ironic that such relatively poor productivity and performance is prevalent at a time when understanding of molecular biology and of physiological and pathological processes has never been better. Failures during development, and the less-than-stellar performance illustrated in Table 3.1 concern materials that probably exhibit excellent potency and specificity in preclinical models. Shortcomings may have been due lack of understanding of the complexity or patient-specificity of many clinical conditions, and of the nuances of the modes of action of drug candidates used to treat them. It is also possible, however, that many medications could be more effective and safer if delivered at a rate, time, and to a location that optimizes performance, rather than "dose dumping" by rapid release in the gastro intestinal tract.

Historical "failures" and shortcomings can often, and with hindsight be attributed to knowledge limitations and paradigms at the time. Efficacy or side effects were probably sought in large genetically and ethnically diverse patient populations. Quantitative and precise measurements of outcome were not possible in many cases. There may, now and in the future be opportunities for modifying release from the dosage form to enhance drug performance as a first intent. Such optimism is based on the following developments and considerations:

- Ever-increasing understanding of the molecular basis of pathological conditions and of drug action.
- Advances in molecular profiling technologies, e.g., proteomic profiling and metabolomic analysis.
- The emergence of diagnostics to measure proteins or other validated biomarkers of the clinical condition, its alleviation, or of drug side effects. More precise determinations of effect can lead to smaller, faster, and adaptive clinical trials, with feedback enabling drug delivery (release) to be optimized with respect to rate or time of delivery in addition to optimum dose.
- Appreciation of the importance of "the time dimension" in treating some clinical conditions (chronotherapeutics) whereby the therapeutic target can vary during the day or even seasonally.

Such developments, allied to better clinical evaluation and better dosage form design can potentially lead to better and earlier indications of performance, less failures in development, and more effective and safer medications.

The increased interest in using biomarkers for surrogate evaluation of the molecular pharmacology of a drug, along with more sophisticated molecular design tools to optimize Absorption, Distribution, Metabolism and Elimination (ADME) characteristics has led to R&D organizations hurrying drug candidates to Phase 1 trials that incorporate such techniques and evaluations. This strategy enables rapid evaluation of a candidate's preclinical safety and surrogate efficacy. Ironically, little has been done to determine the impact of formulation design on performance of novel compounds. Should the drug candidate exhibit favorable clinical and pharmacokinetic behaviors in such trials a deeper evaluation of its PK–PD (pharmacokinetic–pharmacodynamic) relationship may then be warranted to define an optimized delivery profile for further clinical assessment.

Such optimization of "drug delivery" to produce a desired Product Profile may be a better strategy than seeking a backup candidate from Discovery which then must run the gamut of preclinical, Phase 1, safety and efficacy studies and possibly encounter the same or new risks that may have been discharged with the lead candidate. Such "optimization" strategies have been successful in Development programs for products such as Enablex® (Darifenacin, Pfizer-Novartis) and FlowMax® (Tamsulosin, Boehringer Ingelheim).

3.2 Concepts and Principles

Modifying the rate, location, or time of drug delivery may render it more effective, safer, or more convenient than if released and absorbed rapidly following ingestion. Table 3.2 provides a limited list of approaches and potential benefits. Other chapters provide more comprehensive information.

Table 3.2 Strategies for and benefits of modifying drug release

Modification	Objective	Benefit	Example
Sustain or "Slow" release	Prolong plasma residence	Longer duration of action	Once-daily Ca Antagonists (e.g., Nifedipine for hypertension)
	Maintain constant plasma level	Plasma level remains in therapeutic window	Beta blockers e.g., Propanolol
	Obviate tissue irritancy		Slow release potassium chloride
Site-specific release in GI tract	Avoid presystemic degradation or metabolism	Optimize or reduce variability of absorption	Proton pump inhibitors (avoid low pH). Attenuate CYP 450 metabolism (Propanolol)
	Avoid exposure to specific site or tissue	Obviate irritation or toxicity	Aspirin (GI bleeding), Doxycycline (oesophageal/ gastric irritation)
	Target specific area/tissue	Provide "local" effect	Colonic delivery e.g., Mesalazine or Corticosteroids in inflammatory bowel disease (IBD)
Pulsatile release	Plasma profile from OD dosing mimics BD or TID dosing. May also circumvent metabolic barriers	Varying plasma levels prevent tolerance development while providing convenient regimen	Dexamphatamine salts for ADHD
Delay release	Delay onset of effect	Align onset of effect with therapeutic requirement	Theophylline controlled release. H2 receptor antagonists for nocturnal GERD
Prevent release in vitro	Abuse avoidance	No therapeutic benefit but socially useful	Opoid painkillers

Whatever the desired outcome it is imperative, when considering modifying drug delivery to the biosystem that objectives and performance standards are clearly defined at the outset. These include:

- A valid biological, clinical, or patient-specific rationale for controlling delivery. This requires knowledge of the biological effects of the drug that can determine efficacy and side effects.
- A Target *Product* Profile (TPP) that clearly identifies the benefits being sought, viz, improvements over what can be provided by a conventional dosage form.
- A Target *Plasma* Profile for the drug and/or its metabolites that can deliver the desired benefits.

Without clear and measurable targets, and a ready means to monitor them, a program to modify drug release is likely to fail. Furthermore, the above requirements usually need to be augmented with additional information, viz:

- Stability in the milieu of the GI tract (pH, enzymatic).
- Degree of and location of absorption in the GI tract.
- Barriers to systemic absorption (recycling and metabolizing enzymes).
- Route, location, and extent of metabolism and the clinical biology of the metabolites.
- Drug disposition and elimination rate, and route (pharmacokinetics).

Furthermore, the following information may be essential for defining the target plasma profile:

- Dose response with respect to therapeutic effect and side effects.
- Relationship, if any between pharmacokinetics, duration of action, and side effects.

Such multiplicity of requirements has meant that most current modified release products were developed after an immediate release product had been available and widely used over an extended period of time. Performance attributes could only be defined following such extensive clinical experience. However, for the previously mentioned reasons product design to modify delivery and enhance performance of a novel drug candidate may become increasingly feasible at any stage of its preclinical or clinical evaluation. Such possibilities will now be discussed.

3.3 Preclinical Studies

Preclinical formulation-related investigations, as currently practiced mostly concern enhancing absorption by improving solubility or dissolution rates of poorly soluble compounds. Other factors such as poor stability in the gastric milieu or metabolic transformation can also compromise absorption. Such hazards may be predicted by in vitro or animal preclinical studies such that optimization for Phase 1 dosing can be considered.

Table 3.3 Half lives of compounds at low pH and relationship with oral bioavailability

	pH									Oral
Compound	1	1.3	1.5	1.8	2	2.6	3	3.4	4	bioavailability
Amoxicillin	312	540			19 h				177 h	Very good (85–90%)
Ampicillin		660								Good
Benzylpenicillin		3.5		6.4		16		59	172	Poor
Methicillin		2.3								Nonexistent
Phenethicillin		68								Good
Phenoxymethyl Penicillin	160	160								Good
Clavulanate Potassium			2.5		11	21.8		115	255	Good (ca 70%)
Lansoprazole					1.4					Nonexistent
Pantoprazole					3.2					
Omeprazole					1.75					
Metronidazole					2,200 h				∞	Very good (>90%)
Clarithromycin	6				58.9		15.8 h		97 h	ca 50%

3.3.1 *Drug Stability on Ingestion*

If a drug is rapidly degraded during gastric passage there is less available for absorption. Low-pH instability can be obviated by enteric coating to prevent release in the stomach or by other ways of delaying release. Such approaches have been used successfully to boost the bioavailability of Proton Pump Inhibitors such as omeprazole and the antibiotic Erythromycin.

In vitro pH-stability studies may help predict performance in first-time in humans (FTIH) studies. Table 3.3 lists degradation half lives at low pH and 37°C together with bioavailability on oral dosage.

The values in Table 3.3 do not suggest a straightforward pH-stability relationship with absorption. This should not surprise. Other factors may also contribute. Cursory inspection might suggest that short half life at pH values below about 2.0 could lead to significant gastric degradation. However, the values in Table 3.2 were determined in "simple" buffer systems. Other gastric components may also affect stability. The slope of the pH-stability relationship could also be important. For instance, clavulanate half life seems comparable to that for benzylpenicillin at pH values up to about 2.5 but it is more stable at higher pH. This may partly explain its better performance on oral dosage. However, while bioavailability is good in fasted subjects, dosage after food reduces clavulanate bioavailability to about 40%, possibly because of longer gastric residence that may be accompanied by lower pH. In general, it would appear that materials that are unstable at low pH may have poor oral bioavailability but assigning a specific pH/stability value for screening purposes is inadvisable.

Promising technologies for assessing absorption in differing regions and different conditions in the GI tract are now available [http://www.simulations-plus.com (Gastroplus)]. Confidence in predictability is likely to improve with greater experience and as techniques are progressively refined. Studies in animals may also provide useful markers on absorption potential. Nonetheless, uncertainties remain. Hence, for First Time in Human studies it may be prudent to test a nonmodified formulation such as a simple solution, suspension or powder-filled capsules alongside units protected from acidic degradation. Findings could clarify the formulation approach for subsequent development.

3.3.2 Surmounting Enzymatic Barriers

Absorption from the gastro-intestinal (GI) tract can be compromised by P-glycoprotein (Pgp) recycling ("ejecting") drug from the enterocyte or by metabolizing enzymes, such as the Cytochrome P450 (CYP450) family. Such drug liability can be established in vitro. Some excipients are also Pgp inhibitors but these do not come within the scope of this article. CYP450 inhibitors such as quinidine can also enhance the bioavailability of CYP450-susceptible drugs on pre or coadministration [2]. However, this strategy may complicate the coadministration of other medications that are also CYP450-susceptibile; the bioavailability of concurrently administered medication might also be altered.

It is reported that 20–25% of drugs used clinically are metabolized by CYP450 2D6 [3]. Consequently, many organizations screen candidate drugs for CP450 susceptibility, rejecting those that are substrates, because of absorption and drug interaction complications. This raises the barrier to selection and progression. Other features that contribute to absorption, disposition, and activity (such as lipophilicity and basicity) may also render them CYP substrates [4]. Thus it may be difficult, if not impossible in many cases to "design out" CYP450-susceptibility without compromising other desirable features of promising drug candidates. Formulation strategies to mitigate CYP450 liability and provide more consistent absorption may be worth considering in cases where drug candidates have novel and exciting possibilities as therapeutic agents.

CYP liability may be reduced by modifying drug release. Amidon and Leesman [5] used a "delayed release" approach to provide enhanced and more consistent absorption of the beta blocker propranolol. Susceptibility of this drug to CYP450 metabolism (2D6 variant) causes variable systemic levels and nonlinear kinetics. The problem is exacerbated when drug release is prolonged to provide once-daily dosage. Higher CYP450:substrate ratios that probably pertain when drug is released slowly results in greater metabolism and reduced and variable drug plasma levels. The same effect is likely when drug that is rapidly released in the stomach is diluted on passage to and through the small intestine prior to absorption. However, when formulated so that gastric release is avoided but drug is then rapidly released as a "pulse" or pulses in the small intestine the enzyme may be saturated, metabolic lability reduced, and systemic drug levels increased.

The proximal small intestine is CYP450-richer than the distal region. Thus a significant delay in drug release might be beneficial as enzyme:drug complex ratio is likely to be lower. Such "tailored" delay can be accomplished using appropriate pH or time-sensitive release-modifying polymers. The same approach can also be utilized to prolong plasma levels by formulating to deliver separate sequential pulses while transiting the small intestine. However, such strategies require drug to be absorbed from the region(s) where it is released.

"Visualizing" the transit of a modified release dosage form in the GI tract can provide useful insights on site of absorption by associating particle or dosage form disposition with pharmacokinetic parameters. Techniques such as gamma scintigraphy are noninvasive, are readily performed in a Phase 1 environment and could usefully be considered for parallel monitoring along with established practices such as plasma sampling and physiological measurements (Chap. 2).

Another novel strategy for reducing CYP450 liability concerns the HIV Protease Inhibitors, Lopanavir and Ritonavir. These, in combination utilize the principle of "sacrificial substrate." The Ritonavir component, a potent CYP450 3AR Inhibitor blocks the metabolism of Lopinavir, a substrate of the same enzyme, thereby boosting its plasma profile and bioavailability [6].

3.4 Phase 1 Clinical Trials

Phase 1 volunteer studies provide valuable information on absorption, distribution, and excretion (bioavailability, metabolic pathways, and pharmacokinetics). Side effects and their relationship with dose may also become apparent in ascending dose studies. It is also now common to genotype subjects so that genome-associated effects can be ascertained, affording better patient selection for efficacy studies. However, volunteer studies do not usually provide evidence of therapeutic effect. Healthy volunteers (by definition) are not burdened with the specific clinical condition. However, this need not be a constraint in certain therapeutic areas. Dosage form design to control pharmacokinetics may be possible in some instances, however, where relevant efficacy indicators can be defined and monitored.

3.4.1 Anti-Infectives

The pathogen can be considered as the "receptor/target," whether in animals or humans. Appropriate animal models may therefore predict efficacy in humans by indicating optimum plasma/time profiles. Craig states that "animal models can describe the timecourse of in vivo antibiotic therapy and dose response relationships" [7]. Such information could be aligned with findings from Phase 1 concerning attributes such as absorption, pharmacokinetics, protein binding, dose response,

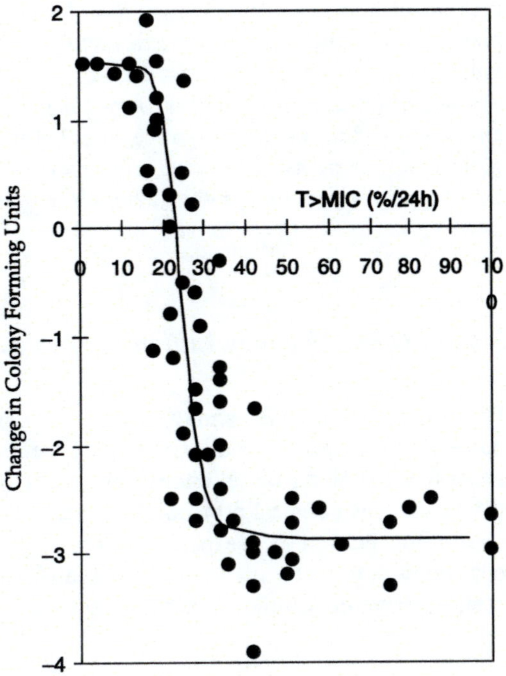

Fig. 3.1 Effect of time above inhibitory concentration (*T*>MIC) of amoxicillin and kill rate for penicillin-resistant *Streptomyces pneumonia.* Reproduced with permission from

metabolism, etc., to define a target plasma profile for dosage form design to deliver the requisite efficacy.

Craig and coworkers also propounded and validated the concept of "time above MIC" (*T*>MIC) as the performance standard for some antibiotics, showing that such time need not span the full dosage interval [8]. Woodnut et al., using a rat infection model showed that inhibitory levels of the antibiotic amoxicillin against resistant *S. pneumonia* should exceed the MIC for about 35% of the dosing interval for optimal kill (Fig. 3.1 and [9]). This enabled the design of a prolonged release dosage form that delivered the requisite plasma profile and was clinically effective [10, 11].

Concepts such as *T*>MIC or "postantibiotic effect" may not apply to all novel anti infectives. Different mechanisms of action may decree otherwise. Nevertheless, it may be beneficial to evaluate the dynamics of activity in preclinical models. A dosage form, providing a plasma profile reflecting the findings could offer benefits such as better efficacy, reduced dose, and consequent reduced cost of goods in a therapeutic area where doses are traditionally high.

The aforementioned microbiological and formulation studies on amoxicillin were performed when it was a mature drug. In the light of today's knowledge, similar studies may be feasible during preclinical evaluation. Too often corporate policies mandate that Discovery Teams quickly pass on a drug candidate to Development groups so that their search for a followup candidate is not delayed. Discovery groups then try to eliminate (in followup compounds) deficiencies identified primarily by preclinical findings. This is high risk. In silico, in vitro, and other preclinical predictions of drug ADME properties remains an inexact science. At the same time, much expertise on the biology, molecular, or otherwise of a novel compound remains in Discovery. Feedback from Phase 1 or even later studies may suggest and warrant additional investigative work before further progression. Development groups may not be resourced for such work. Cooperative Discovery and Development operations, rather than silo cultures can greatly benefit the overall program.

3.4.2 *Other Efficacy Indicators in Phase 1 Trials*

Phase 1 studies in patients may become more prevalent as many areas of interest move from acute to chronic conditions. Genomics-based selection criteria may mean fewer subjects (in all clinical programs) while retaining adequate statistical power. Adaptive clinical protocols can increase flexibility and power as can protocol adjustments in response to emerging findings. Such developments could increase opportunities for evaluating the effect of differing drug delivery profiles where relevant measurements reflect efficacy or side effects.

3.4.2.1 Biomarkers

A Biomarker, as defined by the Biomarker Definition Working Group is "a characteristic that is objectively measured and evaluated as an indicator of normal biological processes, or pharmacological responses to therapeutic intervention." Genomic biomarkers are exciting much interest, particularly in cancer therapy, being considered as offering great potential in patient selection and drug–patient matching (personalized medicine). Proteins and other agents associated with a clinical condition and its treatment are also being studied as potential markers of disease presence, progression, or response to therapy. They have the potential to have an enormous impact on novel drug development, providing earlier indications of efficacy or toxicity. This should facilitate early termination, reduce later stage attrition, and shorten overall development times (http://science.thomsonreuters.com/info/biomarkers) [12–17]. Table 3.4 lists some established and putative biomarkers, their possibilities and constraints.

Biomarkers are hardly new (as the list in Table 3.4 indicates). Well-established and approved surrogate markers of clinical efficacy include LDL cholesterol in atherosclerosis, Hemoglobin A1C (Hb_{A1C}) in Type 2 Diabetes and viral RNA load in HIV. Many others are at the exploratory stage. Some may not fulfill the promise to adequately predict or reflect all facets of efficacy and safety. Panaceas they are not. Monitoring a single putative marker of efficacy or safety is hardly likely to provide a comprehensive snapshot of safety and efficacy. Overall clinical performance will remain the standard.

Table 3.4 Biomarkers or surrogate measurements of disease progression or drug performance

Potential biomarker/diagnostic	Clinical relevance	Comment
C-Reactive protein	AMI propensity (plaque formation) Inflammation e.g., rheumatoid arthritis	Ubiquitous in body so specificity a challenge
Lipoprotein-associated phospholipase A2 (Lp-PLA2)	Cardiac disease	Marker of plaque formation (Atherosclerosis)
Cyclo Oxygenase (COX-2)	Over-expressed in Colon Cancer	COX-2 inhibitors may inhibit development and spread of colorectal cancer
Serum troponins	Myocardial damage e.g., in AMI	Also indicator of drug-induced damage
Interleukin 6	Inflammatory cytokine in rheumatoid arthritis	Implicated many inflammatory processes so specificity an issue
Neopterin	Infectious diseases	Marker of cellular immune system activation (Predict Adverse Events)
Procalcitonin	Sepsis	Effectiveness of antibiotic treatment (may indicate when treatment can be halted)
Hemoglobin A1C (Hb_{A1C})	Type 2 diabetes	3–4 Months therapy required for meaningful values to be generated
Amyloid β, Tau-Protein, P-Tau 181P	Alzheimers	Measured in CSF so requires spinal tap. Furtheromre only "early identifiers" of the condition at this time as no effective treatment that can validate
Serum urate	Gout	High serum urate (hyperuricemia) a marker of disease
Guanisine analogs in CSF	Parkinson's disease	Sampling challenges and poor sensitivity/specificity

Despite these qualifications there is hope, confidence, and some notable successes to suggest that monitoring appropriate proteomic biomarkers can provide valuable insights in many Discovery, Clinical, and other Development programs. Markers, techniques, and kits are becoming increasingly affordable. Further improvements in specificity, robustness, simplicity of operation, and rapidity of data generation will increase usage.

Current interest in nongenomic biomarkers is largely focused on assessment and validation as surrogate markers of efficacy. Biomarkers may also be valuable in determining the influence not just of dose level per se but to determine whether modes of delivery other than conventional "bolus" dosage can elicit or optimize efficacy. It may be possible, with some drugs and clinical conditions to determine dose response, duration of effect, optimum time for dosing, plasma profile (including frequency, amplitude, etc.) by monitoring relevant markers in small populations of volunteers or patients. Such information could then be used to define a target plasma profile to optimize efficacy and guide dosage form design to control release to provide the requisite plasma profile. Such possibilities could be explored at a relevant (early) stage in a clinical evaluation program, depending on available knowledge on the condition and the availability of the relevant markers for safety and efficacy. The outcome would be a better designed dosage form for progression to pivotal studies, a greater chance of overall success and a superior medication.

3.5 Phase 2 Clinical Trials

Phase 2 clinical trials are usually designed to determine whether a drug has the therapeutic benefit predicated by preclinical findings. Trials may be subdivided into sequential 2A and 2B programs, the former possibly designed to provide "proof of concept" results. Phase 2B studies then may focus on a narrower dose range or frequency for optimum efficacy. Some Phase 2 trials on drugs of high clinical need or life-saving drug candidates (e.g., cancer medications) may be designed to provide sufficient information for product registration if warranted by the clinical findings.

Compound attrition rates are usually high in Phase 2 but creative dosage form and study design may increase chances of success, or improve performance in subsequent pivotal trials. Some Phase 2 studies may employ more plasma and tissue sampling and measurements than later larger trials. Data from such monitoring (which could include biomarker measurements as discussed earlier) together with information from Phase 1 trials as well as other insights may suggest that modified release units be included in one or more arms of subsequent studies. Examples could include aligning the timing of delivery to optimize effect, or lowering the incidence of side effects associated with C_{max} or high rates of release within the upper GI tract.

3.5.1 Time, Clinical Efficacy, and Drug Delivery

3.5.1.1 Cardiovascular Conditions

The risk of a cardiovascular emergency is greatest between 6 and 10 a.m. Furthermore the fall in blood pressure between 2 and 3 a.m. that occurs in normotensives does not occur in many hypertensives [18]. If onset and duration of action parallels plasma presence the ideal drug delivery system for such conditions (where the duration of action of the drug is short) might comprise a dosage form, taken at bedtime with

- A delayed release component to make drug available 2–3 h after dosage.
- A followup "pulse" of drug occurring approximately 6 h later, providing cover during the vulnerable early morning period.

Chronotherapeutic delivery is now common for some cardiovascular medications.

3.5.1.2 Inflammatory Conditions

Night-time levels of inflammatory cytokines such as interleukin-6 (IL-6) are elevated in rheumatoid arthritis, leading to pain and joint stiffness after waking. Time of dosing may be critical to optimize control. Prednisone alleviates morning stiffness when administered at around 2 a.m. but this dosing time is not realistic. It has accordingly been formulated to delayed release of the drug about 4 h after bedtime dosage (http://www.skyepharma.com). Another anti-inflammatory, ibuprofen needs to be administered 4–6 h before achieving maximum anti-inflammatory effect. Night time dosage of a delayed release medication would seem warranted if ibuprofen is used to alleviate osteo arthritis-related morning stiffness.

3.5.1.3 Pulmonary Disease

Lung function is lowest in early morning. Such poor functionality is exacerbated in asthmatics. Long-acting bronchodilators would ostensibly seem to be called for but such bronchodilation may present long-term risks. Modified release formulations should therefore make drug available before the critical early morning period.

Innovations in timed delivery have traditionally been introduced after drugs are commercialized and widely used. Clinical findings, accumulated from widespread experience drove the development of such controlled release systems. It is now evident that "chronobiological dynamics" are prevalent in many therapeutic areas, more than hitherto appreciated. Exploratory studies to identify and capitalize on such effects would seem warranted where meaningful markers can be employed and/or modest patient cohorts can provide reliable indications of performance. "Lingering in Phase 1 or 2" to generate such information may identify opportunities to control drug delivery to provide more effective medication.

3.5.2 Dosing Frequency

Phase 2 studies usually explore the effect of dose frequency on drug behavior. Superior activity after frequent dosing may reflect the kinetics and dynamics of the drug. However, "time of dosing" may be masked as a dose frequency finding. In simple terms, evening dosing in a BD regimen may be the determinant of activity. Drug dosed OD, but at some time other than in the evening could appear to be less or not effective. Once daily administration might be efficacious if dosed "at the right time." Detailed exploration of clinical data for individual patients might uncover such possibilities and opportunities for better design for subsequent programs.

Clinical findings in Phase 2 may, however, categorically indicate that frequent dosing is more effective. The possibility of controlling release from a once-daily dosage unit that mimics the plasma profile seen after more frequent dosing in Phase 2 can then be considered. Patient convenience is enhanced when OD dosage is possible

Table 3.5 Drugs with durations of action consistent with plasma presence

Drug	Elimination half life (h)	Duration of action (h)	Clinical effect
Diltiazem	3.1–6.6	ca 6 h	Ca antagonist
Dextromethorphan	1.5–3.9	3–8	Cough suppression
Salbutamol (oral)	1.5	1.5	Bronchodilation in asthma
Theophylline	3–9	6	PDE inhibitor
Nabumetone	24	24	NSAID
Tramadol	5–7	4–6	Opoid analgesic

and convenience usually leads to better compliance and consequently better efficacy. A unit that delivers two "pulses" of drug at time intervals akin to those used with BD dosage should provide the same therapeutic effect. Such a unit should *deliver* an immediate release pulse and one delayed by about 10–12 h. However, there may be formidable barriers to such timed delivery, viz:

- *The GI Tract*: Small intestinal residence time is limited (about 3 h). Overall GI transit can also be variable (due to variable gastric emptying) and influenced by physiological and pathological variables as well as time of day and body position (standing, sitting, sleeping, etc.). These are discussed and exemplified in Chap. 2 and may cause release and absorption to be constrained and inconsistent.
- *Drug Absorption*: Delayed or controlled release to "space out" profiles requires that drug is absorbed along a significant length of the small intestine (significant colonic absorption is rare). However, absorption of some drugs is limited in the distal small intestine and beyond. Gastro-retentive technologies (refer to Chap. 17) that seek to retain the dosage form in the stomach for long periods might be considered in such cases.
- *Physicochemical Properties of the Drug*: Drug solubility may be pH-dependent, or influenced by GI tract contents. This can result in variable release, dissolution rate, and absorption in different regions of the GI tract. It is common to use a form of drug with optimal solubility for conventional (rapidly released) formulations so that absorption is enhanced in the upper GI tract. For modified release systems it might be appropriate in some cases to use a salt, or other form with a less steep pH-solubility profile to ensure more consistent dissolution that may provide less variable absorption throughout the GI tract.

Prolonging plasma residence to sustain an effect requires that duration of action parallels plasma presence (Table 3.5). However, relationships between plasma presence and duration of action can be more complex (Table 3.6). Receptor recovery rather than occupancy may define duration of effect so pharmacological effect may outlast receptor occupancy. Conversely, receptors can be inactivated by long-lasting association with drug, leading to tolerance and attenuation of effect. Soaking a receptor in a solution of the drug may not sustain a response other than possibly where the mechanism of action concerns receptor blockade. There may be little

Table 3.6 Drugs where duration of action exceeds plasma presence

Drug	Elimination half life (h)	Duration of action (h)	Clinical effect
Carvedilol	3.5	9	Beta blockade
Diclofenac	1.2–2	6–8	NSAID
Granisetron	3.5	17	Antiemetic
Methylphenidate	2.1–2.4	3–6	CNS stimulant (in ADD)
Morphine	2–2.5	4	Analgesia
Nifedipine	2–3.4	8	Antianginal and antihypertensive
Omeprazole	0.5–1	ca 24	PPI
Propanolol	3–4	12	Beta blockade
Ibuprofen	2	6–8	NSAID

information in such a "constant signal." The "message" may reflect the timing, frequency, or amplitude of the interaction. Time and timing are essential components of biological processes. Hence, controlling release of rapidly cleared drugs, to prolong plasma presence and compensate for short duration of action is not good strategy without supporting biological evidence for the approach.

Tachyphylaxis or tolerance can also occur where drugs are formulated to deliver "flat" plasma profiles. Gradual receptor desensitization occurs over time and the dose must be increased to achieve the desired effect. This is a major issue with Nitrate drugs used to treat Angina and other cardiovascular diseases. The phenomenon can also be mimicked by enzyme inducers stimulating increased metabolic activity over time against themselves, thus increasing doses are needed over time. For Nitrates and some other agents (e.g., some antihistamines and broncodilators) the effect is at the receptor level whereby the cells react by providing less receptors per cell or open up competing homeostatic physiological or biochemical pathways. Either way, variable plasma profiles over 24 h periods are usually best for such agents, to obviate the need for increasing doses and increased chances of side effects. Such behavior can generally be assessed at the preclinical stage or in Phase 1 and if a once-daily formulation is required, pulsatile delivery may be the best way to avoid tachyphylaxis or other side effects.

Whenever plasma profiles are generated, regardless of stage of development it is good practice to carefully review individual plots and seek association with other measurements or observations. Such diligence may uncover valuable indicators of effects associated with plasma profiles that may not be evident in meaned or otherwise "homogenized" information. Data mining need not be confined to in-house databases. Trials on compounds with similar purported mechanisms of action may be ongoing and listed, as mandated in Clinical Trial Registries (http://clinicaltrials.gov). These do not contain raw data per se but change history or trial design features may provide useful information. Published trial data on compounds with the same mechanism of action may be worth perusing for the same reasons.

3.6 Phase 3 Clinical Trials

Historically, the chance of success for a compound entering Phase 3 clinical trials was considered to be high as such studies invariably followed indications of success in earlier trials. This is no longer the case. Furthermore, product withdrawals after approval and commercialization are now not unusual, following reports of adverse events that only become apparent on widespread use. Regulatory Agencies may request more information during review of new drug applications, including additional efficacy or safety-related studies. Approval may be delayed as a consequence and the extra studies may not guarantee eventual approval. Large R&D-based organizations usually have a followup compound in case of rejection. "Planning for failure" in effect. Planning for success also features prominently throughout organizations during the Phase 3 program. "Partial success" may not feature in the lexicon of risk management. Yet, that is what may emanate from a Phase 3 program.

Late-stage clinical trials have historically involved large patient numbers, spread across many centers but perhaps with less biological sampling than in earlier studies. Data collection and conversion to knowledge and decision making was slow. Pharmaceutical R&D organizations are now adopting innovative approaches to improve trial design and knowledge collection at all stages of clinical programs (http://science.thomsonreuters.com/info/biomarkers). These include:

- Trials aimed at specific patient subsets, where there is a good genomics basis for doing so
- Use of premedication diagnostics (where available) to better target responders and provide "personalized" treatment
- Adaptive clinical protocols to facilitate rapid response to findings

These developments are predicated to increase study power, reduce trial numbers, and provide faster feedback and trial completion. Widespread IT networks have greatly improved remote data collection, transmission, and analysis. There has been a concomitant improvement in response capability to important incidents or findings. The marriage of life sciences information with technology is likely to enhance the process even further. These developments, together with those discussed earlier (e.g., biomarkers) mean that valuable information on many facets of drug activity may be gleaned during and after completion of Phase 3 studies.

Organizations usually have little appetite to redesign the dosage form at this late stage. It would probably mean additional Phase 1 studies, then followup Phase 3 trials; in effect a 1–3 year delay, depending on the clinical area. However, it may be a better option than bringing a followup drug forward, particularly where the putative mechanism of action remains unchanged. At the least there should be a backup plan in place in the event of an Agency request for additional studies to allay a safety or efficacy concern or to limit label claims such that the asset is less attractive commercially. Phase 1 studies could explore redesign possibilities during registration and agency review. Learnings from the Phase 3 studies and any new insights generally on the molecular biology of the drug class or clinical condition could then be incorporated in the redesign brief.

Such contingency operations might even ultimately reduce time to market, where review and approval are held up by clinical or regulatory concerns. A better performing formulation could be available for any Phase 3 studies mandated by agency concerns. Even with successful first time registration an enhanced medication form would be available as a followup product that provides better patient treatment and maximizes asset value. Pharmaceutical development scientists should review clinical data in its granularity, consider the opportunities that may emerge, and explain and propound these to the organization. Plans and evaluations for alternative better formulations should be in place for any drug candidate entering Phase 3 so that asset value is maximized early in its life cycle.

3.7 Drug Repositioning

Development programs on novel drugs cannot cover each and every permutation of dose, dose frequency, or PK profile when seeking safety and efficacy. When a compound progresses to a medicinal product, clinical studies do not stop. Trials to evaluate suitability for patient and disease subsets and comparative studies with other medications are all likely. Observational studies and pharmacovigilance data mining can identify noteworthy patient responses. Nuances of drug behavior, not evident in the precommercial clinical trials may emerge. Concurrently advances in molecular biology, receptor pharmacology, and better understanding of the clinical condition may identify new indications or new opportunities for modifying delivery by reformulation so that the medication is more effective, safer, or more convenient. Such opportunities may concern.

- Time-related effects
- Associations between plasma presence and duration of activity
- Dose response (therapeutic window)
- Unique plasma timecourses for better safety/efficacy
- More convenient dosing
- A side effect that suggests a new indication

The common denominator for capitalizing on such findings is the identification of a target plasma profile to deliver a desired effect. Other opportunities for improving medications concern:

- Changed disease patterns
- New indications
- New treatment Paradigms
- Drug/drug combinations

Enhancements could concern modifying the time, place, or rate of release from the dosage unit as discussed earlier.

3.7.1 *Changed Disease Patterns*

Infectious agents readily elaborate resistance mechanisms against anti infectives. In the absence of novel antibacterials, treatment strategies have included raising the dose (antibiotics) or using combinations of antiviral drugs for HIV treatment. Smarter delivery may also be advantageous. The MIC for the antibacterial amoxicillin was 0.2 μg/ml when first introduced (1972). By 1995 a target MIC of 4 μg/ml was required for organisms such as resistant *S. pneumonia*. An Extended Release tablet comprising amoxicillin with the beta lactamase inhibitor, clavulanic acid delivered the requisite plasma profile, prolonging amoxicillin levels for at least 40% of the dosage interval [10]. It was not necessary to prolong plasma levels of the clavulanate component. Its early presence at the infection site meant that resistant organisms were inhibited so would not readily inactivate the antibiotic. Thus the antibacterial activity of amoxicillin is better sustained during the dosage interval. Clinical studies validated the approach [11]. "Timed" delivery of antiviral combinations might also prolong their usefulness, e.g., "pulsed" dosing of one agent allied to with consistent and persistent plasma presence of another.

3.7.2 *New Indications for Established Drugs*

In concept an "active" drug has no single biological effect. Receptor occupancy or modulation of a mediator cascade can have consequences for other biological processes. Aspirin is widely used as an anti inflammatory and platelet inhibitor. Recently there have been reports that, when dosed chronically it may be beneficial in colorectal cancer. Gastric bleeding is a known consequence of aspirin dosage. Gastro-protective coats have been used for many years to mitigate such gastric irritancy but with limited benefit. Studies to determine the effect of time of and frequency of dosing, as well as frequency and amplitude of plasma profiles on gastric bleeding would seem warranted in the light of known circadian rhythms of inflammatory cytokines.

The cardiovascular liabilities that have been identified in some patients taking COX-2 antagonists have stimulated novel approaches to reduce NSAID erosion and ulceration of the upper GI mucosa. Lecithin-complexed NSAIDs has been proposed as an alternative to Gastro-protective coats, or in combination with Proton Pump inhibitors or H2 antagonists [19]. Such complexes, if combined with capability for site-specific delivery in the GI tract may facilitate the use of NSAIDs in broader indications and at higher doses.

Other reported possibilities, or where concepts have already been proven include:

- Angiotensin Receptor Blockers may be beneficial in Alzheimer's Disease.
- Long-term ibuprofen use may lower the risk of Alzheimers. Ibuprofen has also been shown to be effective in a preclinical model for Colon Cancer.
- Metformin combined with Doxorubicin to prevent or halt progression of Pancreatic and Breast cancer.

- Tamoxifen in Bipolar Disorder.
- Sildenafil for the treatment of Pulmonary Hypertension in neonates.

Novel modes of delivery may warrant consideration in such cases.

3.7.3 New Treatment Paradigms

Maintenance therapy is now eliciting interest for treating various cancers by preventing re-emergence following initial aggressive treatment. Oral agents such as tyrosine kinase inhibitors would be convenient for such chronic dosage. Biomarkers are increasingly being used, have been shown to be useful in cancer therapy and can help identify target plasma profiles for such maintenance treatment.

3.7.4 Drug/Drug Combinations

Combination drug therapy is common for treating cancer, diabetes, cardiovascular conditions, and infectious diseases. In many cases the drugs are combined in the same dosage unit to help convenience and compliance. Such combinations are usually formulated as simple "immediate release" units. However, simultaneous release of both (or more) drugs may not be optimal for absorption or modes of action.

- Statins are more effective when dosed in the evening. Combinations with antihypertensives might be designed so that the statin is readily released after night time dosage with release of the anti hypertensive delayed and prolonged to align with time of maximum cardiovascular need.
- The antiviral Ritonavir inhibits CYP450. Other antivirals are metabolized by the same enzyme. Cocktails of antiretroviral drugs are used to treat HIV; some are metabolized by CYP450. Rapid release of a CYP450 inhibitor like Ritonavir, with delayed release of other drugs that may be CYP450-susceptible may result in prior inhibition of the CYP450 and greater overall bioavailability. This might lead to reduced dosing in an area where drug loading is very high.
- Doxorubicin, dosed in the morning, with Cisplatin dosed in the afternoon/evening enhances survival in ovarian cancer. These drugs are administered parenterally but the mechanism for the synergy, if elucidated may be applicable to combination oral treatments for cancers.

3.8 Conclusions

The successful design of modified release drug delivery systems requires knowledge of the technologies, materials, and release mechanisms used to effect such modifications. Such knowledge ensures that the most appropriate system is adopted for the drug in question. It is also vital for the pharmaceutical development scientist to be familiar with the anatomy and physiology of the GI Tract and of the factors that determine and control drug absorption. Knowledge of the molecular biology of the clinical condition and of that predicating the mode of action of a drug should also be incorporated in dosage form design considerations. Such information may identify the constraints as well as the opportunities for modifying drug release but will also lead to a better decision.

If the promise of Biomarkers is sustained judicious use of these, together with other considerations related to the disease and treatment strategy may mean that dosage forms can be conceived as part of new drug development programs to a far greater extent than has been hitherto possible.

References

1. Spear BS, Heath-Chiozzi M, Huff J (2001) Clinical application of pharmacogenetics. Trends Mol Med 7(5):201–204
2. Kharasch ED et al (2004) Quinidine as a probe for the role of P-Glycoprotein in the intestinal absorption. J Clin Pharmacol 44:224–233
3. Evans WF, Relling MV (1999) Pharmacogenomics: translating functional genomics to rational therapeutics. Science 286:487–491
4. Gibbs J, Hyland R, Youdim K (2006) Minimizing polymorphic metabolism in drug discovery. Drug Metab Dispos 34(9):1516–1522
5. Amidon GL, Leesman GD (1993) US Patent 5,229, 131 (20 July 1993)
6. Cvetkovic RS, Goa KL (2003) Lopinavir/Ritonavir: a review of its use in the management of HIV infection. Drugs 63(8):769–802
7. Andes D, Craig WA (1998) In vivo activities of amoxicillin and amoxicillin-clavulanate against Streptococcus pneumonia: application to breakpoint determinations. Antimicrob Agents Chemother 42(9):2375–2379
8. Craig WA (2003) Pharmacokinetic and pharmacodynamic factors in antimicrobial research. Infect Dis Clin North Am 17:479–501
9. Woodnutt G, Berry V (1999) Efficacy of high-dose amoxicillin-clavulanate against experimental respiratory tract infections caused by strains of Streptococcus pneumonia. Antimicrob Agents Chemother 43:35–40
10. Kaye C et al (2001) The clinical pharmacokinetics of a new pharmaceutically enhanced formulation of amoxicillin/clavulanate. Clin Ther 23(4):578–584
11. Jacobs MR (2005) Extended release amoxicillin/clavulanate: optimizing a product for respiratory infections based on pharmacodynamic principles. Expert Rev Anti Infect Ther 3(3):353–360
12. Peetz D et al (2005) Glycogen phosphorylase BB in acute coronary syndromes. Clin Chem Lab Med 43(12):1351–1358
13. Rabitzsch G et al (1995) Immunoenzymometric assay of human glycogen phosphorylase isoenzyme BB in diagnosis of ischemic myocardial injury. Clin Chem Lab Med 41(7):966–978
14. Rao SP, Miller S, Rosenbaum R, Lakier JB (1999) Cardiac troponin I and cardiac enzymes after electrophysiologic studies, ablations, and defibrillator implantations. Am J Cardiol 84(4):470
15. Nissen NI, Ranlov P, Weis-Fogh J (1965) Use of cardiac markers in the emergency department: eMedicine. Br Heart J 27:520–526. http://emedicine.medscape.com/article/811905-overview

16. Christenson RH et al (1997) Assessment of coronary reperfusion after thrombolysis with a model combining myoglobin, creatine kinase–MB, and clinical variables. Circulation 96(6):1776–1782
17. Lewis GD, Wei R, Liu E et al (2008) Metabolite profiling of blood from individuals undergoing planned myocardial infarction reveals early markers of myocardial injury. J Clin Invest 118(10):3503–3512
18. Anar YA, White WB (1998) Chronotherapeutics for cardiovascular disease. Drugs 55(5):631–643
19. Lichtenberger L (2004) Method and compositions employing formulations of lecithin oils and NSAIDs for protecting the gastro-intestinal tract and providing enhanced therapeutic activity. US Patent 2004/0077604

Chapter 4
Animal Model Systems Suitable for Controlled Release Modeling

Steven C. Sutton and Philip L. Smith

Abstract Studies in animals are far more accessible than human studies and they are less expensive. Setup and data generation are much shorter and animal studies can be attractive as a preclinical option to guide dosage form design of controlled release formulations. However, differences in gastrointestinal (CV) anatomy, environmental conditions in the GI tract and other factors that can affect release, absorption and drug metabolism can lead to confounding, as well as nonpredictive findings. There is far greater potential for baffling findings than in the case of conventional immediate release (IR) formulations.

It is vital therefore to be aware of interspecies differences, as well as human–animal differences so that the most appropriate species is chosen for testing specific drug formulations and that findings are carefully interpreted with respect to their relevance to performance in humans.

This chapter presents and discusses the anatomical, physiological and metabolic differences that can be encountered in a range of animals and the possibilities and limitations of such testing of controlled release formulations.

4.1 Physiological Factors

Successful development of controlled release (CR) dosage forms requires an understanding of the interactions between the dosage form and physiological factors of the gastrointestinal tract (GIT) including: transit, pH, microflora, transepithelial transport mechanisms, and metabolism. Each of these will be discussed with relation to those aspects which are relevant to design and development of a controlled release dosage form.

S.C. Sutton (✉)
Department of Pharmaceutical Sciences, College of Pharmacy,
University of New England, Portland, ME, USA
e-mail: ssutton1@une.edu

C.G. Wilson and P.J. Crowley (eds.), *Controlled Release in Oral Drug Delivery*, Advances in Delivery Science and Technology, DOI 10.1007/978-1-4614-1004-1_4,

4.2 Intestinal Transit

Passage of substances aborally in the GIT is mediated by coordinated contractions of the surrounding smooth muscle layers [1–4]. Gastric smooth muscle is responsible for:

- Ensuring that the volumes of material presented following a meal are accommodated.
- Contracting to provide for optimal mixing with gastric secretions.
- Metering the products of digestion into the small intestine at a rate that provides for optimal interaction with intestinal, pancreatic and biliary secretions and allows for maximal absorption of nutrients.

In general, the stomach is not a site of drug absorption.

Intestinal smooth muscle is arranged in three major layers. The muscularis mucosa is situated closest to the epithelium. Contractions of the muscularis mucosa aid in mixing the luminal contents from the crypts to the villus tips and ensure optimal contact with the surface of the columnar epithelial cells. Circular and longitudinal muscles surround the muscularis mucosa and are responsible for segmentation of the luminal contents and for the longitudinal movement of materials aborally [3]. The small intestine is the site of a variety of transport mechanisms specialized for absorption of many substances including sugars, amino acids, vitamins, minerals, and also therapeutics.

In the colon, absorption of fluid and electrolytes occurs along with some vitamins. Since the colonic epithelium does not contain the variety of specialized nutrient transport mechanisms found in the small intestine, absorption of therapeutics within this segment occurs predominantly by passive mechanisms.

Compared to the human intestine, the GITs of the carnivores (dog and cat) are relatively short and simple [5]. For example, the cat's and dog's small intestine is about 25% and 50%, respectively, of the length of the human small intestine [5]. The colons of the cat, dog, and rhesus monkey are all roughly 40 cm in length – which is about 25% of the human colon [5, 6].

4.2.1 *Gastric pH*

Gastric pH may be the most thoroughly studied GIT parameter. Stomach pH varies as a function of age (older has lower pH), location (anterior pH > posterior pH [5]), migrating myoelectric complex (MMC, housekeeper wave), disease, stress, food, and time. Because of these influences, an individual may have different gastric pH on different days. And, as has been reported for many species, the range of gastric pH may overlap somewhat with humans, for some period of time under many of these conditions.

For example, the fasted stomach pH in humans has been reported as 1.7 (range: 1.4–2.1) [5], 1.5–7.1 [7] and 1.1 (1–3.2) [8]. The maximum fed pH in humans has been reported as 5.0 (4.3–5.4) [5], or 3.6 (2–7) [8]. In the fasted dog, pH has been reported as 1.5 (0.9–2.5) [8], mean ± standard deviation (SD): 2.0±0.6 [9], 6.8 [10, 11]. Fifteen minutes after a meal of 10 or 200 g "dog food," the pH has been recorded as 2.1 [8], and did not increase at any time after the meal [9]. In contrast, 30 min after a meal of 50 g "dog food," the pH was recorded as 4, and fell to pH 2 at 120 min [12]. From their work, Sagawa concluded that pharmacologic intervention may be needed (to increase gastric pH) when the fed dog model is used to predict performance in fed humans for pH sensitive formulations or drug whose solubility is pH-dependent [9].

4.2.2 Gastric Emptying

Before the formulation is emptied from the stomach, various methods may be applied to prolong gastric residence [13]. The impact of a species' pylorus diameter and gastric contractions on an integral tablet's dimension has been discussed [14, 15]. The impact of food on gastric emptying has also been well studied. However, in comparing the effect of food on gastric emptying in different species, the meal composition needs careful consideration. Calories, volume, consistency, fat, protein, carbohydrates all play important roles in gastric emptying, as well as species-dependent physiologic response to the stimulus. Only when the meals are carefully controlled over a wide range can one determine the advantages and disadvantages of a particular animal model. Additionally, the meal may need to be different for an animal model that mimics the impact of food on gastric emptying vs. solubilizing an active pharmaceutical ingredient (API) that has poor wetting and/or aqueous solubility characteristics. Some investigators have attempted to mimic the Food and Drug Administration (FDA) high fat meal by simply blending the entire meal, and then feeding the meal to dogs [16]. This meal corresponds to 1,000 calories (kcal), 50% of which comprises fat.

In one pharmacokinetic project, the lipophilic drug ziprasidone HCl was orally administered to ten dogs in the fasted state and with several meals. Gastric pH and emptying was also recorded in matching studies with a radio-telemetric Heidelberg capsule (Heidelberg International, Electro-Medical Devices, Inc. Norcross, Ga.). The typical meal is fed once a day and consists of 300–350 g dry dog food (e.g. ProLab Canine 1600, PMI Feeds Inc., St. Louis, MO: Crude protein 21%, Crude Fat 8%, Crude Fiber 5.5%, moisture 11%) totaling 1,470 kcal. The standard high fat breakfast was scaled down for a 10 kg dog with a combination of dry dog food and olive oil. Olive oil was selected due to its precedence in food studies [17]. The high-fat meal consisted of 14 g of dry dog food and 8 g of olive oil (9.1 g fat (41%), 128 kcal). After a 40 h fast (water available) the test meal was completely eaten by the dog. Preliminary studies suggested that the usual rations of dry dog food (approximately 300 g per dog) inhibited gastric emptying of the Heidelberg capsule for >12 h. A meal consisting entirely of 50 g dry dog food was selected as a low-fat meal (4 g fat (8%), 210 kcal).

The results of these meals on gastric emptying and pH are shown in Table 4.1. The peak gastric pH was tabulated for fed dogs. Gastric pH in the fasted dog seemed

Table 4.1 Summary of gastric pH and gastric emptying time (GET) values. Meals shown are "Normal rations" (300 g dry dog chow), "Low-fat" (50 g dry dog chow) and "High-fat" (a mixture of 14 g dry dog chow and 8 g olive oil)

		Meal		
Parameter	Fasting	Normal rations	Low-fat	High-fat
pH[a]	2.3	3.5	3.6	2.6
GET (h)	1.0±0.8	12–24	4.2±0.9*	6.5±1.7**
n[b]	10	2	9	4

[a]For the fasted condition, only the average pH is shown; the peak pH values are shown for all fed conditions (see text for details)
[b]Number of studies
* Significantly different from fasted (ANOVA, $p<0.01$)
** Significantly different from the low fat (ANOVA, $p<0.01$)

Table 4.2 Summary pharmacokinetics after ziprasidone HCl administration to dogs under fasted and fed conditions. See Table 4.1 for meal definitions

		Meal	
Parameter	Fasting	Low-fat	High-fat
$AUC_{0-\infty}$ (ng/ml-h)	1,049±420	1,891±452*	2,015±368*
C_{max} (ng/ml)	196±117	282±122*	278±41*
T_{max} (h)	2.1±1.0	3.1±1.5	4.1±2.1*

* Significantly different from fasted (ANOVA, $p<0.01$)

to decrease slowly to a baseline value of pH 1 after capsule administration. The range of gastric emptying times (GET) in the fasted state was 5–140 min, with a mean value of 60 min.

Compared to fasting, the low fat meal significantly ($p<0.01$) increased the GET. The range of GET in dogs fed the low fat meal was 2.4–4.8 h with a mean of 4.2 h. The high fat meal also significantly increased ($p<0.01$) the GET to a value of 7.4 h. The GET after the high fat meal was significantly longer than the GET value after the low fat meal ($p<0.01$).

The GET of the Heidelberg capsule in humans fed a similar high-fat breakfast was 4.8 h [18]. It is not obvious why in the dog the *low fat* meal delayed gastric emptying to a similar extent as the standard *high fat* breakfast delayed gastric emptying in humans. Since fat content in meals has been reported as a major factor for gastric emptying in the dog [19], perhaps the dog required an extraordinary amount of time to digest such meals: once the fat had been digested, gastric emptying of the Heidelberg capsule occurred.

For the pharmacokinetic portion of the study, one capsule containing the commercial blend of 40 mg ziprasidone HCl was administered, followed immediately by an oral gavage of 50 ml of tap water. The peak plasma concentration (C_{max}), time to peak plasma concentration (T_{max}), and area under the plasma concentration/time curve ($AUC_{0-\infty}$) for dogs administered ziprasidone HCl under fasted and fed conditions are listed in Table 4.2. The presence of food increased the average plasma ziprasidone concentration in dogs (Fig. 4.1). Interestingly, both high and low fat meals produced a significant ($p<0.01$) increase in $AUC_{0-\infty}$.

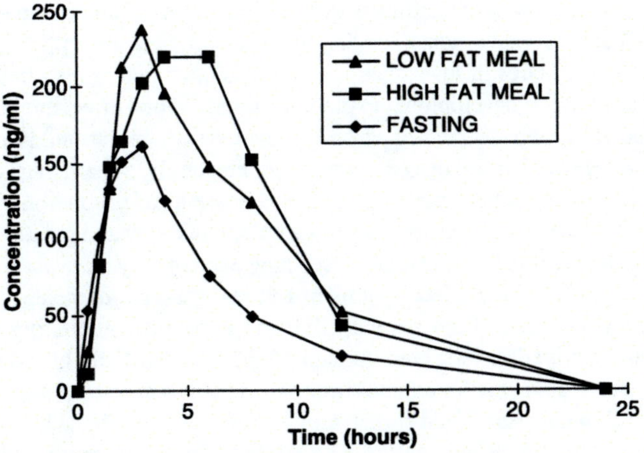

Fig. 4.1 Average ziprasidone concentrations in dogs after a single dose of ziprasidone HCl capsules under fasting, low-fat and high-fat fed conditions

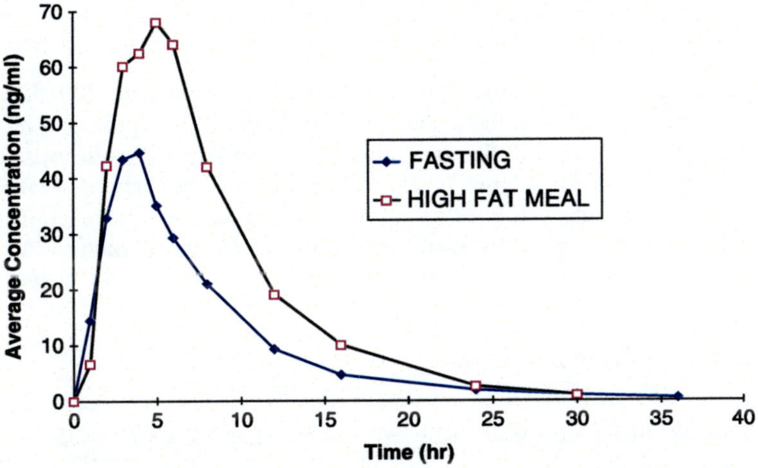

Fig. 4.2 Average ziprasidone concentrations in humans after a single dose of ziprasidone HCl capsules under fasting and high-fat breakfast conditions [20]

A similar food interaction was reported for ziprasidone in healthy volunteers [20]. In that study, the standard high fat breakfast caused a statistically significant increase in $AUC_{0-\infty}$ following 20 mg ziprasidone (1.7-fold, Fig. 4.2). Ziprasidone is a lipophilic drug with a clogP of 3.6 [21]. Increased exposure to lipophilic drugs administered with a meal has been frequently reported [22]. This observation has been explained by:

- Increased bile flow
- Prolonged gastric emptying in response to the meal

Bile contains bile salt and lecithin, which may improve solubilization of lipophilic compounds in fatty food contents, leading to more complete absorption [23].

The pharmacokinetics of investigational drugs is often determined in dogs during the preclinical phase of development to predict whether a food effect might be present. In canine studies, the low-fat meal apparently provided sufficient stimulus for increased bile flow, leading to enhanced drug absorption. Interestingly, a sham meal causes little bile secretion in humans but stimulates a 30% increase in bile secretion in dogs [24]. The dog also secretes a significantly larger volume of bile/day and greater amounts of bile salts than humans [5]. These differences may explain why a *low fat* meal in dogs mimicked the positive food effect in humans following a *high fat* meal.

In summary, because of the similar GET and apparent solubilizing affects, the low fat meal described, or a scaled version of the blended FDA high-fat breakfast may be the preferred meals for canine studies designed to predict a food effect in the clinic [109]. However, until additional studies have been published, it is uncertain how universally this recommendation may apply to various formulations and other lipophillic compounds.

4.2.3 Small Intestine pH

Formulations which rely on the small intestine pH to liberate the API often use enteric polymers. Therefore, the pH at which the polymer is expected to dissolve and the pH of the small intestine is a critical factor in the API's release from an enteric formulation. If the polymer is selected to dissolve at the pH typically reported for humans (duodenum: 4–5.5 [8]), a study in the beagle dog (duodenal pH often reported as 1 pH unit higher) [6] may result in too rapid a release of the API.

4.2.4 Small Intestine Transit

Transit studies of liquids, beads, and larger noneroding tablets and capsules may give different relative values for the small intestine transit time (SITT). For example, the time for 50% of the marker (liquid or beads) to transit the human small intestine was (mean ± SD): 4.1 ± 0.38 h (liquid); 3.3 ± 0.86 h (beads); 3.2 ± 0.30 h (single unit) [15], 180 ± 30 min (single unit) [25]. After the FDA high fat meal, the SITT in humans of the single-unit marker was 3.4 ± 0.5 h [15]. Most studies have shown that the SITT is relatively independent of feeding state, regardless of marker [26]. However, sources of SITT variability are the MMC and *food-induced acceleration*. For example, an MMC may propagate a formulation just "ahead" of the MMC – resulting in a short SITT – while the formulation just "behind" the MMC would have an SITT that was approximately 2 h longer [27]. When food was administered 45 min after the formulation in one study of fasted subjects, the SITT was decreased from 200 to 100 min [28].

The SITT was reported for dog as 1.9 h [29] and 111 ± 17 min [25], 3.3 h for rat [30], 3–4 h for minipig [31], 2.3–2.5 h for cynomolgus monkey [32]. However in the dog, the range of SITT was from 15 to 206 min compared to 180 to 300 min in humans [29]. Gamma scintigraphy imaging was useful for precisely determining the small intestine residence time specific to each study (see below).

4.2.5 *Intestinal Microflora*

More than 400 species of bacteria can be found throughout the GIT. The greatest number is present in the distal small intestine and large intestine in man and rabbit [5, 33]. Gastrointestinal bacteria primarily function to hydrolyze carbohydrates and proteins that are not metabolized in the upper intestine. Another important aspect of bacterial metabolism in the intestine is the hydrolysis of glucuronide conjugates; these are important components of the enterohepatic circulation of compounds [34]. In guinea pig, rat, mouse, monkey, dog and pig not only are there significant numbers of bacteria in the distal small intestine and colon but also in the stomach and proximal small intestine [5]. An additional observation has been the variability in the types of microorganisms present in these segments [5]. Approaches to target the colon by taking advantage of bacterial metabolism present in this intestinal segment have been most successful for treating local disease (e.g. ulcerative colitis) with prodrugs utilizing an azo linkage such as salicyl azosulfapyradine [35]. Studies designed to evaluate delivery of steroid drugs glycosylated with galactose, glucose, or cellobiose (known colonic bacterial substrates) were evaluated by incubation with homogenates from rat stomach, proximal ileum, distal ileum, or cecum [36, 37]. Homogenates from all segments hydrolyzed the conjugates although hydrolysis rate in the cecum was greater than in the other segments. These studies indicated that the specificity of the glycosylated steroids in the colon resulted from their longer residence time along with the relatively more rapid rate of transit and slower rate of hydrolysis in the upper intestine. Subsequent studies to evaluate this colon-specific delivery approach in animal models of inflammatory bowel disease provided encouraging results [38–42]. Despite these promising results in animal models, it was reported that hydrolysis rates of the prodrugs in man were significantly less than in rat [43]. Saffran and coworkers employed rat and dog models to evaluate polymer-coated peptide drugs (insulin, lysine-vasopressin) designed for release by the action of colonic bacteria [44, 45]. Release of peptide drugs can be achieved in the colon, however development of products based on this approach for systemic delivery have not been forthcoming. Basit posits that the potentially carcinogenic hazard of the azo-aromatic linkage on these compounds required for the bacteria activation may explain why no further development of this approach has been reported [46].

In addition to carbohydrates and proteins, gastrointestinal bacteria may also be involved in the metabolism of drugs to alter activity and/or safety [33]. Anaerobic bacteria in the colon have been shown to degrade drugs in humans [33] and in the dog [14, 47]. Thus, for drugs that are substrates for bacterial metabolism, the development of extended-release or colon-specific controlled release systems may not succeed.

4.2.6 Intestinal Transport

Intestinal absorption requires the active component of a dosage form to first be released into solution in the lumen of the GIT followed by passage across the intestinal epithelial barrier into the circulation. Passage across the GIT can occur by a number of pathways including: passive paracellular or passive transepithelial transport, carrier-mediated transport facilitated by a specific membrane carrier either passive or active and low capacity endocytic pathways [48, 49]. These transporters can be generally classed as solute carriers (SLC) [50]. In addition, there is another set of transporters responsible for drug efflux which consists of an ATP-binding cassette and are classified as ABC transporters [50]. For compounds that are absorbed by carrier-mediated mechanisms, there may be regional specificity (e.g. carrier-mediated transport of antibiotics in the small intestine) whereas passive transcellular or paracellular transport may occur along the entire length of the intestine [51]. In addition, it has been shown that there are regional differences in the distribution of both SLC and ABC transporters along the length of the intestine and across different species [50, 52–54]. In the development of CR dosage forms, it is important to know the sites and mechanisms of absorption to allow a dosage form to be tailored to meet these specific requirements. Prior to the evaluation of CR dosage forms in vivo, it is often helpful to evaluate mechanisms of transport using in vitro cell culture or tissue preparations [55, 56].

4.2.7 Intestinal Metabolism

In addition to species differences in general metabolism, species differences in gut metabolism may confuse the interpretation of PK results with CR formulations [88, 89]. Whenever dog studies are completed, polymorphic metabolism should be considered and animals should be characterized for phenotype [110].

4.3 Animal Models

4.3.1 Rat/Mouse

Based on its size and minimal requirements for drug substance, the rat has been employed in a myriad of physiological, formulation development, and safety assessment studies [57]. Drug absorption after oral administration to the rat has been reported to be predictive of human [58]. However, because of the size of the rat GI tract, adjustments must be made to orally administered nondisintegrating formulations. For example, solid formulations are usually administered as a powder or as small coated granules, directly into the stomach [59] or intestinal segments [60].

The rat has been reported to have some predictive potential for a certain gastric mucosa adhesion formulation [61]. In one study, the release of furosemide in non-adhesive and adhesive microspheres was examined. The adhesive polymers were cross-linked polyacrylic acid derivatives (carboxyvinyl polymer) dispersed in a matrix of polyglycerol esters of fatty acids. The (dry) microspheres were administered via an intubation tube into the stomach, followed by 1 ml of water. Rats administered the adhesive formulation demonstrated prolonged, elevated plasma furosemide concentrations compared to the nonadhesive formulation. This was predictive of the subsequent clinical findings with the same two formulations [62]. While this result appeared to support the rat model, the reader is cautioned not to apply these findings to all mucoadhesive formulations [63].

Enteric-coated microspheres have long been administered to rats. Cellulose acetate phthalate (CAP)-coated microspheres administered into the throat of rats resulted in the expected protective benefit of enteric formulation for an acid-labile vaccination of an enterotoxin [64]. More recently, an enteric polymer incorporated into nanospheres was compared to traditional enteric-coated microparticles in the rat. Interestingly, the colon targeting property of the nanospheres was superior to that of the enteric microspheres, resulting in a "superior therapeutics efficacy" in a rat colitis model [59].

The rat has been reported to be useful to evaluate colonic targeting formulations using pH, time and enzyme release mechanisms of action. According to Smrdel et al., "the colon-specific delivery of the drug using microcapsules relies on the combination of pH (outer gastro-resistant coating), time (inner retard coating of Eudragit® RS and RL,) and enzyme (pectin core) controlled drug release mechanisms." The results showed improved performance in the ulcerative colitis rat model over an orally or rectally administered suspension of the API [65].

Recently, a hydrogel CR formulation has been shown to release theophylline in rats. The mechanism of action for this formulation was polyionic complex based on xanthan–chitosan. While no data on humans were presented, the formulation was easily administered to rats and results were promising [66].

4.3.2 Dog

There are numerous examples in the literature of pharmacokinetic evaluation of human dosage forms using the dog [14, 61–75]. The beagle dog has been the most common breed employed for these models, although some studies with mongrel and large dogs have also been reported [67]. Various inconsistencies between dog and human pharmacokinetic studies for single unit CR dosage forms are attributed to the following:

- Size of pylorus aperture [12, 68]
- Stomach contractions [69]
- Hydrodynamic flow [70]
- Food content [16]
- GI fluid volume [5]
- Metabolism and transport [87–89]

While these factors have been problematic in predicting size-controlled gastric retention [71], the dog has been reported to have some predictive potential for a certain gastric mucosa adhesion formulation. Although these reports were not imaging studies, pharmacokinetic results were impressive; the area under the plasma theophylline vs time curve ($AUC_{0-8\,h}$) doubled for the adhesive microspheres [61]. Such improvement was similar to that reported for these formulations in humans [62].

Most published CR pharmacokinetic canine studies have examined the pharmacokinetics of an orally administered API which exhibited good solubility and permeability in humans. When osmotic formulations were administered, the in vivo performance of the formulation closely predicted the performance in humans [72, 73].

For matrix tablets administered to dogs, the results were usually predictive of performance in humans when the underlying mechanism for release was primarily diffusion [74, 75]. If erosion was the primary mechanism of release, the formulation appeared to have an accelerated release compared to humans. This was attributed to the stronger contraction of the canine stomach [69] and faster hydrodynamic flow [70]. In a pharmacoscintigraphic study where 6 or 18 h diffusion-controlled matrix tablets were administered to fed and fasted beagle dogs, the variability caused by the above factors obfuscated the conclusions [16]. When accelerated tablet erosion in the stomach was completed within 2 h, there was no correlation with in vitro dissolution results. However as will be discussed in a later section, scintigraphic imaging permitted the identification of subjects without accelerated tablet erosion. For example, as in another study, the dogs without accelerated tablet erosion – as determined by scintigraphy – correlated with its in vitro release [14]. Consistent with the human studies (unpublished data), the fed dog studies suggested that the robustness of the formulation needed to be improved.

When the API had a low permeability in humans, correlation with dog studies was variable; relative bioavailability (RBA) could be over or under-estimated. Regardless, the shorter dog SITT must be reckoned with when evaluating the performance of long-duration formulations. Such formulations may appear to have a truncated release in dogs, yet perform as designed in humans [76]. In these cases, scintigraphy may again permit the proper selection of only those animals with longer SITT for prediction of the formulation performance in humans.

The small intestine in the dog has been reported to be more permeable than human to certain APIs, resulting in overprediction of the bioavailability of certain compounds [77]. This species difference may argue in favor of the dog model for evaluation of the in vivo performance of a CR formulation of a compound with poor permeability. Such a study may be predictive for the *short-duration* release (CR) formulation in dogs, but less so for formulations of longer duration of release.

When the API had poor aqueous solubility or wetting characteristics, the dog also provided some critical formulation performance information. However, the interpretation for prediction of the RBA in humans was again difficult. The dog colonoscopy model was an early attempt to overcome some of the GI physiology differences between the dog and human [78]. This model consisted of an intubation of the API (solution or suspension formulation) via the rectum, directly into the ascending colon. Eleven compounds – including examples in each of the four Biopharmaceutics Classification System (BCS) cases were studied in this model and in humans [79]. The correlation between the RBA in dogs and humans was 0.92, suggesting that the dog was a good model for predicting this relative absorption in humans.

A further development of the dog colonoscopy model was the "tablet insertion method" (TIM). In this model, a CR tablet that had been "primed" in saline for a specified period of time was inserted directly into the dog colon using a colonoscope. Serial blood sampling from the dog provided a measure of how well the formulation performed in situ. In one osmotic tablet formulation, a solubility-enhancing formulation was delivered, and the TIM provided an in situ evaluation of how well the formulation enhanced the solubility and subsequent absorption of the API. The TIM dog model indicated a 400% improvement in colon absorption for the "solubility-enhanced" formulation compared to a standard CR formulation [80].

Interest in colonic drug delivery for treatment of ulcerative colitis and cancer has been steadily increasing [46, 81]. Strategies for colonic targeting include pH, transit time, pressure, and bacteria [46]. While in some dogs the gastric pH is variable [9], the pH of the colon is ≈6.5 in both dog [5] and human [81]. Furthermore, colon transit time in both dogs and humans is sensitive to diet, and colon transit times overlap [14, 82–84]. Finally, the similar performance of a pressure-activated dosage form in dog [85] and human [86] suggested that this parameter was also similar in both species [46].

4.3.3 *Cat*

Perhaps because of the potentially fractious nature of cats, their use has not been as popular as the dog for studies to predict the performance of CR formulations in human. However, there may be some circumstances where the cat should be considered. For example, nicotine and coumarin are two drugs metabolized by cytochrome p450-2A [87], and the cat (not the dog) presents with cytochrome p450-2A [88, 89]. It is possible that for some APIs, the intestinal first pass extraction of the API is saturated after oral administration of the immediate release (IR) formulation, but not after the CR formulation. In this case, the formulation release profile may have to be optimized to account for such metabolizing enzymes. If the in vivo release parameters are calculated from the deconvolution of pharmacokinetic data (instead of scintigraphy), evaluation of the CR formulation in the cat would probably be more predictive of human performance than a dog study for nicotine and coumarin.

A few studies have examined the performance of enteric-coated formulations in cats. In one study an enteric-coated formulation was orally administered to cats to treat a feline acquired immune deficiency disease. The AUC with an enteric formulation of thiophoscarnet was 50% better than oral gavage [90]. In another study, CR tablet of methimazole was shown to provide sustained blood levels over a 24 h period. Twelve hours after the conventional (IR) tablet, plasma methimazole concentrations (C_{12}) averaged about 0.20 μg/ml, while after the CR tablet, $C_{12} \approx 0.60$ μg/ml [91].

4.3.4 Pig/Minipig

Although the GI physiology of the pig/minipig has some similarities with humans [92], the 10 h fasted GET for particulates makes it an unlikely candidate as a model of *orally* administered single-unit formulations [5]. Furthermore, the half-life for the gastric emptying of liquids was reported as 1.4 h (cf. humans 11 min) [31]. One possible "work around" for such a problem is to bypass the stomach. This has been accomplished in the dog using a Thiry-Vella fistula [93], and a modified "vascular access port" [94–96].

A few reports have demonstrated the utility of the pig to prove the in vivo performance of CR formulations [97–99], but no comments on their predictive value as a model for humans were made.

4.3.5 Monkey

Of the species discussed so far, the monkey is genetically closest to human, however there are some GI physiology differences that may impact the predictive utility of this model. For example, the cynomolgus monkey appears not to have predictive potential for a certain gastric mucosa adhesion formulation [61]. Despite having a duodenum which is only about 20% the length of human [6], the fraction absorbed in "rhesus and cynomolgus macaques" appeared to predict the F_a for humans [100]. Enteric-coated beads, tablets, and CR tablets have been administered to the cynomolgus monkey [101]. The performance of a CR ethylcellulose-coated multiparticulate formulation was examined in vitro and in the cynomolgus monkey [102]. The formulation appeared to perform in the monkey as intended, according to in vitro dissolution data. Note that the short GIT in the cynomolgus monkey was a clear disadvantage to an attempt at an IVIVC for the 8-h release diffusion-controlled matrix tablet formulation [103].

4.4 Pharmacoscintigraphic Studies

The science of imaging has been useful for gaining insights into the mechanisms of formulations and for troubleshooting difficult drug delivery problems.

In one project [14], a matrix formulation of poorly water soluble API was orally administered to dogs. A few milligrams of samarium oxide ($^{152}Sm_2O_3$) were incorporated in the formulation, which was compressed into tablets with standard equipment and then sent out to a research reactor for neutron activation. Activation changed the stable samarium 152 isotope to the gamma emitting radioactive samarium 153 isotope (≈47 h half life). The gamma-labeled formulations were then shipped back to the study facility and were orally administered to dogs in a pharmacokinetic study. Following administration, images were collected to determine dosage form location in the GIT. The purpose of the study was to determine whether the in vivo release of the API from the matrix tablet was at the rate designed from in vitro dissolution experiments. This was determined using the deconvolution method of calculation [104]. The gamma scintigraphy imaging provided a visual confirmation of GI tract location during the study. Findings indicated that, while the formulation was in the small intestine, in vivo and in vitro release closely correlated. Implicit in this study design is the assumption that liberated API would rapidly wet, dissolve, and be absorbed. Therefore, the lag between release and appearance in the systemic circulation could be ignored.

In another study [80], the API did not rapidly wet and dissolve, leading to a lag in both absorption, and appearance in the systemic circulation. In this case, the poorly soluble (but otherwise well absorbed) API was formulated as a single CR unit. Although the API was delivered from the osmotic-driven formulation, the purpose of the study was to determine whether a solubility-enhancing formulation provided sustained supersaturation of the API, with subsequent absorption in the colon. A pharmacoscintigraphic dog study again provided some insight into in vivo performance. Taking into account the point at which the dosage form entered the colon (arrows, Fig. 4.3), it was observed that the bioavailability of the API from the solubility-enhanced formulation was at best only about 20% better than the traditional formulation. A subsequent clinical pharmacoscintigraphic study in humans revealed that colonic absorption of the solubility-enhanced formulation was also similar to that from the traditional formulation ("cap," in Fig. 4.4). The authors concluded that the pharmacoscintigraphic studies provided critical additional information to allow for a better prediction of the in vivo performance of the formulation in humans.

With the power of imaging studies comes the confidence to cull "outliers" based on artifacts of GI physiology. In the pharmacoscintigraphic study discussed earlier where 6 or 18 h diffusion-controlled matrix tablets were administered to fed beagle dogs, erosion of some tablets was accelerated in the stomach [16]. In dogs where tablet erosion was not excessive before gastric emptying, the in vivo performance predicted performance in humans. In another project, the in vivo release of sertraline HCl from a matrix tablet was highly variable in dogs [14]. Since sertraline is

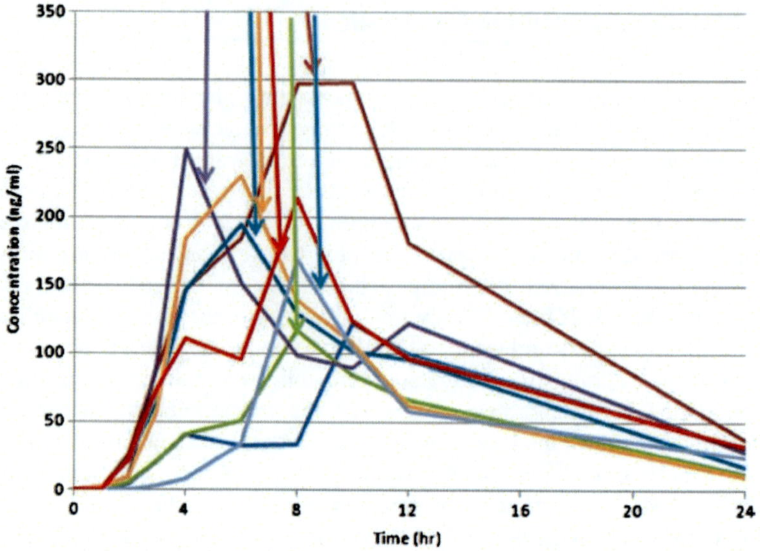

Fig. 4.3 Plasma concentrations in individual dogs administered an osmotic formulation in a pharmacoscintigraphic study. *Arrow heads* indicate the time at which each tablet entered the colon [80]

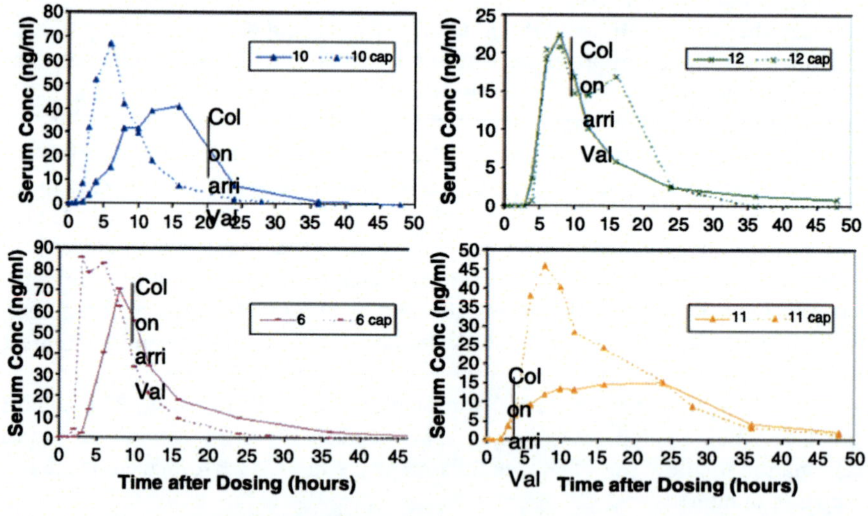

Fig. 4.4 Plasma concentrations in individual subjects administered traditional ("cap") and "solubility enhanced" osmotic formulations in a pharmacoscintigraphic study. *Vertical lines* indicate the time at which the formulation entered the colon [80]

poorly water soluble, its release in the colon resulted in low blood levels, contrasting with release in the small intestine which resulted in high blood levels. Therefore, when the location of the eroding formulation was matched with the deconvolution analysis, there was a satisfactory in vivo–in vitro correlation (IVIVC).

These examples emphasized the value of pharmacoscintigraphic studies during the preclinical development of CR formulations.

4.5 Conclusions

From the preceding discussion, it is apparent that there is currently insufficient information to identify one animal model for use in evaluation of all CR dosage forms. This conclusion results from the observations that in terms of physiology, anatomy, location and density of transporters and enzymes and location and number of bacteria, there can be significant differences between human and animal models. Despite these differences, it is often possible to rank the relative effectiveness of CR dosage forms within an animal model thereby allowing a reduction in the number of studies to be conducted in man. Additionally, our enhanced understanding of the mechanisms involved in absorption across the GIT allows a more judicious selection of therapeutics and CR dosage forms for further development [105, 106]. Finally, the advent of new methods – such as *in situ* digital imaging – for evaluating CR dosage form performance may provide further support for development of a limited number of animal models [107, 108].

References

1. Meyer JH (1987) Motility of the stomach and gastroduodenal junction. In: Johnson L (ed) Physiology of the gastrointestinal tract, Chap. 19, 2nd edn. Raven, NY, USA, pp 613–629
2. Smith M, Morton D (2001) The digestive system. Elsevier, London
3. Weisbrodt NW (1987) In: Johnson L (ed) Physiology of the gastrointestinal tract. Raven, New York, pp 631–663
4. Christensen J (1987) In: Johnson LR (ed) Physiology of the gastrointestinal tract. Raven, New York, pp 665–693
5. Kararli T (1995) Comparison of the gastrointestinal anatomy, physiology, and biochemistry of humans and commonly used laboratory animals. Biopharm Drug Dispos 16:351–380
6. Dressman J, Yamada K (1991) In: Welling P, Tse F (eds) Pharmaceutical bioequivalence. Dekker, New York, pp 235–266
7. Simmons D (1993) Characterization of the precipitates of poorly water soluble drugs and determination of the solubilities in human gastro-intestinal fluids by microscopy. Drug Dev Ind Pharm 19(10):1103–1112
8. Youngberg CA, Wlodyga BA, Schmaltz S, Dressman JB (1985) Radiotelemetric determination of gastrointestinal pH in four healthy Beagles. Am J Vet Res 46(7):1516–1521
9. Sagawa K, Li F, Liese R, Sutton S (2009) Fed and fasted gastric pH and gastric residence time in conscious beagle dogs. J Pharm Sci 98(7):2494–2500
10. Akimoto MN, Furuya N, Fukushima A, Higuchi K, Shohei F, Suwa T (2000) Gastric pH profiles of Beagle dogs and their use as an alternative to human testing. Eur J Pharm Biopharm 49:99–102
11. Rong Z, Paul M, Christopher H, Lu Z, Mathias N, Teresa NF, Doris AW, Munir AH, Ronald LS, Duxin S (2005) pH-dependent dissolution in vitro and absorption in vivo of weakly basic drugs: development of a canine model. Pharm Res 22(2):188–192
12. Itoh T, Higuchi T, Gardner C, Caldwell L (1986) Effect of particle size and food on gastric residence time of non-disintegrating solids in beagle dogs. J Pharm Pharmacol 38:801–806

13. Waterman KC (2007) A critical review of gastric retentive controlled drug delivery. Pharm Dev Technol 12(1):1–10
14. Sutton SC (2004) Companion animal physiology and dosage form performance. Adv Drug Deliv Rev 56:1383–1398
15. Davis SS, Hardy JG, Fara JW (1986) Transit of pharmaceutical dosage forms through the small intestine. Gut 27:886–892
16. McInnes F, Clear N, Humphrey M, Stevens HN, McInnes F, Clear N, Humphrey M, Stevens HNE (2008) In vivo performance of an oral MR matrix tablet formulation in the beagle dog in the fed and fasted state: assessment of mechanical weakness. Pharm Res 25(5):1075–1084
17. Inskeep PB, Davis KM, Reed AE (1995) Pharmacokinetics of the acyl coenzyme A:cholesterol acyl transferase inhibitor CP-105,191 in dogs–the effect of food and sesame oil on systemic exposure following oral dosing. J Pharm Sci 84(2):131–133
18. Mojaverian P, Ferguson RK, Vlasses PH, Rocci ML Jr, Oren A, Fix JA, Caldwell LJ, Gardner C (1985) Estimation of gastric residence time of the Heidelberg capsule in humans: effect of varying food composition. Gastroenterology 89(2):392–397
19. Mizuta H, Kawazoe Y, Haga K, Ogawa K (1990) Effects of meals on gastric emptying and small intestinal transit times of a suspension in the beagle dog assessed using acetaminophen and salicylazosulfapyridine as markers. Chem Pharm Bull 38:2224–2227
20. Hamelin BA, Allard S, Laplante L, Miceli J, Wilner KD, Tremblay J, Lebel M (1998) The effect of timing of a standard meal on the pharmacokinetics and pharmacodynamics of the novel atypical antipsychotic agent ziprasidone. Pharmacotherapy 18(1):9–15
21. Friesen DT, Shanker R, Crew M, Smithey DT, Curatolo WJ, Nightingale JA, Friesen DT, Shanker R, Crew M, Smithey DT, Nightingale JAS (2008) Hydroxypropyl methylcellulose acetate succinate-based spray-dried dispersions: an overview. Mol Pharm 5(6):1003–1019
22. Welling PG, Huang H, Hewitt PF, Lyons LL (1978) Bioavailability of erythromycin stearate: influence of food and fluid volume. J Pharm Sci 67(6):764–766
23. Gibaldi M, Feldman S (1970) Mechanisms of surfactant effects on drug absorption. J Pharm Sci 59:579–589
24. Calabrese E (1991) Principles of animal extrapolation. Lewis, Chelsea
25. Davis SS, Wilding EA, Wilding IR (1993) Gastrointestinal transit of a matrix tablet formulation: comparison of canine and human data. Int J Pharm 94:235–238
26. Davis SS, Hardy JG, Taylor MJ, Whalley DR, Wilson CG (1984) A comparative study of the gastrointestinal transit of a pellet and tablet formulation. Int J Pharm 21(2):167–177
27. Summers RW, Helm J, Christensen J (1976) Intestinal propulsion in the dog. Its relation to food intake and the migratory myoelectric complex. Gastroenterology 70(5 pt.1):753–758
28. Fadda H, McConnell E, Short M, Basit A (2009) Meal-induced acceleration of tablet transit through the human small intestine. Pharm Res 26(3):740–740
29. Youngberg C (1984) Radiotelemetric determination of GI pH in Man and Dog. Pharmaceutical Chemistry. University of Michigan, Ann Arbor
30. Sawamoto T, Haruta S, Kurosaki Y, Higaki K (1997) Prediction of the plasma concentration profiles of orally administered drugs in rats on the basis of gastrointestinal transit kinetics and absorbability. J Pharm Pharmacol 49:450–457
31. Davis S, Illum L, Hinchcliffe M (2001) Gastrointestinal transit of dosage forms in the pig. J Pharm Pharmacol 53:33–39
32. Kondo H, Takahashi Y, Watanabe T, Yokohama S, Watanabe J (2003) Gastrointestinal transit of liquids in unfed cynomolgus monkeys. Biopharm Drug Dispos 24(3):131–140
33. Basit AW, Lacey LF (2001) Colonic metabolism of ranitidine: implications for its delivery and absorption. Int J Pharm 227(1–2):157–165
34. Shamat MA (1993) The role of the gastrointestinal microflora in the metabolism of drugs. Int J Pharm 97:1–13
35. Peppercorn MA, Goldman P (1972) The role of intestinal bacteria in the metabolism of salicylazosulfapyridine. J Pharmacol Exp Ther 181:555–562
36. Friend DR, Chang GW (1984) A colon-specific drug-delivery system based on drug glycosides and the glycosidases of colonic bacteria. J Med Chem 27:261–266

37. Friend DR, Chang GWJ (1985) Drug glycosides: potential prodrugs for colon-specific drug delivery. J Med Chem 28:51–57
38. Friend D, Phillips S, McLeod A, Tozer T (1991) Relative anti-inflammatory effect of oral dexmamethasone-b-D-glucoside and dexamethasone in experimental inflammatory bowel disease in guinea pig. J Pharm Pharmacol 43:353–355
39. McCleod A, Fedorak R, Friend D, Tozer T, Cui N (1994) A glucocorticoid prodrug facilitates normal mucosal function in rat colitis without adrenal suppression. Gastroenterology 106: 405–413
40. Cui N, Friend D, Fedorak R (1994) A budesonide prodrug accelerates treatment of colitis in rats. Gut 35:1439–1446
41. Fedorak R, Haeberlin B, Empey L (1995) Colonic drug delivery of dexamethasone from a prodrug accelerates healing of colitis in rat without adrenal suppression. Gastroenterology 108:1688–1699
42. Nolen HI, Fedorak R, Friend D (1997) Steady-state pharmacokinetics of corticosteroid delivery from glucuronide prodrugs in normal and colitic rat. Biopharm Drug Dispos 18:681–695
43. Nolen H, Fedorak R, Friend D (1995) Budesonide-beta-D-glucuronide: a potential prodrug for treatment of ulcerative colitis. J Pharm Sci 84:677–681
44. Saffran M, Kumar GS, Savariar C, Burnham JC, Williams F, Neckers DC (1986) A new approach to the oral administration of insulin and other peptide drugs. Science 233:1081–1084
45. Saffran M, Bedra C, Kumar GS, Neckers DC (1988) Vasopressin: a model for the study of effects of additives on the oral and rectal administration of peptide drugs. J Pharm Sci 77:33–38
46. Basit AW (2005) Advances in colonic drug delivery. Drugs 65(14):1991–2007
47. Evans LAF, Sutton SC (2003) Canine model to predict colonic microflora degradation and non-specific adsorption of compounds. AAPS Northeastern Regional Discussion Group Meeting. AAPS, Rocky Hill, CT
48. Smith PL, Wall DA, Gochoco CH, Wilson G (1992) Oral absorption of peptides and proteins. Routes of delivery: case studies. Adv Drug Deliv Rev 8:253–290
49. Mori K, Duh E, Gehlbach P, Ando A, Takahashi K, Pearlman J, Mori K, Yang HS, Zack DJ, Ettyreddy D, Brough DE, Wei LL, Campochiaro PA (2001) Pigment epithelium-derived factor inhibits retinal and choroidal neovascularization. J Cell Physiol 188(2):253–263
50. Anderle P (2009) C. Nielsen. In: van de Waterbeemd H, Testa B (eds) Drug bioavailability: estimation of solubility, permeability, absorption and bioavailability. Wiley, Weinheim, pp 223–276
51. Ghandehari H, Smith PL, Ellens H, Yeh P-Y, Kopecek J (1997) Size-dependent permeability of hydrophilic probes across rabbit colonic epithelium. J Pharmacol Exp Ther 280:747–753
52. Stephens R, Tanianis J, Higgs N, Humphrey M, Warhurst G (2002) Region-dependent modulation of intestinal permeability by drug efflux transporters: In vitro studies in mdr1a (–/–) mouse intestine. JPET 303:1095–1101
53. Hilgendorf C, Ahlin G, Seithel A, Artursson P, Ungell A-L, Karlsson J (2007) Expression of thirty-six drug transporter genes in human intestine, liver, kidney, and organotypic cell lines. Drug Metab Dispos 35:1333–1340
54. Gutmann H, Hruz P, Zimmermann C, Beglinger C, Drewe J (2005) Distribution of breast cancer resistance protein (BCRP/ABCG2) mRNA expression along the human GI tract. Biochem Pharmacol 70:695–699
55. Smith PL (1996) In: Borchardt R, Smith P, Wilson G (eds) Models for assessing drug absorption and metabolism. Plenum, New York, pp 13–34
56. Hidalgo IJ (1996) In: Borchardt RT, Smith PL, Wilson G (eds) Models for assessing drug absorption and metabolism. Plenum, New York, pp 35–50
57. Mashimo T, Serikawa T (2009) Rat resources in biomedical research. Curr Pharm Biotechnol 10(2):214–220
58. Chiou WL, Barve A (1998) Linear correlation of the fraction of oral dose absorbed of 64 drugs between humans and rats. Pharm Res 15(11):1792–1795
59. Makhlof A, Tozuka Y, Takeuchi H (2009) pH-sensitive nanospheres for colon-specific drug delivery in experimentally induced colitis rat model. Eur J Pharm Biopharm 72(1):1–8

60. LeRoy S, Sutton S, Beall P (1988) A method for studying small intestinal transit in the rat. Ann NY Acad Sci 529:131–134
61. Nagahara N, Akiyama Y, Higaki K, Kimura T (2007) Animal models for predicting potency of oral sustained-release adhesive microspheres in humans. Int J Pharm 331(1):46–53
62. Akiyama Y, Nagahara N, Nara E, Kitano M (1998) Evaluation of oral mucoadhesive microspheres in man on the basis of the pharmacokinetics of furosemide and riboflavin, compounds with limited gastrointestinal absorption sites. J Pharm Pharmacol 50(2):159–166
63. Streubel A, Siepmann J, Bodmeier R (2006) Gastroretentive drug delivery systems. Expert Opin Drug Deliv 3(2):217–233
64. Klipstein FA, Engert RF, Sherman WT (1983) Peroral immunization of rats with Escherichia coli heat-labile enterotoxin delivered by microspheres. Infect Immun 39(2):1000–1003
65. Smrdel P, Cerne M, Bogataj M, Urleb U, Mrhar A (2010) Enhanced therapeutic effect of LK-423 in treating experimentally induced colitis in rats when administered in colon delivery microcapsules. J Microencapsul 27(7):572–582
66. Popa N, Novac O, Profire L, Lupusoru C, Popa M (2010) Hydrogels based on chitosan–xanthan for controlled release of theophylline. J Mater Sci Mater Med 21(4):1241–1248
67. Iwanaga Y, Wen J, Thollander MS, Kost LJ, Thomforde GM, Allen RG, Phillips SF (1998) Scintigraphic measurement of regional gastrointestinal transit in the dog. Am J Physiol Gastrointest Liver Physiol 275(5):G904–G910
68. Coupe A, Davis S, Evans D, Wilding I (1991) Correlation of the gastric emptying of nondisintegrating tablets with gastrointestinal motility. Pharm Res 8(10):1281–1285
69. Kamba M, Seta Y, Kusai A, Nishimura K (2002) Comparison of the mechanical destructive force in the small intestine of dog and human. Int J Pharm 237:139–149
70. Katori N, Aoyagi N, Terao T (1995) Estimation of agitation intensity in the GI tract in humans and dogs based on in Vitro/in Vivo correlation. Pharm Res 12(2):237–243
71. Fix JA, Cargill R, Engle K (1993) Controlled gastric emptying III. Gastric residence time of a nondisintegrating geometric shape in human volunteers. Pharm Res 10:1087–1089
72. McClelland GA, Stubbs RJ, Fix JA, Pogany SA, Zentner GM (1991) Enhancement of 3-hydroxy-3-methylglutaryl-coenzyme A (HMG-CoA) reductase inhibitor efficacy through administration of a controlled-porosity osmotic pump dosage form. Pharm Res 8(7):873–876
73. McClelland GA, Sutton SC, Engle K, Zentner GM (1991) The solubility modulated osmotic pump: in vitro/in vivo release of diltiazem hydrochloride. Pharm Res 8:88–92
74. Hayashi T, Kanbe H, Okada M, Kawase I, Ikeda Y, Onuki Y, Kaneko T, Sonobe T (2007) In vitro and in vivo sustained-release characteristics of theophylline matrix tablets and novel cluster tablets. Int J Pharm 341(1–2):105–113
75. Thombre AG, Appel LE, Chidlaw MB, Daugherity PD, Dumont F, Evans LAF, Sutton SC (2004) Osmotic drug delivery using swellable-core technology. J Control Release 94:75–89
76. Yamada K, Furuya A, Akimoto M, Maki T, Suwa T, Ogata H (1995) Evaluation of gastrointestinal transit controlled-beagle dog as a suitable animal model for bioavailability testing of sustained-release acetaminophen dosage form. Int J Pharm 119(1):1–10
77. Chiou WL, Jeong H, Chung S (2000) Evaluation of using the dog as an animal model to study the fraction of oral dose absorbed of 43 drugs in humans. Pharm Res 17:135–140
78. Sutton SC, Evans L, Fortner J, McCarthy J, Sweeney K (2006) Dog colonoscopy model for predicting human colon absorption. Pharm Res 23(7):1554–1563
79. Sutton S (2009) The use of gastrointestinal intubation studies for controlled release development. Br J Clin Pharmacol 68(3):342–354
80. Sutton SC (2005) Biopharmaceutical properties that impact the successful prediction of in vivo performance of oral dosage forms. AAPS Annual Meeting and Exposition. AAPS, Nashville, TN
81. Vertzoni M, Goumas K, Söderlind E, Abrahamsson B, Dressman J, Poulou A, Reppa C (2010) Characterization of the ascending colon fluids in ulcerative colitis. Pharm Res 27(8): 1620–1626
82. Bouchoucha M, Thomas SR (2000) Error analysis of classic colonic transit time estimates. Am J Physiol Gastrointest Liver Physiol 279(3):G520–G527

83. Veysey MJ, Thomas LA, Mallet AI, Jenkins PJ, Besser GM, Wass JAH, Murphy GM, Dowling RH (1999) Prolonged large bowel transit increases serum deoxycholic acid: a risk factor for octreotide induced gallstones. Gut 44(5):675–681
84. Veysey M, Thomas L, Mallet A, Jenkins P, Besser G, Murphy G, Dowling R (2001) Colonic transit influences deoxycholic acid kinetics. Gastroenterology 121(4):812–22
85. Yassin AEB, Alsarra IA, Alanazi FK, Al-Mohizea AM, Al-Robayan AA, Al-Obeed OA (2010) New targeted-colon delivery system: in vitro and in vivo evaluation using X-ray imaging. J Drug Target 18(1):59–66
86. Hu Z, Mawatari S, Shimokawa T, Kimura GO, Yoshikawa Y, Shibata N, Takada K (2000) Colon delivery efficiencies of intestinal pressure-controlled colon delivery capsules prepared by a coating machine in human subjects. J Pharm Pharmacol 52(10):1187–1193
87. DeVore NM, Smith BD, Urban MJ, Scott EE (2008) Key residues controlling phenacetin metabolism by human cytochrome P450 2A enzymes. Drug Metab Dispos 36(12):2582–2590
88. Tanaka A, Inoue A, Takeguchi A, Washizu T, Bonkobara M, Arai T (2005) Comparison of expression of glucokinase gene and activities of enzymes related to glucose metabolism in livers between dog and cat. Vet Res Commun 29(6):477–485
89. Chauret N, Gauthier A, Martin J, Nicoll-Griffith DA (1997) In vitro comparison of cytochrome p450-mediated metabolic activities in human, dog, cat, and horse. Drug Metab Dispos 25(10):1130–1136
90. Straw JA, Loo TL, de Vera CC, Nelson PD, Tompkins WA, Bai SA (1992) Pharmacokinetics of potential anti-AIDS agents thiofoscarnet and foscarnet in the cat. J Acquir Immune Defic Syndr 5(9):936–942
91. FrÉNais R, Burgaud S, Horspool LJI (2008) Pharmacokinetics of controlled-release carbimazole tablets support once daily dosing in cats. J Vet Pharmacol Ther 31(3):213–219
92. van der Laan JW, Brightwell J, McAnulty P, Ratky J, Stark C (2010) Regulatory acceptability of the minipig in the development of pharmaceuticals, chemicals and other products. J Pharmacol Toxicol Methods 62(3):184–195
93. Fix J, Engle K, Porter P, Leppert P, Selk S, Gardner C, Alexander J (1986) Acylcarnitines: drug absorption-enhancing agents in the gastrointestinal tract. Am J Physiol 250:G332–G340
94. Kim J, Oberle R, Krummel D, Dressman J, Fleisher D (1994) Absorption of ACE inhibitors from small intestine and colon. J Pharm Sci 83:1350–1356
95. Meunier L, Kissinger J, Marcello J, Nichols A, Smith P (1993) A chronic access port model for direct delivery of drugs into the intestine of conscious dogs. Lab Animal Sci 43:466–470
96. Dobson G, Owusu-Ababio G, Furlong M, Rager J, Khan M, Ij H (2004) Establishment of an intraduodenal fistula dog model for biopharmaceutics studies. AAPS National Metting. AAPS, Nashville, TN
97. Bernkop-Schnürch A, Pinter Y, Guggi D, Kahlbacher H, Schöffmann G, Schuh M, Schmerold I, Del Curto M, D'Antonio M, Esposito P, Huck C (2005) The use of thiolated polymers as carrier matrix in oral peptide delivery–Proof of concept. J Control Release 106(1–2):26–33
98. Kesisoglou F, Panmai S, Wu Y (2007) Nanosizing – Oral formulation development and biopharmaceutical evaluation. Adv Drug Deliv Rev 59(7):631–644
99. Shiu GK, Sager AO, Velagapudi RB, Prasad VK, Skelly JP (1988) The effect of food on the absorption of controlled-release theophylline in mini-swine. Pharm Res 5(1):48–52
100. Chiou WL, Buehler PW (2002) Comparison of oral absorption and bioavailability of drugs between monkey and human. Pharm Res 19:868–874
101. Ikegami K, Tagawa K, Narisawa S, Osawa T (2003) Suitability of the cynomolgus monkeys as an animal model for drug absorption studies of oral dosage forms from the viewpoint of gastrointestinal physiology. Biol Pharm Bull 26(10):1442–1447
102. Ikegami K, Tagawa K, Osawa T (2006) Bioavailability and in vivo release behavior of controlled-release multiple-unit theophylline dosage forms in beagle dogs, cynomolgus monkeys, and göttingen minipigs. J Pharm Sci 95(9):1888–1895
103. Scheubel E, Adamy L, Hoffart V, Cardot JM (2010) Selection of the most suitable dissolution method for an extended release formulation based on IVIVC level A obtained on cynomolgus monkey. Drug Dev Ind Pharm 36(11):1320–1329

104. Gillespie W (1997) In: Young D, Devane J, Butler J (eds) In vitro in vivo correlations, vol 423. Plenum, New York, pp 53–65
105. Pang SK, Weiss M, Macheras P (2007) Advanced pharmacokinetic models based on organ clearance, circulatory, and fractal concepts. AAPS J 9(2):E268–E283
106. The International Transporter Consortium (2010) Membrane transporters in drug development. Nat Rev Drug Discov 9(3):215–236
107. K. Baggott (2008) Remote control for pill cameras. Technology review. MIT, Cambridge
108. Pahwa R, Dutt H, Kumar V, Sharma P (2010) In vivo evaluation using gamma scintigraphy. Pharmaceutical Technology, Pharmtech.com
109. Lentz KA, Quitko M, Morgan DG, Grace JE, Gleason Jr. C, Marathe PH (2007) Development and validation of a preclinical food effect model. J Pharm Sci 96(2):459–472
110. Mealey KL, Bentjen SA, Gay JM, Cantor GH (2001) Ivermectin sensitivity in collies is associated with a deletion mutation of the mdr1 gene. Pharmacogenetics 11(8):727–733

Chapter 5
In Vitro Testing of Controlled Release Dosage Forms During Development and Manufacture

Michael J. Rathbone and James M. Butler

Abstract This chapter provides the reader with some basic considerations relating to the purpose, aims, goals, development and validation of an in vitro drug release test for a modified release oral product in order to maximize its successful research, optimization and development.

5.1 Introduction

Dissolution is defined as the process by which a solid substance enters in the solvent to yield a solution and is controlled by the affinity between the solid substance and the solvent [1]. It is an important property of a dosage form that is a necessary prerequisite to drug absorption and one that contributes to the rate and extent of drug availability to the body [2]. The dissolution properties of a dosage form are assessed using an in vitro dissolution test which is developed to determine the dissolution characteristics of an immediate release dosage form. In contrast, drug release is that process by which a drug is released under the control of its formulation in a manner that modifies (extends or hastens) its release compared to an immediate release preparation.

Drug release from a modified release dosage form is controlled by various methods, depending upon the formulation of the product, including diffusion, dissolution partitioning or osmotic pressure. Release is assessed using an in vitro drug release test which is developed to determine the drug release characteristics of a modified release dosage form. The same equipment and apparatus is used to

J.M. Butler (✉)
GlaxoSmithKline R&D, Harlow, Essex, UK
e-mail: james.m.butler@gsk.com

C.G. Wilson and P.J. Crowley (eds.), *Controlled Release in Oral Drug Delivery*, Advances in Delivery Science and Technology, DOI 10.1007/978-1-4614-1004-1_5,

undertake a dissolution test or an in vitro drug release test. However, different pharmacopeial monographs apply to each method and may contain different instructions for development and operation. The equipment is simple and readily available commercially. However, it is this simplicity, together with the learnings from a long and informative history of the field that results in the need to spend considerable time and effort to develop and validate a methodology that is meaningful for a particular modified release product. The development of an in vitro drug release test for a modified release product is undertaken on each product on a case-by-case basis, with each product demanding the same detailed attention in time, effort and procedure.

A well-developed method can contribute to the successful research, optimization and development of a product and shorten the time between innovation and marketing. A poorly ill developed (or meaningless) method in contrast can lead to false conclusions and send investigators down a wrong track, resulting in inappropriate recommendations on the formulation. Each modified release product requires the same arduous process to be undertaken and is likely to be associated with one or more special consideration(s), e.g., a particular condition of the test; a slight modification of the apparatus; and a particular release media composition. Such considerations depend upon the physicochemical properties of the drug, the formulation components and the physical attributes of the delivery system.

In an in vitro drug release test, the apparatus generally comprises a vessel, drug release media, a lid and some method of agitation, e.g., a stirrer or basket. Supporting this is the equipment that maintains the temperature of the drug release media constant at a predefined value within narrow limits, and rotates the stirrer at a fixed and constant rate, again within narrow limits, while maintaining it in a preset position and avoiding wobble.

Before an investigator embarks on the process of developing an in vitro drug release test, they must first familiarize themselves of the historical developments in this area as this provides the relevant science that underpins all the decisions made to develop a meaningful methodology [2]. Those underlying scientific principles must be understood, appreciated and applied in order to develop a sensitive, robust, discriminative, meaningful test that does what it purports to do. The fundamental principles that underpin any in vitro drug release test include mass transfer, hydrodynamic theories, boundary layer theories, mechanisms of drug release from dosage forms, mathematical modeling, amongst others. They explain why the shape of the vessel, shape and size of the stirrers, positions of the stirrers, locations of sampling points, etc., need to be precisely defined and why such narrow limits on their tolerances are documented and must be adhered to [3, 4]. They also explain why any modification to existing apparatus, or the adoption of any new apparatus, must be precisely defined, meet predefined specifications and tolerances and why all new apparatus must undergo qualification before use.

It is the purpose of this chapter to provide the reader with a brief treatise on what useful considerations and techniques can be employed for evaluating performance during oral modified release product development to maximize the chances of success with a clinical product.

5.2 The Basics

5.2.1 What Is In Vitro Drug Release Testing?

In vitro drug release testing is an experimental methodology in which the drug release properties of a modified release dosage form are characterized, documented and/or compared.

5.2.2 How Is In Vitro Drug Release Testing Conducted?

In vitro drug release testing is conducted in precisely defined apparatus under equally precisely defined conditions. This is because the shape, size and overall geometry of the apparatus can affect the hydrodynamics of the drug release media, and subsequently influence drug release from the delivery system. History has shown that slight changes in vessel design can significantly affect drug release [3, 4]. This is why the shape and dimensions of the apparatus are precisely defined (together with narrow tolerances) in pharmacopeias. Ultimately the test conditions should meet those defined in dedicated monographs of various countries pharmacopeias; however, this does not need to be the case in the early stages of the development of a drug release test. It is feasible to initially use a test method very different to that described in a pharmacopeia in order to maximize the chances of correlation at the beginning of the process.

5.2.3 What Are the Aims and Goals of In Vitro Drug Release Testing?

The aim of an in vitro drug release test is to reproducibly provide an accurate estimate of the rate of release of the incorporated drug from a modified release dosage form. It should aim to be sensitive enough to detect differences in product formulation or manufacturing changes, but robust enough to detect such differences only when they are biologically relevant.

The ultimate goal of the test is to generate information that can provide an insight into the mechanism by which the drug is being released from the dosage form and provide data to facilitate the rational and rapid research, optimization and development of a modified release dosage form.

5.2.4 What Is an In Vitro Test Used for?

An in vitro drug release test can be used for many purposes. These include:

- Product research and development
- Stability testing
- Final product QC
- Process monitoring, process characterization and in-house process control monitoring
- Prediction of biological performance

An in vitro drug release test is an invaluable tool in product research and development. During this phase of the product life cycle, an in vitro drug release can change and evolve as more information comes to hand. However, at each stage the test changes should be clearly documented and experiments performed which bridge one test to the other, thereby providing continuity and relevance between tests that allow the rational development of the product to be clearly seen. This allows regulatory product development reports to define a clear and logical development progression of the product.

Using an in vitro drug release test to monitor the physical stability of the product over time has become an essential component of any stability trial. All pharmacopeias specify the need for including and undertaking an in vitro drug release test in any stability trial conducted for regulatory purposes. Such a test need not be the final product test (although it usually is), and its limits need not be the same as those of the final product test (they may be narrower than the final product specifications). The test must be able to detect the influence of a physical change in the product that affects the release characteristics of the product to assure the investigator that a product stored over the shelf life of the product will perform with the same efficacy and safety and is of equivalent quality as a freshly made product.

Use of an in vitro drug release test as a final product quality control test has become an essential requirement for product release. Such a test is the culmination of years of investigation on the developmental and final products and should reflect biological performance (see later). Much time and effort go into both the development and validation of the final product QC test as well as the setting of the release specifications.

In vitro drug release tests are sometimes used to monitor studies to characterize the manufacturing process and/or are adopted as an in-house process control monitor. Use for this purpose depends on whether some critical manufacturing parameter that specifically affects drug release is key to manufacture. An in vitro drug release test will certainly be utilized as part of manufacturing process validation.

For an in vitro drug release test to have any value, it must exhibit some ability to relate to the biological performance of the product. This relevance can range from being simply indicative of some biological indicator, to being able to predict a given biological indicator, to being able to demonstrate a valid in vivo–in vitro correlation.

5.2.5 In Vitro Drug Release Testing of a Modified Release Oral Dosage Form Versus Dissolution Testing of an Immediate Release Oral Product

The in vitro drug release test evolved from the dissolution test. As such, much can be learnt and extrapolated between the studies, outcomes and learnings between the two tests. The same apparatus and similar conditions of use are employed in both tests. However, some differences occur and are obvious. Many of the test conditions are the same between the two procedures and final method conditions are chosen on a case-by-case basis after undertaking an extensive development process that involves a critical examination of all equipments, apparatus and media-related factors that could possibly affect drug release.

The major difference between an in vitro dissolution test and an in vitro drug release test is that a dissolution test lasts for less than an hour, whereas an in vitro drug release test can last many hours. Hanson and Banakar list a number of external variables that affect drug dissolution [3, 4] including eccentricity of the stirrer, vibration of the apparatus, alignment and centering of the stirrer, agitation rate, dissolved gasses, media pH, media composition, evaporation, temperature, flow pattern of the media arising, for example, from probes, sampling position, blockage of filters, interference of drug detection methods and sorption of drug onto the equipment. This list applies equally well to in vitro drug release test methodologies.

The significance and relevance of each of these external factors to drug release (and subsequently on the design of the final methodology) varies on a case-by-case basis and is evaluated in the development phase. The lid used in an in vitro dissolution test is specifically designed to allow ease of access for sampling. It is designed with a wide access groove in order to facilitate rapid sampling. The size of the groove exposes a large surface area of the dissolution media to the air (and potentially evaporative losses); however, the brevity of the test (less than 1 h) precludes the problem of evaporation over that time. The same lid cannot be used for an in vitro drug release test and one must be used that retards evaporative losses of the release media to prevent excessive evaporation between samples. Hanson lists evaporation as a key variable that must be controlled during a test and recommends that its acceptable limits are "None", i.e., no evaporation of the dissolution media is acceptable if the test is to be considered valid [4]. Banakar concurred with this statement and calculated that even 2% inadvertent evaporative losses would result in an invalid methodology [3] since evaporation of the release media can result in violation of sink conditions or induce other effects on a case-by-case basis. Because during evaporation only water molecules leave the surface and dissipate into the atmosphere (leaving behind solutes) this can result in changes in pH, ionic strength, viscosity, surface active agent concentration, etc. as the changes (reduction) in volume concentrates the buffer salts, ions, hydrophilic polymers/surface active agents, etc., that may be present in the receptor medium. In addition, changes in volume can affect the hydrodynamics of the release media resulting in changes in diffusional distances. Besides, the fundamental theories that apply to drug diffusion (Ficks Laws of Diffusion) assume that the volume of the receptor medium remains constant throughout the diffusive process since concentration changes in the receptor media must reflect changes in the amount of drug entering the media following release and not be influenced by volume changes ($C = a/v$; where C is the concentration of drug in the receptor media, a is the amount of drug in the receptor media and v is the volume of the receptor media). This assumption must be adhered to in the design of the drug release test in order to ensure that drug release equations such as the Higuchi "Square Root of Time" equation holds true. Fortunately in the case of drug release testing, prevention of evaporative losses is a simple matter of using an appropriately designed lid that prevents evaporation occurring.

5.2.6 What the Concentration Versus Time Profile Depends upon

Evaluation of a modified release dosage form in the apparatus and media described in a pharmacopeia will inevitably result in the appearance of the drug in the release media. That does not, however, mean that that the resultant release profile is meaningful, or that the shape and duration of the observed profile is controlled by release of drug from the delivery system. A variety of factors affect the appearance of drug in the release media. These include:

- System-dependant parameters (apparatus design and dimensions)
- Environmental conditions (stirring speed, receptor phase composition, etc.)
- Dosage form type
- Dosage form composition (formulation)
- Manufacturing process

Each of these factors influences the shape and duration of the observed profile. The latter three factors are of primary interest to the pharmaceutical scientist, and the method is developed accordingly to detect differences in either dosage form, dosage form components or changes in the manufacturing process, in order to monitor or optimize product performance. Development of the method is through systematic changes in system-dependant parameters and changes in environmental conditions. However, if an incomplete development phase is undertaken, a system-dependant parameter or an environmental condition may result in a parameter external to the delivery system controlling the shape and duration of the profile. A simple example is that a prolonged release profile may be observed if the stirring speed is reduced to 10 rpm. Such a profile may not, however, be due to retardation of release from the delivery system, but rather from excessively large diffusive boundary layers adjacent to the surface of the dosage form. Under less obvious circumstances an investigator may believe that a prolonged observed concentration versus time profile is a reflection of the formulation, whereas in actuality, it is being defined and controlled by a factor external to the delivery system. Thus system-dependant parameters and environmental conditions must be thoroughly investigated during a development phase and adjusted in order to establish that the observed profile is a result of, and arises from, drug release from the dosage form, and is not a result of some external influence. Some system-dependant parameters and environmental conditions that could be investigated include:

- pH
- Ionic strength
- Stirring speed
- Aqueous phase composition (surface active agents, alcohol)
- Stirrer height/position
- Sampling point
- Degassing
- Volume
- Paddle wobble
- Vibration

In addition, several supporting experiments should be undertaken to assure the investigator that the correct conditions have been chosen in the final method. These include:

- Solubility of the drug(s) in the various release media
- Development and validation of a suitable analytical assay
- Physical and chemical changes in the delivery system during release
- Experiments that provide an insight into the mechanism of release of the drug from the dosage form

5.2.7 *Importance of Development and Validation*

It is essential that an extensive and complete development process is undertaken to define the final method. There are several reasons for this. First, the development process provides the developer with a broad insight into those factors that potentially affect drug release in the system and to what extent they affect drug release. These can be different between products depending upon their formulation. Knowledge of those factors that affect drug release can then be used by the developer to define how much control must be impelled on those factors within appropriate tolerances or limits in the final test. For example, if the pH of the release media is around the pK_a of the drug, then control of the pH of the release media would need to be adhered to very narrow tolerances, both initially and during the test procedure, compared to the situation where the pH is 2.5 or 3 pH units above or below the pK_a of the drug. Secondly, an extensive evaluation of those factors that affect release provides knowledge for the developer to draw upon as the test method develops or if the formulation changes. That knowledge can be used to modify the test method without the need for further extensive investigations. Thirdly, extensive developmental knowledge allows the operator to modify the method accordingly as new data on the product comes to hand. This is particularly useful when clinical performance data are generated and the method needs to be adjusted in order to provide some in vivo relevance to the test.

Validation is the process where experiments are undertaken that demonstrate that the method does what it claims to do. The method should be shown to be reproducible, precise, accurate and robust.

5.3 Noncompendial Methods

In some cases the uniqueness of the dosage form requires the investigator to consider some modification to the standard apparatus or conditions. If this is to occur, whatever is done must be considered under the constraint that it is undertaken in accordance with the scientific principles inherent within the pharmacopeia. For example, new apparatus must be exactly designed, its manufacture adhere to the design drawings and it must be qualified before use to demonstrate that its physical parameters meet those of the design drawing and fall within the predefined tolerances.

5.4 Novel Experimental Methods

The in vivo environment of the gastrointestinal (GI) tract is more complex than can be created using compendial apparatus. There has always been a conflict between the need for a simple, reproducible dissolution or drug release test for routine batch release, and a desire to more fully mimic the gastrointestinal conditions. One emerging strategy to deal with this conflict is to separate these two aims, and have drug release tests that are complex, but are more likely to be prospectively predictive for human performance and therefore suited for the early stages of formulation development, in particular at the formulation selection and optimization stage. A separate test for batch release testing and quality control (QC) would then be developed later or perhaps in parallel. Although the QC test may still be able to incorporate some key features of the more complex predictive test, the primary aim is to provide a robust test method that can be used on a routine basis. It is also possible that a more complex method, once developed, might still be used later in the product development cycle for the purposes of an in vitro–in vivo correlation (IVIVC) if this could not be adequately achieved with the simpler test method.

The desire for greater in vivo relevance, particularly during early development, and for a greater general understanding of formulation performance in vivo in the context of quality by design (QbD) has led to the emergence of technologies that mimic functional aspects of the GI tract to a greater extent than does conventional dissolution or drug release test.

5.4.1 Conventional Apparatus, Unconventional Media

There has long been an awareness of the role of solubilization by the mixed micellar systems containing bile salts in the GI tract [5, 6]. However, these materials have some inherent variability in their impurity profile, solutions of these materials have a short stability shelf life and solution preparation methods have historically been relatively complex, so they tend to be avoided for routine work.

The use of biorelevant media in in vitro drug release tests has its roots in research conducted using dissolution tests designed to evaluate the dissolution of poorly soluble immediate release formulations. Systematic work to identify simulated media that were convenient for routine experimental laboratory use in dissolution tests to mimic the human intestine was first published in the mid-to-late 1990s by Dressman et al. [7, 8]. The recommended media recipes contained mixed micelles consisting of taurocholic acid (a bile salt) and lecithin (an endogenous phospholipid) at levels relevant to those found in the human intestine both in the fasted and fed state.

This combination of components means that both the wetting and solubilization properties of GI fluids can be mimicked relatively easily in the laboratory. This enables scientists developing new oral formulations to run tests in media, much closer in nature to GI fluids, even if it does not result in a test that provides sink conditions for high dose/low solubility drugs. More recently, the range of biorelevant media has been expanded to include bio-relevant gastric media [9], and further modifications have been recommended to better match the concentrations of the ingredients to that observed for actual GI fluids [10]. The availability of pre-prepared powder blends to overcome solution preparation complexity has also helped in the adoption of these media for more routine use [11].

As pointed out above, initially, the focus of the application of biorelevant media was on the dissolution of poorly soluble immediate release formulations. This enabled dissolution to be used in a more prospective and prognostic manner for the prediction of oral drug absorption. Application to extended release formulations has followed more recently. Here, the use of biorelevant media may be combined with the reciprocating cylinder or flow through cell dissolution apparatus. Both types of apparatus enable multiple media change to mimic transit through the GI tract, which is more difficult to achieve with the more commonly used paddle or basket methods [12–14].

5.4.2 *Novel, Noncompendial Methods for the Investigation of Drug Release*

Obtaining an adequate prediction of in vivo performance may require a degree of biorelevance beyond that possible with a standard compendial test even when used in conjunction with appropriate biorelevant dissolution media. Apparatus described in a pharmacopeia are necessarily kept as simple as possible and are carefully specified in terms of equipment design, and will only mimic some aspects of the in vivo environment (e.g., temperature). With regard to other aspects (such as hydrodynamics, digestion or dynamic fluid mixing) a test using the standard apparatus will lack biorelevance, mainly because they are more difficult to mimic within the constraints of the equipment design. For instance, a rotating paddle or basket can hardly realistically mimic the motion or movement of fluids in the GI tract.

Once again, recent research in the dissolution test arena may provide direction for those scientists interested in the development of more realistic in vitro drug release tests. Recently, holistic dissolution systems have begun to be used in the development of oral dosage forms that aim to mimic many of the relevant dynamic and digestive features of the GI tract that might impact on in vivo performance. These systems, which include the TNO Intestinal Models TIM-1 [15] and TIM-2 [16], and the Institute of Food Research's Dynamic Gastric Model (DGM) [17], offer a significant improvement in the ability of the in vitro dissolution test to imitate the dynamics, fluid input and removal, food, enzyme-dependent digestive processes, and even, in the case of TIM-2, gut microflora in the GI tract that may impact dissolution (and stability) in vivo. They can be seen as attempts to recreate multifunctional aspects of the GI tract in the laboratory. Initially designed for use by the food industry, they offer the possibility of a more reliable, although not necessarily less expensive in vitro alternative to the use of in vivo animal models for the prediction of oral drug formulation performance in humans. However, sample throughput, cost and obtaining a significant number of replicates for improved confidence remain challenges for the widespread application of these systems, and the number of applications to pharmaceuticals reported in the literature is currently limited.

For the prediction of the performance of oral modified release dosage forms, the TNO-TIM-1, a gastric and small intestinal model (Fig. 5.1), has been shown to provide an improved prediction of the behavior of theophylline matrix tablets, whereas simpler methods needed retrospective adjustment of the agitation intensity in order to obtain a good IVIVC [18]. The DGM is a dual chamber gastric model (Fig. 5.2) designed to mimic both the fundus (lower agitation "storage" region of the stomach) and the antrum (higher agitation region). It has been used to characterize the differences between two different oral nifedipine products that use contrasting release mechanisms – an osmotic pump formulation and a matrix tablet formulation. These differences may explain the known differences between the in vivo robustness of these formulations [19].

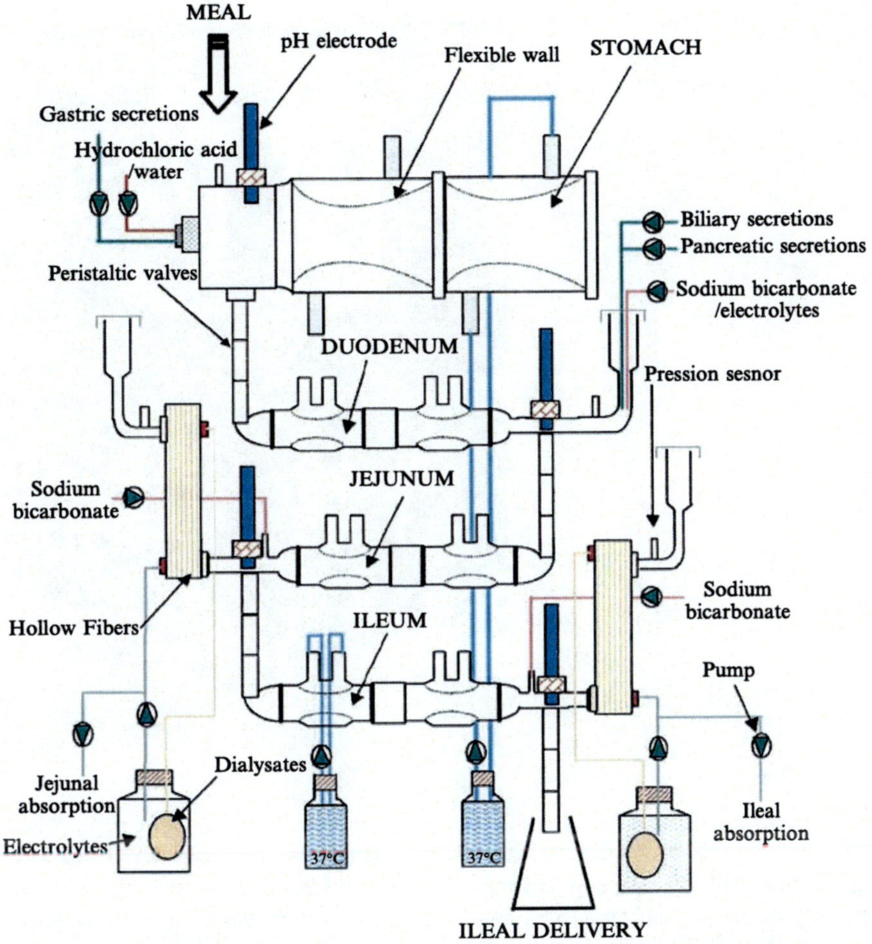

Fig. 5.1 Schematic representation of TNO's "TIM 1." From S. Blanquet, J. P. Meunier, M. Minekus, S. Marol-Bonnin, and M. Alric. Applied and Environmental Microbiology, May 2003, Vol. 69, No. 5, pp 2884–2892. Recombinant Saccharomyces cerevisiae Expressing P450 in Artificial Digestive Systems: a Model for Biodetoxication in the Human Digestive Environment. http://aem.asm.org/cgi/reprint/69/5/2884. See also TNO website: http://www.tno.nl/content.cfm?context=thema&content=markt_product&laag1=891&laag2=180&laag3=297&item_id=874&Taal=2

In addition to these "GI tract in a laboratory" systems, other dynamic systems of intermediate complexity that attempt to improve the biorelevance of key variables that are inadequately mimicked in more established dissolution tests have been developed in recent years [20–25]. These systems focus on factors such as dynamic fluid mixing [20, 21], improved mimicking of the impact of GI motion on

Fig. 5.2 The dynamic gastric model (DGM). From [26]. See http://pubs.acs.org/doi/pdfplus/10.1021/mp1001203

dosage form [22, 23], lipid digestion [24] or drug removal from the GI tract by absorption [25]. By focusing on improving the biorelevance of a specific aspect of the test, these methods tend to retain at least some of the simplicity of a pharmacopeia dissolution method such as the ability to test multiple replicates, and can be used to target and improve the biorelevance of the test for specific types of drug and formulation.

A fuller description of these emerging dynamic dissolution testing systems has recently been the topic of a useful and comprehensive literature review [26].

Dissolution apparatus attempting to improve the mimicking of the impact of GI motion have particular application to modified release dosage forms, especially when developing matrix tablets which may be vulnerable to some loss of their controlled release characteristics when subjected to the physical stresses associated with GI transit, such as during gastric empting and illeo-cecal transit. The physical stresses associated with gastric emptying, where the dosage form may be subjected to a short period of very high agitation, particularly upon passage through the pyloric sphincter, may be critical to the integrity of some modified release dosage forms [22]. These stresses are likely to be a significant contributory factor to differences in pharmacokinetics seen between the fasted

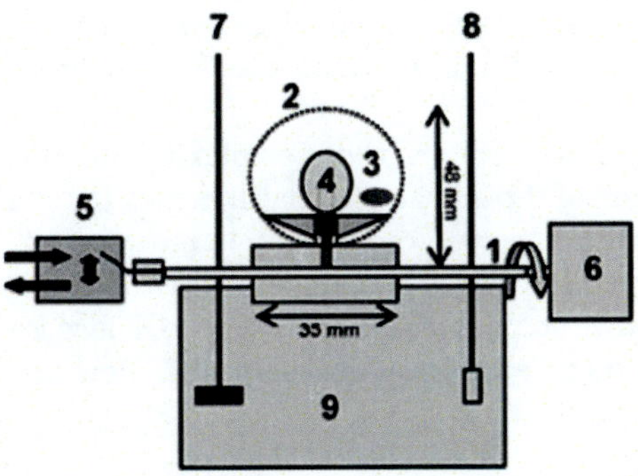

Fig. 5.3 Schematic representation of the dissolution stress test device described in [22]

Fig. 5.4 Representation of a modified paddle apparatus for mimicking gastric dynamics (paddle not shown for visual clarity)

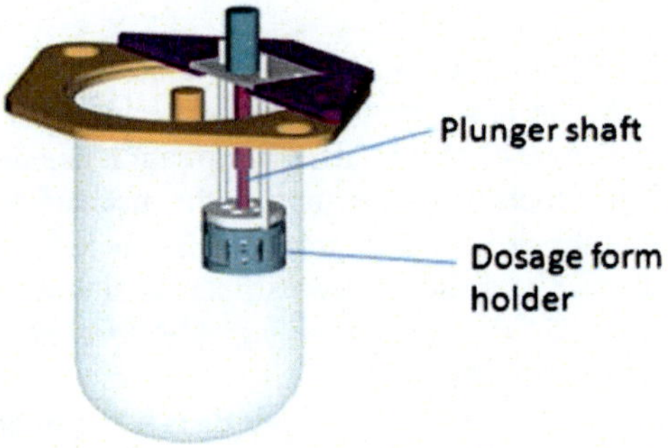

and fed state in vivo with some modified release matrix tablets, which rely on erosion for drug release.

Garbacz et al. [22] have demonstrated that the irregular pharmacokinetic profiles seen with a Diclofenac modified release tablet can be mimicked using a novel dissolution test apparatus (Fig. 5.3) that uses both a rotary/dipping motion and an intermittent squeezing motion applied via an inflatable balloon. The system was devised to mimic the physical forces acting upon a matrix tablet, determined from pressure measurements exerted on a telemetric capsule [22]. The test apparatus was also used to examine modified release matrix nifedipine formulations marketed in Europe. Matrix formulations showed a higher susceptibility to variable and more extensive drug release during the applied biorelevant stress test than an osmotic pump formulation, particularly during the simulated high stress events associated with gastric empting [27].

Another device, also using two different means of agitation within the same apparatus and devised to mimic the physical forces in vivo in a more realistic way than established dissolution methods, is described by Burke et al. [28] (see Fig. 5.4).

This apparatus uses a plunger to apply stress to matrix formulations placed within a stationary basket. The basket is placed within a standard paddle dissolution apparatus.

These various approaches, although of different levels of complexity, offer the hope of greater accuracy in the prediction of the human pharmacokinetics for oral modified release dosage forms, and a much improved opportunity to prospectively predict likely in vivo profiles at the very earliest stages of modified release formulation development. If these approaches prove successful, they may help to reduce the development cycle times for new oral modified release dosage forms. However, more confirmatory work is needed.

5.4.3 Mechanistic Tools to Understand Factors Controlling In Vitro Drug Release

In addition to the various novel apparatus described above, there are also novel tools available to improve the characterization of the dosage form behavior during the in vitro test, so that more information is gained than simply a drug release profile. These tools therefore improve the mechanistic understanding of the factors controlling the drug release profile. They include the use of novel techniques to measure the size of disintegrated particles in situ in a dissolution vessel [29], and the use of magnetic resonance imaging (MRI) to better characterize the mechanism of drug release [30, 31].

The MRI imaging technique is now possible with both high-field equipment for high-quality image analysis [32], and low-field equipment that may be located in a standard laboratory [33]. The use of a flow through cell, identical in design to that specified in the main pharmacopeias, is advantageous as the dosage form can be held in position, and the width of the dissolution cell means the magnet and dosage form can be in close proximity in a small fluid volume [34]. MRI imaging is of particular value for modified release formulations, as gaining an understanding of the changes occurring within the dosage form itself during the in vitro drug release test can improve the formulation design process, and identify underlying mechanisms contributing to drug release. A set of sample images for an osmotic pump modified release formulation is shown in Fig. 5.5. Changes in the hydration within the two compartments within the formulation can be detected, as can the plume of material that is forced out of the drilled orifice in the outer coat. Quantitative analysis of the hydration rate and erosion rate of modified release formulations are also possible via the use of image analysis software.

Fourier Transform Infra Red (FTIR) spectroscopy has also been reported as of value in the mechanistic understanding of dissolution. This technique has the advantage of providing a degree of spatial understanding as to where different components such as the drug and polymer are located during the dissolution process [35].

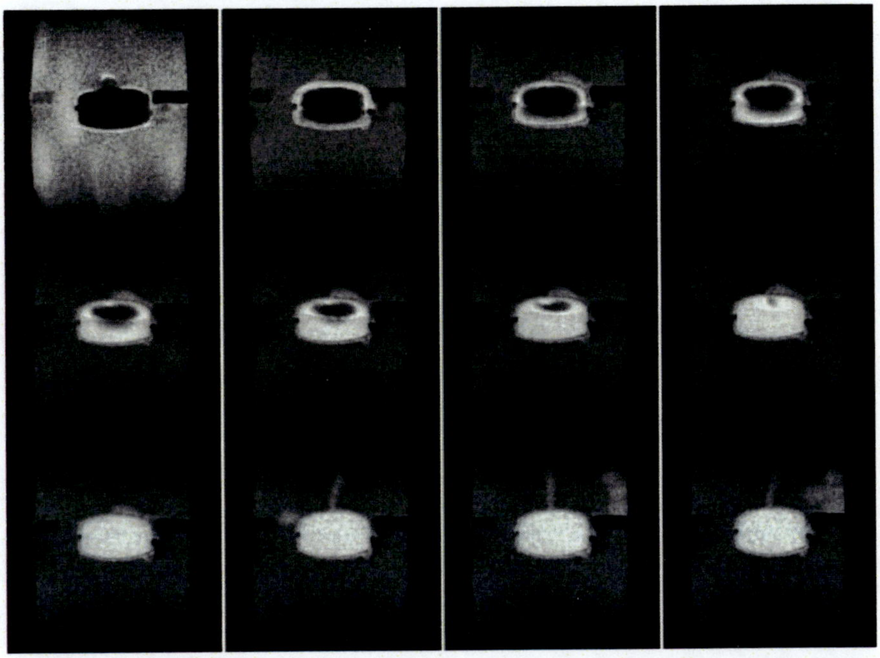

Fig. 5.5 A series of images (over about 12 h) for a nifedipine osmotic pump formulation taken with a Bench-top MRI. The hydration of the bilayer and the extrusion of the hydrated material out of the single orifice can be visualized

5.5 Regulatory Expectations

Extensive guidance on the development and use of drug release tests for modified drug release for regulatory submissions is available [36, 37], and reference to these during drug release test development and validation is essential to ensure regulatory expectations are met with respect to both the product and the test method.

It is worth noting that for an oral modified release product there is an accepted regulatory expectation that a stability-indicating drug release test method with multiple time points documenting the drug release profile is developed.

5.6 Concluding Remarks

An in vitro drug release test is an invaluable tool in the development, characterization and quality control of drug products. To be meaningful, the test must be appropriately developed and fully validated to demonstrate that it does what it claims to do. Pharmacopeias precisely define the apparatus and suggest appropriate release media composition that reflects long historical development and experience. Pharmacopeia methods are underpinned with inherent science and expectations relating to the test conditions, apparatus, tolerances, precision of the apparatus, etc. It is up to the investigator to develop a test through prudent adjustment of apparatus design (occasionally) and experimental conditions (usually). The investigator has the responsibility to develop the test so that the release profiles reflect drug release from the dosage form in a manner that provides useful information on the research, optimization and development of the modified release dosage form. It is also the investigator's responsibility to ensure that any modification in apparatus design meets the same inherent principles of design principles, and undertakes post manufacture qualification to the same level as that of the standard equipment and apparatus that is supplied commercially.

Ultimately, in vitro drug release test conditions should meet those defined in dedicated monographs of various countries' pharmacopeias. However, this does not need to be the case in the early stages of the development of an in vitro drug release test. It is feasible to initially use a test method very different to that described in a pharmacopeia in order to maximize the chances of correlation at the beginning of the process in order to increase the opportunity for a more rapid and relevant product research and development process.

Workers in the field of in vitro drug release testing should have an active interest in the developments occurring in the closely related field of dissolution testing.

References

1. Banakar UV (1991) Introduction, historical highlights, and the need for dissolution testing. In: Banakar UV (ed) Pharmaceutical dissolution testing, Chap 1. Marcel Dekkar, New York, pp 1–18
2. Banakar UV (1991) Pharmaceutical dissolution testing. Marcel Dekkar, New York
3. Banakar UV (1991) Factors that influence dissolution testing. In: Banakar UV (ed) Pharmaceutical dissolution testing, Chap 5. Marcel Dekkar, New York, pp 133–187
4. Hanson WA (1982) Handbook of dissolution testing. Pharmaceutical Technology, Springfield
5. Bates T, Gibaldi M, Kanig J (1966) Solubilising properties of bile salt solutions. J Pharm Sci 55:191–199
6. Bates TR, Lin SL, Gibaldi M (1967) Solubilization and rate of dissolution of drugs in the presence of physiologic concentrations of lysolecithin. J Pharm Sci 56(11):1492–1495
7. Dressman JB, Amidon GL, Reppas C, Shah VP (1998) Dissolution testing as a prognostic tool for oral drug absorption: immediate release dosage forms. Pharm Res 15:11–22
8. Galia E, Nicolaides E, Horter D, Lobenberg R, Reppas C, Dressman JB (1998) Evaluation of various dissolution media for predicting in vivo performance of class I and II drugs. Pharm Res 15:698–705
9. Vertzoni M, Dressman JB, Butler JM, Hempenstall J, Reppas C (2005) Simulation of fasting gastric conditions and its importance for the in vivo dissolution of lipophilic compounds. Eur J Pharm Biopharm 60:413–417
10. Jantratid E, Janssen N, Reppas C, Dressman JB (2008) Dissolution media simulating conditions in the proximal human gastrointestinal tract: an update. Pharm Res 25:1663–1676
11. Boni JE, Brickl RS, Dressman J, Pfefferle M (2009) Instant FaSSIF and FeSSIF – Biorelevance meets practicality. Dissolution Technologies, August 2009, pp 41–45
12. Klein S, Stein J, Dressman J (2005) Site-specific release of anti-inflammatory drugs in the gastrointestinal tract: an in vitro model. J Pharm Pharmacol 57:709–720

13. Fotaki N, Symillides M, Reppas C (2005) In vitro vs. canine data for predicting input profiles of isosorbide-5-mononitrate from oral extended release products on a confidence interval basis. Eur J Pharm Sci 24:115–122
14. Fotaki N, Aivaliotis A, Butler J, Dressman J, Fischbach M, Hempenstall J, Klein S, Reppas C (2009) A comparative study of different release apparatus in generating in vitro–in vivo correlations for extended release formulations. Eur J Pharm Biopharm 73(1):115–120
15. Blanquet S, Zeijdner E, Beyssac E, Meunier J-P, Denis S, Havenaar R, Alric M (2004) A dynamic artificial gastrointestinal system for studying the behavior of orally administered drug dosage forms under various physiological conditions. Pharm Res 21(4):585–591
16. Kamm MA, Sandborn WJ, Gassull M, Schreiber S, Jackowski L, Butler T, Lyne A, Stephenson D, Palmen M, Joseph RE (2007) Once-daily, high-concentration MMX mesalamine in active ulcerative colitis. Gastroenterology 132(1):66–75
17. Mercuri A, Lo Curto A, Wickham MSJ, Craig DQM (2008) Dynamic gastric model (DGM): a novel in vitro apparatus to assess the impact of gastric digestion on the droplet size of self-emulsifying drug-delivery systems. J Pharm Pharmacol 60(Suppl. 1):A-2
18. Souliman S, Beyssac E, Cardot JM, Denis S, Alric M (2007) Investigation of the biopharmaceutical behaviour of theophylline hydrophilic matrix tablets using USP methods and an artificial digestive system. Drug Dev Ind Pharm 33(4):475–83
19. Mercuri A (2010) The dynamic gastric model (DGM) a new in vitro model for biorelevant dissolution and delivery assessment of oral dosage forms. UKICRS Newsletter July 2010, pp 28–30
20. Gu C-H, Rao D, Gandhi RB, Hilden J, Raghavan K (2005) Using a novel multicompartment dissolution system to predict the effect of gastric pH on the oral absorption of weak bases with poor intrinsic solubility. J Pharm Sci 94(1):199–208
21. Carino SR, Sperry DC, Hawley M (2005) Relative bioavailability estimation of carbamazepine crystal forms using and artificial stomach-duodenum model. J Pharm Sci 95(1):116–125
22. Garbacz G, Wedemeyer R-S, Nagel S, Giessmann T, Moennikes H, Wilson CG, Siegmund W, Weitschies W (2008) Irregular absorption profiles observed from diclofenac extended release tablets can be predicted using a dissolution test apparatus that mimics in vivo physical stresses. Eur J Pharm Biopharm 70(2):421–428
23. Burke M, Kalantzi L, Parr A (2010) Pharmaceutical analysis apparatus and method. US Patent Application US20100126287 A1
24. Fatouros D, Mullertz A (2008) In vitro lipid digestion models in design of drug delivery systems for enhancing oral bioavailability. Expert Opin Drug Metab Toxicol 4(1):65–76
25. Buch P, Langguth P, Kataoka M, Yamashita S (2009) IVIVC in oral absorption for fenofibrate immediate release tablets using a dissolution/permeation system. J Pharm Sci 98(6):2001–2009
26. McAllister M (2010) Dynamic dissolution: a step closer to predictive dissolution testing? Mol Pharm 7(5):1374–1387
27. Garbacz G, Golke B, Wedemeyer R-S, Axell M, Söderlind E, Abrahamsson B, Weitschies W (2009) Comparison of dissolution profiles obtained from nifedipine extended release once a day products using different dissolution test apparatuses. Eur J Pharm Sci 38:147–155
28. Burke M, Kalantzi L, Parr A (2010) Pharmaceutical analysis apparatus and method. USA Patent Application US20100126287 A1
29. Coutant CA, Skibic MJ, Doddridge GD, Kemp CA, Sperry DC (2010) In vitro monitoring of dissolution of an immediate release tablet by focused beam reflectance measurement. Mol Pharm 7(5):1508–1515
30. Richardson JC, Bowtell RW, Mader K, Melia CD (2005) Pharmaceutical applications of magnetic resonance imaging (MRI). Adv Drug Deliv Rev 57:1191–1209
31. Nott KP (2010) Magnetic resonance imaging of tablet dissolution. Eur J Pharm Biopharm 74:78–83
32. Chen YY, Hughes LP, Gladden LF, Mantle MD (2010) Quantitative ultra-fast MRI of HPMC swelling and dissolution. J Pharm Sci 99(8):3462–3472
33. Butler JM, Nott KP (2010) Using low-field MRI to improve tablet dissolution testing. Tablets Capsules 8(1):25–29

34. Fyfe CA, Grondey H, Blazek-Welsh AI, Chopra SK, Fahie BJ (2000) NMR imaging investigations of drug delivery devices using a flow-through USP dissolution apparatus. J Control Release 68(1):73–83
35. van der Weerd J, Kazarian SG (2004) Combined approach of FTIR imaging and conventional dissolution tests applied to drug release. J Control Release 98:295–305
36. CDER (1997) Guidance for industry – extended release oral dosage forms: development, evaluation, and application of in vitro/in vivo correlations. Food and Drug Administration, Center for Drug Evaluation and Research (CDER)
37. CDER (1997) Guidance for industry SUPAC-MR: modified release solid oral dosage forms. scale-up and post-approval changes: chemistry, manufacturing, and controls; in vitro dissolution testing and in vivo bioequivalence documentation food and drug administration. Center for Drug Evaluation and Research (CDER)

Chapter 6
Oral Controlled Delivery Mechanisms and Technologies

Hossein Omidian, Shahin Fesharaki, and Kinam Park

Abstract This chapter reviews release mechanisms and corresponding technologies used to control drug release from oral dosage forms. Products utilizing such approaches range from simple matrix systems to those employing more complex osmotic delivery technologies. Technologies including Oros™, gastroretentive devices, TIMERx™, Contramid™, Geomatrix™, and SODAS™ are discussed. These approaches are generally utilized to provide once-daily administration but some can also be used for differential release of more than one drug from the same dosage form or timed release of drug to align with time of clinical need.

6.1 Introduction and General Principles

Interaction between a drug and a polymeric material generally forms the basis of controlled oral drug delivery. Drug in solution exhibits random Brownian motion to equilibrate concentration, where concentration gradients exist. A polymer at certain concentration in such a solution imposes mandatory pathways for drug diffusion. Thus, polymers that dissolve in or otherwise hydrate in aqueous media can alter the drug diffusion process in a time-dependent manner. For example, hydroxypropyl methylcellulose (HPMC or hypromellose), which is water soluble, behaves as a swellable absorptive polymer in the limited volumes of aqueous media in the gastrointestinal (GI) tract. Drug dispersed in this polymer, as in monolithic tablets, diffuses through the viscous hydrated polymer at a rate dependent on the movement kinetics of the polymer chains. The faster these relax, the faster the diffusion rate.

H. Omidian (✉)
College of Pharmacy, Nova Southeastern University, Fort Lauderdale, FL, USA
e-mail: omidian@nova.edu

C.G. Wilson and P.J. Crowley (eds.), *Controlled Release in Oral Drug Delivery*,
Advances in Delivery Science and Technology, DOI 10.1007/978-1-4614-1004-1_6,

Hydrophilic polymers like HPMC may also control drug release by erosion mechanisms. After consumption of the dosage form, the GI tract fluid encounters the dosage unit, causing the polymer to hydrate and swell. Weakened mechanical properties in the swollen state may cause the hydrated polymer to break away from the prime particle (compact or pellet). Drug release may therefore be controlled by a combination of diffusion and erosion. Such release mechanisms can apply to systems where drug is dispersed in or coated with polymer.

Delivery to specific regions of the GI tract may be achieved using polymers with pH-dependent solubilities. These include enteric polymers with carboxylic acid functional groups; their pH-dependent solubility determines location for release. Drug can be released at different segments in the GI tract by using enteric coating polymers that dissolve at different pHs, e.g., Eudragit L100-55 (soluble at pH>5.5), Eudragit L100 (soluble at pH>6.0), and Eudragit FS 30D (soluble at pH>7.0) (http://eudragit.evonik.com/) or combinations of these. Water insoluble polymers can extend or prolong drug release. These include methacrylate- or acrylate-based polymers, with little or low permeability (e.g., Eudragits NE 30D, NM 30D, and NE 40D) (http://eudragit.evonik.com/). Addition of hydrophilic functional groups such as trimethylaminoethyl methacrylate can improve permeability and swellability in water (e.g., Eudragits RL and RS series) thereby altering release behaviors.

Technologies have been developed to exploit diffusion, erosion, and other physicochemical mechanisms and provide drug and disease-specific release profiles. Some are based on the nature of the release-modifying material(s), others on the design of the dosage form:

- TIMERx™ technology controls drug release, consequent to interaction between the two hydrocolloids, xanthan gum and locust bean gum.
- Release from a Contramid™ tablet is controlled by the degree of crosslinking of high amylase starch.
- Alza's Oros™ and Duros™ technologies are based on osmosis-driven release.
- Release from Jago Pharma's Geomatrix™ technology is based on the surface area available for drug release.

These mechanisms and technologies are discussed and exemplified in this chapter. Appendix 1 lists examples of commercial products where release has been modified to enhance performance, safety, or patient convenience.

6.2 Diffusion-Controlled Drug Release

When a matrix comprising drug and a hydrophilic polymer is exposed to GI fluids, it may, depending on composition, break up (disintegrate) or simply hydrate. Disintegration leads to dispersion and dissolution of drug. If the unit retains its structure (does not disintegrate), GI fluid permeates the core and the polymer is hydrated, becoming a viscous mass. Drug must then diffuse through this hydrated matrix at a rate depending on drug solubility and matrix permeability before release from the unit. Diffusion can also be influenced by membrane porosity if the unit is coated or by the presence of a pore-inducing filler in the coat. In such cases, the porosity of the coating membrane (pore size, shape, and distribution), and filler tortuosity can determine diffusion properties, and hence the drug release profile. Examples of matrix systems for water soluble and water insoluble drugs are given in Appendix 2. Biomedical materials may also exhibit matrix-type release profiles. Malcolm et al. developed a crosslinked silicone-based device to release metronidazole [1].

Matrix systems have also been utilized to provide pH-independent release of weakly basic drugs. In such a context, Streubel et al. showed that verapamil hydrochloride tablets exhibited pH-independent behavior when formulated in matrices containing ethylcellulose or HPMC with organic acids such as fumaric, succinic, or adipic acid [2].

6.3 Osmotic-Controlled Drug Release

The osmotic-controlled drug release (OROS™) concept for controlling delivery is based on dissolved drug being transported in a controlled manner from the dosage form to the external media under the influence of osmotic pressure. Figure 6.1 illustrates how a solution containing dissolved solute "attracts" water from an adjacent chamber, separated by a semipermeable membrane. Permeation rate depends on

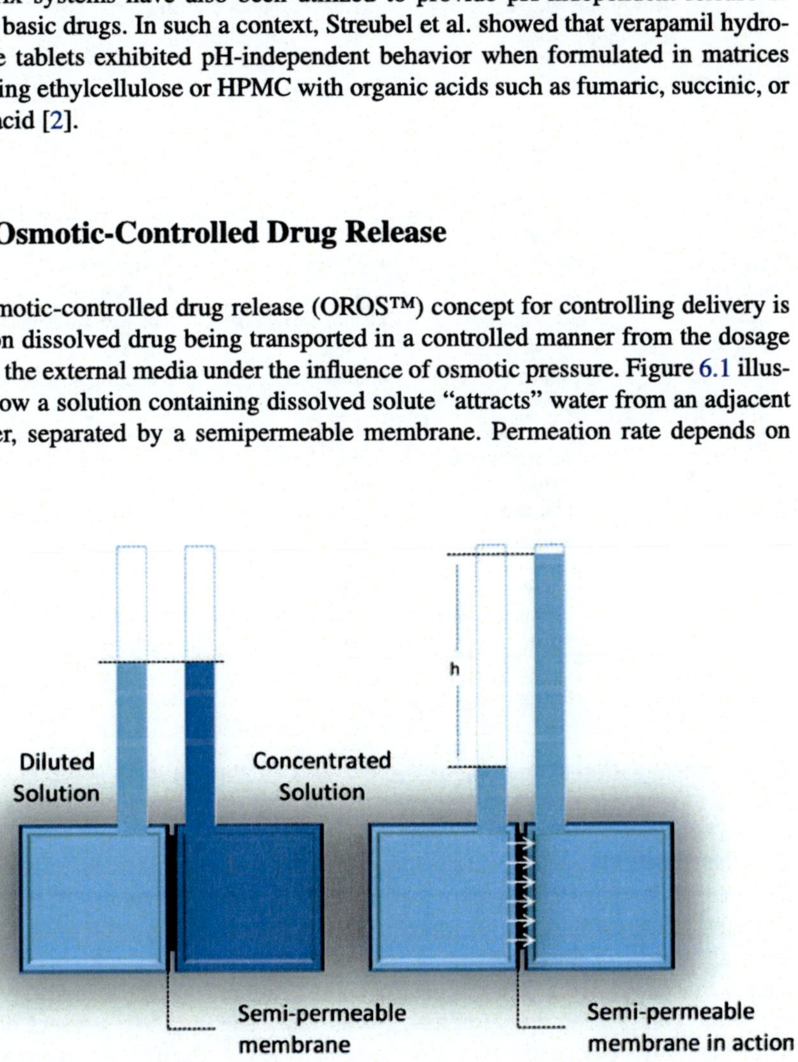

Fig. 6.1 The concept of osmosis

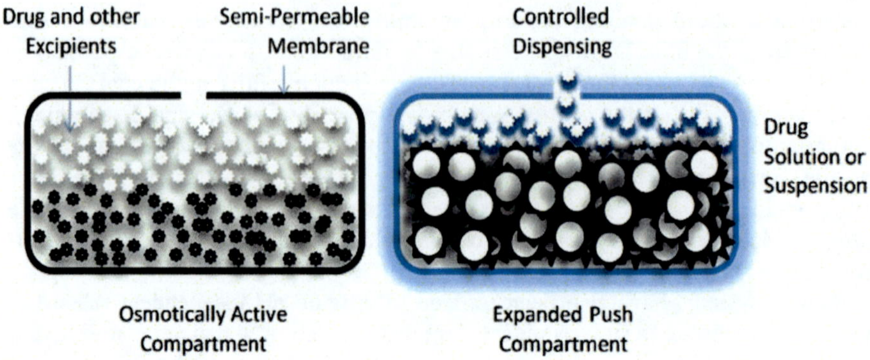

Fig. 6.2 OROS system

Table 6.1 Selected OROS products

Brand	Ditropan XL™	DynaCirc CR™	Covera HS™
Drug	Oxybutynin	Isradipine	Verapamil
Manufacturer	Alza	Novartis	G.D.Searle
Function	Antispasmodic	Calcium antagonist	Calcium antagonist
Solubility in water	Readily soluble	Insoluble	Soluble
Semipermeable membrane	Cellulose acetate	Cellulose acetate	Cellulose acetate
Water-soluble excipients	Hypromellose, poly(ethylene oxide), sodium chloride, poly(ethylene glycol) (http://www.rxlist.com/ditropan-xl-drug.htm)	Hydroxypropyl methylcellulose, poly(ethylene oxide), sodium chloride, poly(ethylene glycol) (http://www.rxlist.com/dynacirc-cr-drug.htm)	Poly(ethylene oxide), sodium chloride, hydroxyethyl cellulose, hydroxypropyl cellulose, hypromellose, poly(ethylene glycol), povidone (http://www.rxlist.com/covera-hs-drug.htm)

solute concentration ("number of molecules") in the receptor solution. Hence, materials that are ionizable and/or very soluble, and of low molecular mass are the most effective osmotic agents.

If the receptor chamber contains drug and excipients that are osmotically active the hydrostatic or hydraulic pressure exerted by the increased volume of fluid drives ("pushes") drug through one or more orifices in the dosage unit at a relatively constant rate (Fig. 6.2). Such delivery is believed to function independently of environmental conditions in the GI tract (pH, regional location, etc). The technology can be applied to compressed units (tablets) or capsules. The OROS™ concept or variations thereof has been applied to many medicinal agents for delivery in a controlled manner over an extended period of time (Tables 6.1 and 6.2).

Table 6.2 Examples of OROS products with two release ports

Brand	Fortamet ER™	Altocor™
Drug	Metformin	Lovastatin
Manufacturer	Andrx/Watson	Andrx/Watson
Function	Antihyperglycemic	Cholesterol-lowering agent
Solubility in water	Freely soluble	Insoluble
Semipermeable membrane	Cellulose acetate	Cellulose acetate
Excipients	Hypromellose, PEG 400, PEG 8000, Povidone (http://www.rxlist.com/fortamet-drug.htm)	Hydroxypropyl methylcellulose, hypromellose phthalate, methacrylic acid copolymer, PEG 400, PEG 8000, poly(ethylene oxide), propylene glycol, sodium chloride (http://www.rxlist.com/altocor-drug.htm)

6.3.1 Conventional OROS™

Glucotrol XL™ (Novartis) delivers 2.5, 5, or 10 mg of the sulphonylurea, glipizide in a sustained manner to enable once-daily dosage. Release modifying components are poly(ethylene oxide), HPMC, cellulose acetate, with sodium chloride as the prime osmotic pressure inducer (http://www.rxlist.com/glucotrol-xl-drug.htm). The bilayer tablet core, containing drug and the osmotically active agents (in separate layers) is coated with cellulose acetate, which is permeable to water but not to the drug or the osmotic agent (Fig. 6.2). On ingestion, GI fluid diffuses through the semipermeable membrane into the tablet core. Dissolved drug from the drug-containing layer is driven through a laser-drilled orifice due to the osmotic pressure buildup in the layer containing sodium chloride (so-called push layer). Delivery rate is independent of pH and gastric motility, thereby extending drug release throughout the GI tract. Neither do other variables, such as posture or diet state have an effect. Since delivery is driven by osmotic pressure, drug is released at a constant rate so long as the osmotic gradient between the drug layer and the GI tract fluid is maintained. The osmotic gradient eventually decreases due to release of osmotic agent and drug, with delivery eventually tailing off (http://www.rxlist.com/glucotrol-xl-drug.htm). Water soluble as well as insoluble drugs have been formulated in such a system as given in Table 6.1.

The bilayer Oros™ oxybutynin tablet also provides extended release over 24 h based on essentially the same release mechanism as Glucotrol XL™. Following administration oxybutinin plasma concentration rises slowly over 4–6 h, followed by a relatively constant plasma level for up to 24 h. A study by Goldenberg showed that such dosage was well tolerated, and as clinically effective as its 5 mg "immediate release counterpart" [3].

Duan et al. prepared an osmotic tablet containing isosorbide-5-mononitrate (5-ISMN). Tablet composition, size, and location of the orifice, and membrane properties affected the drug release. Based on pharmacokinetics and bioavailability data

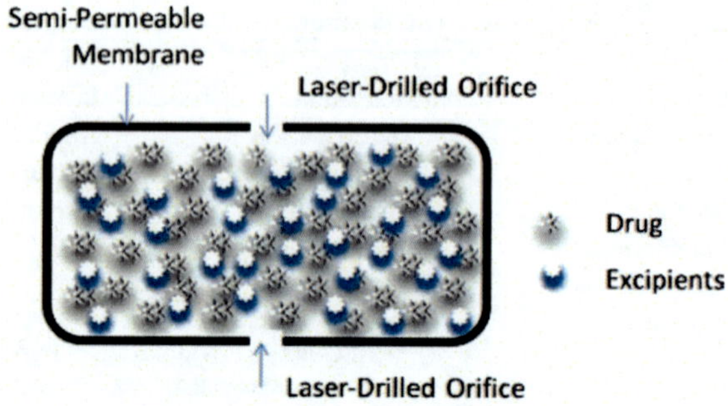

Fig. 6.3 SCOT system

in Beagle dogs, the osmotic tablet was considered to be a more suitable long-acting preparation than a 5-ISMN SR matrix tablet for once-daily dosage [4].

Liu et al. studied factors affecting in vitro and in vivo behavior of an osmotic tablet containing nifedipine. Membrane and orifice size significantly affected release but release per se was independent of dissolution medium variables. The pharmacokinetic data suggested that the formulation was capable of sustaining plasma levels to enable once-daily dosing [5].

Swellable core technology (SCT) partners the drug with a water-swellable excipient. Thombre et al. studied in vitro and in vivo release of tenidap and sildenafil from SCT formulations with different core configurations (single layer, bilayer, and trilayer). Release rates were independent of core configuration, and the in vivo pharmacokinetic parameters in beagle dogs were consistent with in vitro performance [6]. Wagstaff et al. studied drug release from an extended release osmotic tablet containing the biguanide, metformin hydrochloride in patients with type 2 diabetes. The extended release formulation exhibited prolonged T_{max} but overall bioavailability (area under curve) was comparable for both presentations, as were GI adverse effects [7]. Waterman et al. developed an extrudable system for osmotic delivery of poorly soluble drugs. The dosage form comprises a monolith core coated with a semipermeable membrane. The core contains hydroxyethyl cellulose along with sugar as the osmotic agent [8].

6.3.2 *OROS™ with Twin Orifices*

The *SCOT*™ (Single Composition Osmotic Tablet) system shown in Fig. 6.3 has the osmotic agent and drug in a single layer, contrasting with systems comprising two layers. This single layer technology utilizes various osmotic modulating agents and polymeric coating to provide zero-order drug release. The system claims to effect substantially complete release of drug possibly due to twin exit ports on either side of the tablet.

The core comprises primarily drug with low levels of excipient. The coat is permeable to water, but not to higher molecular weight components in biological fluids. Table 6.2 lists examples of drugs utilizing this technology.

Table 6.3 Examples of trilayer OROS™ products

Brand	Concerta™	Invega™
Drug	Methylphenidate	Paliperidone
Manufacturer	Johnson & Johnson	Johnson & Johnson
Function	CNS stimulant	psychotropic agent
Solubility in water	Freely soluble	Insoluble
Semipermeable membrane	Cellulose acetate	Cellulose acetate
Water-soluble excipients	Hypromellose, phosphoric acid, poloxamer, poly(ethylene glycol), poly(ethylene oxide), povidone, propylene glycol, sodium chloride (http://www.rxlist.com/concerta-drug.htm)	Hypromellose, hydroxyethyl cellulose, poly(ethylene glycol), poly(ethylene oxide), povidone, propylene glycol, sodium chloride (http://www.rxlist.com/invega-drug.htm)

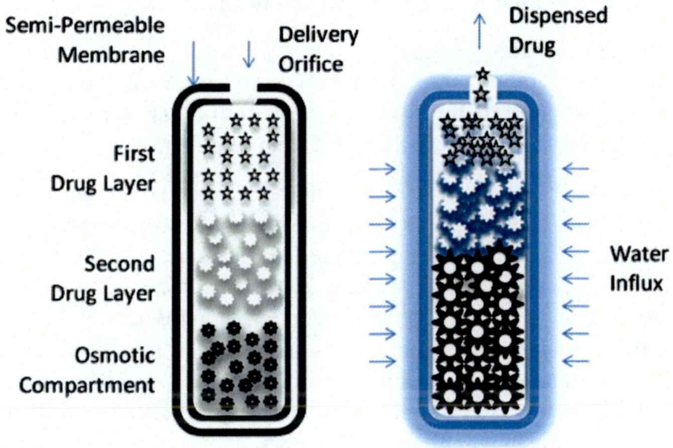

Fig. 6.4 Trilayer OROS system

6.3.3 *Trilayer OROS™*

Trilayer OROS™ further advances the osmotic concept, being applied to brands such as Concerta™ and Invega™ (Table 6.3). Tablets comprise a core and a semipermeable membrane. The core is composed of three layers, i.e., two drug layers and one osmotically active compartment (Fig. 6.4). The drug layers may contain two drugs or the same drug at two concentrations. The Concerta™ presentation has an "immediate-release" overcoat to deliver a rapid initial dose fraction. An orifice in the first layer along with the semipermeable membrane controls subsequent drug release.

The Invega™ tablet technology, while similar to that for Concerta™ is equipped with laser-drilled orifices on the dome of the drug layer. Tablets are coated with a water-dispersible polymer, which is quickly eroded on exposure to GI Tract media (http://www.rxlist.com/concerta-drug.htm). It is claimed that Invega 12 mg once-daily OROS™ tablet provides steady-state plasma concentrations over 6 days (http://www.janssencns.com/invega/schizoaffective-disorder/dosing-and-administration/oros-technology).

6.3.4 *Osmodex™*

Laser-drill technology is used in the Osmodex™ family of "platform" technologies in combination with a variety of single and multiple drug delivery approaches. The technologies are classified as (http://www.osmoticausa.com/):

- Osmodex SD™ for soluble drugs.
- Osmodex™ IR/CR (combined instant and controlled release of one or two drugs).
- Duodex Double CR™ (delivery of two drugs with different release patterns).
- Osmodex Triplet™ (to provide three different release rates).

Allegra-D 24HOUR™ utilizes the Osmodex™ IR/CR technology. It is designed to provide an immediate release of the antihistamine, fexofenadine (180 mg) combined with extended release (240 mg) of the decongestant, pseudoephedrine hydrochloride for 24-h cover of nasal allergy. Excipients facilitating such prolonged delivery are sodium chloride (osmotic agent), poly(ethylene glycol), povidone, hypromellose, croscarmellose sodium, and copovidone (release modifiers). The tablet has cellulose acetate-based coat as a semipermeable membrane, and the coated tablet is then covered with an immediate release drug layer (http://www.rxlist.com/allegra-d-24-hour-drug.htm). Figure 6.5 shows an Osmodex™ system with an immediate release coating and an extended release core.

6.3.5 *OROS™ Safety and Clinical Aspects*

Benefits of osmosis-driven technology relate to its capability to:

- Release drug at a constant (zero order) rate.
- Provide consistent release regardless of drug or environment, i.e., independent of drug, patient physiology, or food effect [9].
- Sustain delivery over a significant time period.
- Pulse and/or delay delivery to align with patient needs or mode of drug action.

Treatment tolerability and patient compliance may improve with some medications as a consequence of such delivery. OROS™ technology, with its capability to provide lower peak plasma levels and "smoother" plasma profiles

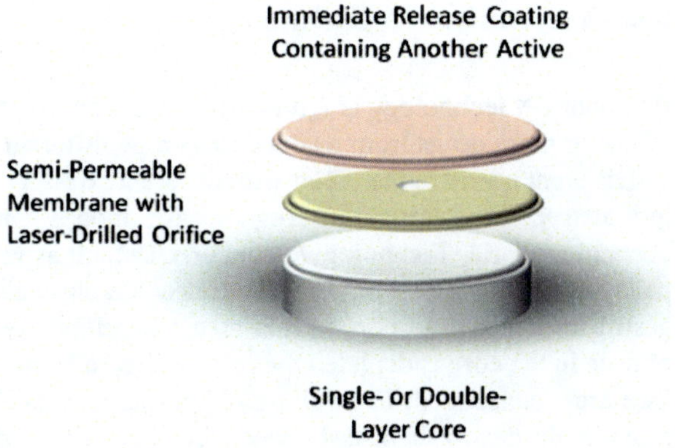

Fig. 6.5 OSMODEX system

may also result in reduced plasma level-related side effects [10]. Wonnemann et al. compared the bioavailability of nifedipine from two commercial modified release nifedipine products, viz., Adalat OROS 30™ and Nifedipine Retard 30™. A significant food interaction effect was noted with Nifedipine Retard 30™. In contrast, food intake did not have a significant effect on release from the OROS™ dosage form, based on pharmacokinetic parameters. Variable and unpredictable plasma levels, suggestive of inconsistent delivery were also noted with the Nifedipine Retard 30™ formulation. Such findings illustrate the hazard of switching medications that might ostensibly seem to have comparable efficacy and safety [11].

Sathyan et al. reported that side effects associated with immediate-release oxybutinin can be alleviated using an extended release OROS™ presentation of the drug [12]. Paliperidone is an oral psychotropic agent for treating schizophrenia. It undergoes limited hepatic metabolism if formulated utilizing OROS technology. Davidson et al. studied safety and efficacy of once-daily paliperidone in acute schizophrenia. All doses of extended-release paliperidone were well tolerated and shown to improve personal and social functioning [13]. In another clinical study in acute schizophrenia, symptoms were improved significantly in patients who used paliperidone ER [14].

Potential drawbacks associated with the OROS™ platforms include the high costs of manufacture (laser drilling is required). Dose dumping is also a potential issue if the semipermeable coat is compromised as the entire daily dose being contained in a single unit. Bass et al. comprehensively reviewed safety aspects of tablets based on OROS™ technology. Long-term safety data indicated a low incidence of clinically significant GI tract side effects including intestinal, gastric, and esophageal irritation, injury, and obstruction. The general experience indicates that for some drugs OROS™-based products can provide substantial therapeutic and convenience benefits without delivery-related risks [15].

6.4 Geomatrix™ Technology

Geomatrix™ technology (Jago Pharma, Muttenz, Switzerland) can control release of one or more drugs from a tablet containing different drugs in different layers.

Different layers in the tablet with different swelling, gelling, and erosion behaviors can provide separate drug release modes (http://www.skyepharma.com/Technology/Oral_Technology/Geomatrix/Default.aspx?id=62). In general, hydrophilic polymers progressively swell on encountering aqueous media thereby increasing gastric residence time, core surface area, and diffusivity for release. Thus, as amount of drug in the core is depleted the rate of release is increased due to the greater surface area, consequent to swelling. Appropriate matching of drug and polymer(s) provides the desired balance between drug depletion and increased diffusion through the matrix to deliver a steady flux.

Various release mechanisms can be achieved using the Geomatrix™ technique. These include:

- Zero order (constant rate over time).
- Binary (release of two drugs at different rates and times).
- Biphasic release (combination of slow and fast release for a same drug).

Biphasic delivery can be further subgrouped as "quick–slow" release and "slow–quick release." With the former, a burst release of a drug is followed by extended release over time (e.g., Zyflo CR™) (http://www.skyepharma.com/Technology/Oral_Technology/Geomatrix/Default.aspx?id=62). Zyflo CR™ (Cornerstone Therapeutics) is an extended release tablet containing the antiasthmatic, zileuton. The triple layer provides an immediate release dose fraction, and a middle layer to regulate drug release from an extended release layer to prolong drug release and effect.

HPMC is one of the most common release modifiers in matrix tablets. Maggi et al. studied poly(ethylene oxide) as an alternative for Geomatrix-based products. HPMC provided more controllable and slower release rates in multiple layer Geomatrix™ systems [16].

Conte et al. reviewed Geomatrix™ technology in terms of efficacy, reproducibility, and technological characteristics [17].

Geomatrix™ technology is primarily intended for water soluble drugs and the release rate is significantly reduced if drug has poor aqueous solubility [18]. It has been successfully applied to drugs for a wide spectrum of clinical conditions. Examples include:

- Dilacor XR™ (Watson Labs) is an extended release capsule providing 24 h release of Diltiazem HCl and prolonged control of hypertension. Ethylcellulose and hypromellose act as release modifiers for this water-soluble drug.
- The antidepressant Paroxetine (Paxil CR™ GSK) is an enteric-coated tablet, controlling drug dissolution over 5 h for gradual release in the small intestine. The enteric coat delays release until tablet leaves the stomach (http://us.gsk.com/products/assets/us_paxilcr.pdf). Drug solubility is about 5 mg/ml and the release

modifiers comprise hypromellose, poly(vinyl pyrrolidone), glyceryl behenate, and methacrylic acid copolymer type C. One tablet layer comprises a degradable barrier, and the other layer contains the active in a hydrophilic matrix.
- Ropinirole HCl is formulated with sodium carboxymethylcellulose, glyceryl behenate, hypromellose, and povidone as a controlled release tablet for Parkinsonism (Requip XL™) (http://www.rxlist.com/requip-xl-drug.htm). It is a triple layer tablet, with the active in the center layer, laminated between two placebo layers, which control the surface area available for the drug release. The relatively low dosage of this drug facilitates the rather complex dosage form design.
- El-Nabarawi utilized Geomatrix technology to prolong duration of action of the anti-inflammatory, tenoxicam from a bilayer unit. The drug-containing layer (drug and HPMC) is welded to a drug-free layer containing HPMC and ethyl cellulose (EC) using a casting/solvent evaporation technique. Study showed that the addition of the drug-free layer, its composition and thickness could change the release profile [19].
- Wilding et al. used gamma scintigraphy and pharmaco-scintigraphy to evaluate the effect of fed/fasting state on GI transit and drug release behavior of a Diltiazem Geomatrix™ tablet formulation. Pharmacokinetic data showed ready absorption in the colon in the fasted state, the tablet remaining intact for almost 17 h. Food slightly increased the overall extent of absorption without changing the release characteristics [20].
- Goutte et al. proposed an experimental design for developing and preparing a Geomatrix™ system for cost effectiveness and time saving (a typical Geomatrix™ system requires one compression and three granulation processes) [21].

6.5 TIMERx™

The TIMERx™ technology developed by Pennwest offers the following modes of release:

- First order, i.e., the release rate decreases over time.
- Zero order release (constant rate over time).
- Combinations of immediate release and controlled release.

The technology is based on a customized, agglomerated hydrophilic complex that forms a matrix on compression. The matrix comprises two polysaccharides, viz., xanthan gum and locust bean gum. Interactions between these in an aqueous environment result in formation of a viscous gel with a slowly eroding core. Such synergy between xanthan and galactomannans was first reported in 1971, when researchers observed formation of a thermoreversible gel in xanthan gum: locust bean gum mixtures. Further studies showed that total polysaccharide concentration, not gum ratio, influenced the setting and melting temperatures of the gel [22].

Tobyn et al. found that interactions between these gums were synergistic in aqueous media when a third component such as dextrose is present [23]. Staniforth et al. reviewed the physicochemical interactions between the two gums, and how they could influence release [24]. Dosage forms utilizing TIMERx™ technology can prolong release over 4 h [25], either as zero-order or chronotherapeutic release modes by manipulating the gum interactions [24]. Tobyn et al. showed that electron spin resonance (ESR) was a useful indicator of interactions between hydrocolloids and drugs, intended for presentation in TIMERx™ systems [26].

The analgesic, *oxymorphone* hydrochloride, has been formulated as an extended release tablet using TIMERx™ technology (Opana ER™; Endo Pharma).

Variants of the TIMERx™ platform comprise:

- Geminex™ technology enabling release of different actives independent of each other.
- SyncroDose™ is designed to deliver the drug at a desired site and time in the body.

6.6 Gastroretention

Retention of a dosage form in the stomach is an attractive concept for prolonging release and absorption from controlled release dosage forms containing the following categories of drug:

- Drugs with narrow absorption windows.
- Drugs locally active in the stomach.
- Drugs unstable in the colon or distal small intestine.
- Drugs with low solubility at high pH.

Discussion in this chapter is limited to the mechanisms that may be considered to provide gastroretention. A separate chapter provides more detailed discussion.

Gastroretention can conceptually be achieved through floating, size expansion (swelling or unfolding), mucoadhesion, sinking, and magnetic attraction as outlined in Fig. 6.6 [27]. The extent of gastric retention, however, depends on various factors, and clinically effective gastric retentive devices are yet to be developed.

6.6.1 Floating Systems

A dosage unit with lower relative density than gastric contents is less disposed to be propelled towards the pyloric sphincter but to reside in the fundus or body of the stomach [27–29]. Units can be either monoliths or multiparticulate. Such buoyancy might be effected by a number of techniques, viz.

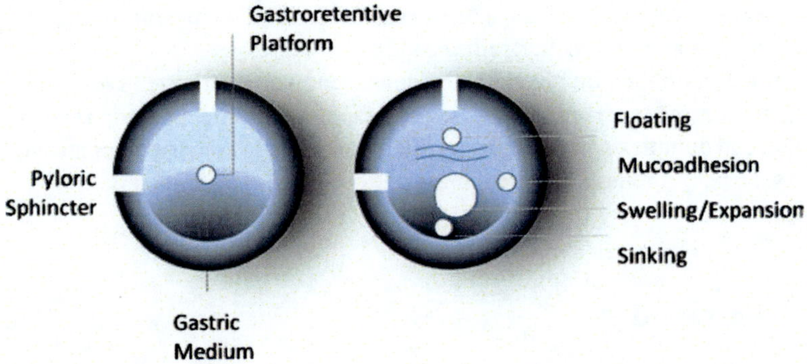

Fig. 6.6 Gastro-retentive systems

- Gel formation: Hydrocolloids (xanthan gum), polysaccharides (HPMC), synthetic polymers (poly(ethylene oxide), carbopol), and natural gums (alginates, guar gum) have been used to prepare flotation platforms [30–34].
- Effervescent systems: Inclusion of an effervescent couple, e.g., sodium bicarbonate, citric acid in the unit leads to interaction and evolution of carbon dioxide in the gastric medium. Gas entrapment by the hydrophilic polymers reduces density, enhances buoyancy, flotation, and gastroretention [31].

6.6.2 Size Expansion

Size expansion strategies comprise enlargement of the dosage unit on exposure to the gastric environment such that passage through the pyloric sphincter is constrained. Swelling systems incorporate polymer capable of absorbing gastric fluid while folding systems are designed to unravel on hydration. Unit size must be sufficiently small to afford oral administration but expansion must be sufficient for gastric retention. Collagen sponge has been utilized to confer unfolding properties [35, 36].

Gastric retention may also be induced by changing the unit rigidity by judicious choice of materials. Drug depletion and physical breakdown can combine to reduce unit size and allow passage to the duodenum [37].

6.6.3 Mucoadhesion

Gastric residency can conceptually be prolonged, by incorporating in the dosage form synthetic or natural polymers with an affinity for gastric mucosa. Drug is then released in a controlled manner for prolonged absorption in the intestine.

Mucoadhesive polymers that have been evaluated include chitosan and its thiolated derivative, carbopol and methylcellulose [38].

Information on the effectiveness and consistency of gastroretentive technologies in humans is rather sparse at this time. The factors that affect gastric residence are manifold and complex and present formidable barriers to consistent, controlled drug delivery. Hence, it remains in the realms of "promise" than delivery.

6.7 Contramid™

Contramid™ technology (Labopharm) utilizes crosslinked, high-amylase starch to control drug release. Release is essentially dependent on unit swelling, with degree of starch crosslinking being the rate controlling factor. When a unit is placed in an aqueous medium, the starch forms a hard gel and displays sponge-like viscoelastic behaviors. X-ray tomography reveals a uniform membrane at the gel surface that controls drug release [39].

- Ryzolt™ (Purdue Pharma) is an extended release tablet containing the antiarthritic, tramadol. It comprises a dual matrix with immediate and extended release components. Release modifying ingredients include pregelatinized modified starch, poly(vinyl acetate), povidone, and xanthan gum.
- Oleptro™ (Labopharm) is an extended release tablet containing the antidepressant, trazodone, which releases drug over a 24-h period. From an absorption perspective, Oleptro™ 300 mg tablets exhibit a T_{max} of about 9 h, postdose under fasting condition. The tablet contains hydroxypropyl distarch phosphate and polymeric ingredients such as hypromellose, poly(ethylene glycol) 3350, and poly(vinyl alcohol) (http://www.rxlist.com/oleptro-drug.htm).
- The in vitro release of sodium diclofenac from a Contramid™ system was studied by Rahmouni et al. Factors such as pH, ionic strength of the medium, and enzyme concentration were studied, which could affect the enzymatic hydrolysis of the crosslinked high amylase starch, and hence the drug release. Excipients such as HPMC and PEO also influenced tablet erosion. In vitro studies to determine the effect of low and high amylase concentrations revealed that the release mechanism is changed from diffusion to a combined diffusion and erosion mechanisms [40].
- Rioux et al. studied the effect of crosslink density of high amylase starch on various mechanical properties of Contramid™ films. Young's modulus, elongation at break, tensile strength, permeability to water and oxygen were all affected by level of crosslinking and environmental humidity [41].

Contramid™-based products have been shown to be safe and effective in clinical studies in patients:

- Once-daily tramadol (Contramid OAD™) was safe and effective for pain management [42]. Contramid OAD™ has also been compared with BID tramadol.

The Contramid presentation provided sustained analgesia throughout the dosing interval [43].

- Sheehan et al. evaluated once-daily trazodone (Contramid™) in major depressive disorder. The extended release formula was well tolerated and more effective than placebo [44].

6.8 Multiparticualte Systems

Multiple Unit Pellet Systems are discussed in a separate chapter in this book. Mechanism-related facets are considered here.

6.8.1 Micropump™ Technology

Flamel's Micropump™ technology has been designed to extend small intestine residence time of appropriately sized small particles that become "lodged" in intestinal villi, thereby prolonging small intestinal retention. Controlling drug release from such particles may sustain absorption. The concept is illustrated in Fig. 6.7.

The approach has been used to provide a prolonged release formulation of the antiviral, acyclovir, for twice daily administration [45]. It may be appropriate for drugs with short half-lives that are absorbed primarily in the small intestine. Coreg CR™ (GSK) also utilizes Micropump technology and comprises three kinds of microparticles, viz.

- Uncoated microparticles that release a fraction of the dose rapidly providing early onset of action.
- Coated microparticles that delay release of another drug fraction.
- A second population of coated microparticles that release drug even later in the small intestine, thereby sustaining absorption and duration of action.

The product provides once-daily therapy for congestive heart failure.

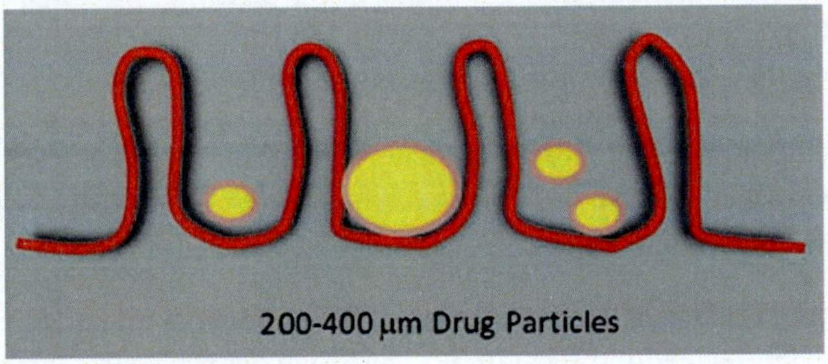

Fig. 6.7 Micropump delivery system

6.8.2 Spheroidal Oral Drug Absorption System

Capsules utilizing Elan's Spheroidal Oral Drug Absorption System (SODAS™) technology contain spherical beads, 1–2 mm in diameter. The beads contain a drug core plus excipient as well as a coating of controlled release polymer. Once ingested, the water-soluble polymers of the coating layer are dissolved, which leaves a porous layer through which the active can diffuse out at a controlled rate. Depending on the drug's physicochemical properties, the polymer composition of the membrane can be different.

- SODAS™ technology is employed to provide once-daily dosage of methylphenidate hydrochloride for treating Attention Deficit Hyperactivity Disorder (ADHD). Bimodal plasma profiles comparable to those obtained after twice daily dosage of immediate release units are obtained with both products (Ritalin LA™ and Focalin XR™).
- Kowalik et al. and McGough et al. reviewed a SODAS™ dosage form containing dexmethylphenidate. The dosage form provided immediate release followed by delayed release after 4 h. The product was shown to be clinically effective over a 12-h period [46, 47].
- Aragon et al. studied the pharmacokinetics of an oral morphine formulation containing both immediate and extended release components. The SODAS™ technology sustained plasma concentrations over 24 h. However, clinical benefit was considered to be limited due to low drug plasma concentration and high variability [48].

Avinza™ (morphine sulfate extended release capsules) has also been formulated utilizing SODAS™ technology, and offers a bimodal release of morphine sulfate. It provides an instant release fraction and an additional fraction to give a sustained pain management over 24 h [49]. Hilleman and Banakar showed that SODAS™ formulations were less vulnerable to food and pH effects than wax-matrix systems as the latter systems may display dose dumping at low pH [50].

6.9 Conclusions

Many concepts and technologies are available for delaying, prolonging, or otherwise modifying drug release. While a relatively limited number of excipients (mainly polymeric materials) are available it is possible, by judicious combinations of these, to design a release profile, suited to specific therapeutic agents. It is important that such design reflects the physicochemical, absorption characteristics, pharmacokinetic behaviors, and dose of drug. Knowledge of the relationship between plasma presence (and plasma concentration) and drug action (duration, onset dose response, viz., a clear target plasma profile) is also a prerequisite for success. Consequently, focusing on a single "platform technology" is undesirable. The variability of the GI tract, with respect to local environment and transit rates also needs to be considered when deciding on a strategy for dosage form design.

Appendix 1. Commercial Products and Their Corresponding Release Mechanisms

Brand name	Active ingredient	Release mechanism	Company
Nexium[a]	Esomeprazole magnesium	Multiparticulate	AstraZeneca
Effexor XR[b]	Venlafaxine HCl	Multiparticulate	Wyeth
Cymbalta[c]	Duloxetine HCl	Multiparticulate	Lilly
Adderall XR[d]	Dextroamphetamine sulfate, dextroamphetamine saccharate, amphetamine aspartate H_2O, and amphetamine sulfate	Multiparticulate	Shire
Flomax[e]	Tamsulosin HCl	Multiparticulate	Boehringer Ingelheim
Detrol LA[f]	Tolterodine tartarate	Multiparticulate	Pfizer
Focalin XR[g]	Dexmethylphenidate HCl	Multiparticulate bimodal release (rapid and delayed)	Novartis
Coreg CR[h]	Carvedilol phosphate	Multiparticulate Micro Pump	GSK
Kadian[i]	Morphine sulfate	Multiparticulate	Actavis
Avinza[j]	Morphine sulfate	Multiparticulate	King
Ultram ER[k]	Tramadol HCl	Diffusion Controlled Tablet	Ortho-McNeil-Janssen
Wellbutrin XL[l]	Bupropion HCl	Diffusion Controlled Tablet	GSK
Ambien CR[m]	Zolpidem tartrate	Matrix Tablet	Sanofi-Aventis
Depakote ER[n]	Divalproex sodium	Matrix Tablet	Abbott
Budeprion XL[o]	Bupropion HCl	Matrix Tablet	Teva
Asacol[p]	Mesalamine	Colonic Delivery Tablet	Proctor & Gamble
Solodyn[q]	Minocycline HCl	Matrix Tablet	Medicis
Allegra-D 12 Hour[r]	Fexofenadine HCl/ pseudoephedrine HCl	Matrix Tablet	Sanofi-Aventis
Enablex[s]	Darifenacin	Matrix Tablet	Novartis
Opana ER[t]	Oxymorphone HCl	Matrix Tablet TIMERx	Endo
Allegra-D 24 Hour[u]	Fexofenadine HCl/ pseudoephedrine HCl	Osmotic	Sanofi-Aventis
Concerta[v]	Methylphenidate HCl	Advanced Osmotic	Ortho-McNeil-Janssen

[a]http://www.nexiumtouchpoints.com/nexium-information/dosing/#Delayed-Release
[b]http://www.effexorxr.com/
[c]http://www.cymbalta.com/
[d]http://www.adderallxr.com/
[e]http://www.4flomax.com/

(continued)

Appendix 1 (continued)

[f]http://www.detrolla.com/
[g]http://www.focalinxr.com/
[h]http://www.coregcr.com/
[i]http://www.kadian.com/
[j]http://www.avinza.com/
[k]http://www.ultram-er.com/
[l]http://www.wellbutrin.com/
[m]http://www.ambiencr.com/
[n]http://www.depakoteer.com/
[o]http://www.rxlist.com/budeprion-xl-drug.htm
[p]http://www.asacol.com/
[q]http://www.solodyn.com/
[r]http://www.allegra.com/allegra-D/allegra-D_12hour.aspx
[s]http://www.enablex.com/
[t]http://www.opana.com/
[u]http://www.allegra.com/allegra-D/allegra-D_24hour.aspx
[v]http://www.concerta.net/

Appendix 2. Conventional Matrix System for Selected Soluble/ Insoluble Drug Products and Their Corresponding Release Modifier(s)

Drugs that are sparingly soluble or insoluble in water		
Lovastatin (antihyperlipidemic)	Advicor™ – a niacin combination (Abbott)	Hypromellose, povidone[a]
Simvastatin (antihyperlipidemic)	Simcor™ – a niacin combination (Abbott)	Hypromellose, povidone[b]
Clarithromycin (antibiotic)	Biaxin XL™ (Abbott)	Cellulosic polymers[c]
Carbamazepine (anticonvulsant)	Tegretol XR™ (Novartis)	Cellulose compounds[d]
Zolpidem tartartae (Insomnia)	Ambien CR™ (Sanofi)	Hypromellose, PEG, sodium starch glycolate[e]
Alprazolam (Panic disorder)	Xanax XR™ (Pharmacia)	Hypromellose[f]
Fluvoxamine (Anxiety disorder)	Luvox CR™ (Elan)	Ammonio methacrylate copolymer[g]
Guanfacine (ADHD)	Intuniv™ (Shire)	Hypromellose, methacrylic acid copolymer, povidone, crospovidone, glyceryl behenate[h]

(continued)

Appendix 2 (continued)

Drugs that are freely soluble, very soluble, or highly soluble in water		
Niacin (antihyperlipidemic)	Niaspan™ (Abbott)	Hypromellose, povidone[i]
Metformin (antihyperglycemic for type 2 diabetes)	Glucophage XR™ (Bristol Myer Squibb)	Sodium carboxymethyl cellulose, hypromellose[j]
Bupropion HCl (Major depressive disorder)	Wellbutrin SR™ (GSK)	Hypromellose[k]
Levetiracetam (Seizure)	Keppra XR™ (UCB)	Hypromellose, PEG6000, partially hydrolyzed polyvinyl alcohol[l]

[a]http://www.rxlist.com/advicor-drug.htm
[b]http://www.rxlist.com/simcor-drug.htm
[c]http://www.rxlist.com/biaxin-drug.htm
[d]http://www.rxlist.com/tegretol-drug.htm
[e]http://www.rxlist.com/ambien-cr-drug.htm
[f]http://www.rxlist.com/xanax-xr-drug.htm
[g]http://www.rxlist.com/luvox-cr-drug.htm
[h]http://www.rxlist.com/intuniv-drug.htm
[i]http://www.rxlist.com/niaspan-drug.htm
[j]http://www.rxlist.com/glucophage-drug.htm
[k]http://www.rxlist.com/wellbutrin-sr-drug.htm
[l]http://www.rxlist.com/keppra-xr-drug.htm

References

1. Malcolm RK et al (2004) Controlled release of a model antibacterial drug from a novel self-lubricating silicone biomaterial. J Control Release 97(2):313–320
2. Streubel A et al (2000) pH-independent release of a weakly basic drug from water-insoluble and -soluble matrix tablets. J Control Release 67(1):101–110
3. Goldenberg MM (1999) An extended-release formulation of oxybutynin chloride for the treatment of overactive urinary bladder. Clin Ther 21(4):634–642
4. Duan XY et al (2009) Development of monolithic osmotic pump tablet system for isosorbide-5-mononitrate delivery and evaluation of it in vitro and in vivo. Drug Dev Ind Pharm 35(4):499–507
5. Liu X, Chen DW, Zhang RH (2003) Evaluation of monolithic osmotic tablet system for nifedipine delivery in vitro and in vivo. Drug Dev Ind Pharm 29(7):813–819
6. Thombre AG et al (2004) Osmotic drug delivery using swellable-core technology. J Control Release 94(1):75–89
7. Wagstaff AJ, Figgitt DP (2004) Extended-release metformin hydrochloride. Single-composition osmotic tablet formulation. Treat Endocrinol 3(5):327–332
8. Waterman KC, MacDonald BC, Roy MC (2009) Extrudable core system: development of a single-layer osmotic controlled-release tablet. J Control Release 134(3):201–206
9. Malaterre V et al (2009) Oral osmotically driven systems: 30 years of development and clinical use. Eur J Pharm Biopharm 73(3):311–323
10. Lautenschlager M, Heinz A (2008) Paliperidone-ER: first atypical antipsychotic with oral extended release formulation. Expert Rev Neurother 8(2):193–200

11. Wonnemann M et al (2006) Significant food interactions observed with a nifedipine modified-release formulation marketed in the European Union. Int J Clin Pharmacol Ther 44(1):38–48
12. Sathyan G, Chancellor MB, Gupta SK (2001) Effect of OROS (R) controlled-release delivery on the pharmacokinetics and pharmacodynamics of oxybutynin chloride. Br J Clin Pharmacol 52(4):409–417
13. Davidson M et al (2007) Efficacy, safety and early response of paliperidone extended-release tablets (paliperidone ER): results of a 6-week, randomized, placebo-controlled study. Schizophr Res 93(1–3):117–130
14. Kane J et al (2007) Treatment of schizophrenia with paliperidone extended-release tablets: a 6-week placebo-controlled trial. Schizophr Res 90(1–3):147–161
15. Bass DM, Prevo M, Waxman DS (2002) Gastrointestinal safety of an extended-release, nondeformable, oral dosage form (OROS (R))(1) – a retrospective study. Drug Saf 25(14):1021–1033
16. Maggi L, Bruni R, Conte U (2000) High molecular weight polyethylene oxides (PEOs) as an alternative to HPMC in controlled release dosage forms. Int J Pharm 195(1–2):229–238
17. Conte U et al (1993) Multilayered hydrophilic matrices as constant release devices (Geonatrix systems). J Control Release 26(1):39–47
18. Conte U, Maggi L (1996) Modulation of the dissolution profiles from Geomatrix(R) multilayer matrix tablets containing drugs of different solubility. Biomaterials 17(9):889–896
19. El-Nabarawi MA (2005) Modulation of tenoxicam release from hydrophilic matrix: modulator membrane versus rate-controlling membrane. Chem Pharm Bull 53(9):1083–1087
20. Wilding IR et al (1995) Pharmacoscintigraphic evaluation of a modified release (Geomatrix) diltiazem formulation. J Control Release 33(1):89–97
21. Goutte F et al (2002) Power of experimental design studies for the validation of pharmaceutical processes: case study of a multilayer tablet manufacturing process. Drug Dev Ind Pharm 28(7):841–848
22. Stephen AM (1995) Food polysaccharides and their applications. Marcel Dekker, New York, p 478
23. Tobyn MJ et al (1996) Prediction of physical properties of a novel polysaccharide controlled release system. Int J Pharm 128(1–2):113–122
24. Staniforth JN, Baichwal AR (2005) TIMERx: novel polysaccharide composites for controlled/programmed release of drugs in the gastrointestinal tract. Expert Opin Drug Deliv 2(3):587–595
25. Endo/Penwest (2003) Oxymorphone--Endo/Penwest EN 3202, EN 3203. Drugs R D, 4(3):204–206
26. Tobyn MJ et al (1996) Investigations of the interactions between a novel polysaccharide controlled release matrix and model compounds using ESR. J Control Release 40(3):147–155
27. Streubel A, Siepmann J, Bodmeier R (2006) Drug delivery to the upper small intestine window using gastroretentive technologies. Curr Opin Pharmacol 6(5):501–508
28. Murphy CS et al (2009) Gastroretentive drug delivery systems: current developments in novel system design and evaluation. Curr Drug Deliv 6(5):451–460
29. Arza RAK, Gonugunta CSR, Veerareddy PR (2009) Formulation and evaluation of swellable and floating gastroretentive ciprofloxacin hydrochloride tablets. AAPS PharmSciTech 10(1):220–226
30. Badhan AC et al (2009) Development and evaluation of sustained release gastroretentive minimatrices for effective treatment of *H. pylori* infection. AAPS PharmSciTech 10(2):459–467
31. Basak SC, Rahman J, Ramalingam M (2007) Design and in vitro testing of a floatable gastroretentive tablet of metformin hydrochloride. Pharmazie 62(2):145–148
32. Boldhane SP, Kuchekar BS (2009) Gastroretentive drug delivery of metformin hydrochloride: formulation and in vitro evaluation using 3(2) full factorial design. Curr Drug Deliv 6(5):477–485
33. Bomma R et al (2009) Development and evaluation of gastroretentive norfloxacin floating tablets. Acta Pharm 59(2):211–221
34. Singh S et al (2007) Gastroretentive drug delivery system of metoclopramide hydrochloride: formulation and in vitro evaluation. Curr Drug Deliv 4(4):269–275

35. Groning R et al (2007) Compressed collagen sponges as gastroretentive dosage forms: in vitro and in vivo studies. Eur J Pharm Sci 30(1):1–6
36. Groning R, Cloer C, Muller RS (2006) Development and in vitro evaluation of expandable gastroretentive dosage forms based on compressed collagen sponges. Pharmazie 61(7):608–612
37. Klausner EA et al (2003) Expandable gastroretentive dosage forms. J Control Release 90(2):143–162
38. Dhaliwal S et al (2008) Mucoadhesive microspheres for gastroretentive delivery of acyclovir: in vitro and in vivo evaluation. AAPS J 10(2):322–330
39. Ravenelle F, Rahmouni M (2006) Contramid (R): High-amylose starch for controlled drug delivery. Polysaccharides Drug Deliv Pharm Appl 934:79–104
40. Rahmouni M et al (2001) Enzymatic degradation of cross-linked high amylose starch tablets and its effect on in vitro release of sodium diclofenac. Eur J Pharm Biopharm 51(3):191–198
41. Rioux B et al (2002) Structure-properties relationship in cross-linked high amylose starch cast films. Carbohydr Polym 50(4):371–378
42. Burch F et al (2007) A comparison of the analgesic efficacy of tramadol contramid OAD versus placebo in patients with pain due to osteoarthritis. J Pain Symptom Manag 34:328–338
43. Mongin G et al (2004) Efficacy and safety assessment of a novel once-daily tablet formulation of tramadol – a randomised, controlled study versus twice-daily tramadol in patients with osteoarthritis of the knee. Clin Drug Invest 24(9):545–558
44. Sheehan DV et al (2009) Extended-release trazodone in major depressive disorder: a randomized, double-blind, placebo-controlled study. Psychiatry (Edgmont) 6(5):20–33
45. Barnard DL (2001) Genvir (Flamel Technologies). Curr Opin Investig Drugs 2(5):622–623
46. Kowalik S, Minami H, Silva R (2006) Dexmethylphenidate extended-release capsules for the treatment of attention deficit hyperactivity disorder. Expert Opin Pharmacother 7(18):2547–2557
47. McGough JJ, Pataki CS, Suddath R (2005) Dexmethylphenidate extended-release capsules for attention deficit hyperactivity disorder. Expert Rev Neurother 5(4):437–441
48. Aragon CL et al (2009) Pharmacokinetics of an immediate and extended release oral morphine formulation utilizing the spheroidal oral drug absorption system in dogs. J Vet Pharmacol Ther 32(2):129–136
49. Semenchuk MR (2002) Avinza Elan. Curr Opin Investig Drugs 3(9):1369–1372
50. Hilleman DE, Banakar UV (1992) Issues in contemporary drug delivery. Part VI: Advanced cardiac drug formulations. J Pharm Technol 8(5):203–211

Chapter 7
Drug–Polymer Matrices for Extended Release

Sandip B. Tiwari, James DiNunzio, and Ali Rajabi-Siahboomi

Abstract Extending drug release from a dosage form can prolong its action, attenuate peak plasma levels, thereby obviating concentration-related side effects or optimize efficacy by matching systemic presence with other time-related effects. Such modifications can be affected by embedding the drug in a matrix that prevents immediate release but delivers at a rate consistent with absorption or disposition requirements. Various polymeric and other materials can be used to design the most appropriate release profile and provide a viable and consistent mode of manufacture. Such materials, their properties, and modes of release modification are presented, reviewed, and discussed in this chapter.

7.1 Introduction

An extended-release (ER) formulation of a drug may enhance its therapeutic benefits, minimize side effects, and boost patient compliance [1–3], thereby improving the management of the disease. Table 7.1 outlines the potential, limitations, and an idealized development path for an ER product [1–3]. Such presentations now comprise a significant number of new product filings in the USA [4]. In a historical context, the first commercial oral ER formulations were pellet-filled capsules (Spansules®), introduced in the 1950s by Smith, Kline, and French [5]. Since then, a number of strategies have been developed to modify drug release, including simple matrix tablets, pellets, or more complex technologies. Matrix-based systems offer many advantages if applied to suitable drugs. A combination of matrix former and other materials can help match the release characteristics to the requirements of the drug and clinical condition.

S.B. Tiwari (✉)
Colorcon Inc. Global Headquarters, Harleysville, PA, USA
e-mail: STiwari@colorcon.com

C.G. Wilson and P.J. Crowley (eds.), *Controlled Release in Oral Drug Delivery*,
Advances in Delivery Science and Technology, DOI 10.1007/978-1-4614-1004-1_7,

Table 7.1 Development path for extended release formulations; their advantages and limitations

Typical development path for extended release formulations
New chemical entity compound approval for development
Preformulation and immediate release formulation development
First human trial with immediate release formulation
Development of extended release formulation based on clinical/commercial triggers
Advantages of extended release formulations
Clinical advantages
To reduce frequency of drug administration (associated with short compound half life)
Improve patient compliance
Reduce drug toxicity (local or systemic associated with high peak exposure)
Reduce drug level fluctuation in blood (avoids over- or under-medication for periods of time)
To stabilize medical condition (because of more uniform drug levels)
Reduce drug accumulation with chronic therapy
Improve bioavailability of some drugs because of spatial control
Reduce total drug usage when compared with conventional therapy
Commercial/industrial advantages
Economical to the health care providers and the patient
Illustration of innovation and technological leadership
Product life-cycle management
Product differentiation and branding
Market expansion
Patent extension
Potential limitations of extended release formulations
Delay in onset of drug action
Possibility of dose dumping in the case of a poorly formulated dosage form
Increased potential for first-pass metabolism
Greater dependence on GI transit time of the dosage form
Cost per unit dose is higher when compared with conventional doses
Not all drugs are suitable for formulating into ER dosage form

Successful development of ER dosage forms requires consideration of factors such as:

- The physicochemical properties of the drug.
- Drug dose, stability, and its solubility in the different regions of the gastrointestinal (GI) tract.
- Dose response.
- Mode, site, and extent of absorption.
- Pharmacokinetic properties such as metabolism, distribution, and elimination.
- Relationship between plasma presence and onset/duration of action.

Dosage form design programs also need to address choice of excipients, modes of manufacture, and equipment availability and capability. These considerations are reviewed in this chapter for matrix systems used to modify drug release from oral dosage units.

The theoretical basis of ER strategies and their relevance to clinical performance have been extensively reported [2, 6–19]. Other chapters in this book deal with specific modes of modifying release and with the associated technologies.

Our aim therefore is not to duplicate such information but focus on practical considerations for design and manufacture of matrix systems, particularly on the characteristics of polymers and other excipients employed to modulate such drug release.

7.2 Dosage Form Design Strategies

The criteria listed in Table 7.2 have been used for preliminary assessment of the suitability of a drug for development in ER form [20].

Matrix-based systems incorporate one or more of the following approaches:

1. Monolithic matrix systems (*mono* meaning single, *lith* is stone or block of material) comprise a release-rate controlling polymer matrix containing dispersed or dissolved drug. Two types may be employed, depending on the release-modifying polymer.

   (a) *Hydrophilic matrix systems*: drug particles are dispersed or dissolved in a water-soluble polymeric matrix, drug becoming available for release as the matrix hydrates, swells (forms a gel), and dissolves. Hydrophilic matrices have the capability to provide desired release profiles for a wide range of drugs using established and well-characterized excipients. Processes for manufacture are also robust.

   (b) Inert or Insoluble Matrix Systems: An inert matrix system contains drug embedded in a polymer that is not soluble in GI fluids [1, 21]. Such matrices can be excreted intact ("exhausted ghosts") following drug release in the GI tract. Drug release occurs by liquid penetration through the polymer or via channels formed by including pore formers or wetting agents in the matrix to enhance fluid permeation, leading to dissolution of the embedded drug and its diffusion through the matrix. Such systems are not suitable for high drug loads, the polymer content probably being insufficient to form a matrix. Neither is it suited to poorly soluble drugs as dissolution in the matrix would be release rate limiting. Release from insoluble matrices is not affected by volume of dissolution medium, pH, enzyme content, and other attributes of the digestive fluids, unless drug solubility is pH dependent. Thus, inert matrices are less susceptible to hydrodynamic or food effects.

2. Reservoir (coated) systems comprise a drug-containing core enclosed within a polymer barrier coat. Two types of reservoir systems can be used:

   (c) Simple diffusion/erosion systems where a drug-containing core is contained within hydrophilic and/or water-insoluble polymer coatings. Drug release is achieved by diffusion of drug through the coatings or following coat erosion.

   (d) Osmotic systems where the drug core is contained within a semipermeable polymer membrane with a mechanical/laser drilled hole for drug release, driven by osmotic pressure generated within the tablet core.

   Reservoir-based systems are discussed elsewhere in this book.

Table 7.2 Initial criteria in controlled release feasibility assessments (reprinted with permission from [20])

		Comments
Physicochemical factors		
Dose	<1 mg	Developmental complexity (potential drug content uniformity issue)
	10–250 mg	Average degree of difficulty
	>>250–300 mg	Could need more than one tablet to accommodate the drug load
Dose/solubility ratio[a]	<1 ml	Several technology options exists for CR development
	1–100 ml	Average degree of difficulty
	>1,000 ml	Need solubilization – CR development will be difficult
	>10,000 ml	CR development practically impossible
Stability	Generally stable as a solid or solution and with common CR excipients	Predict average degree of difficulty
	Compound shows or is predicted to have significant degradation	Predict higher degree of difficulty
Biopharmaceutical factors		
Absorption mechanism	Transcellular passive diffusion	Average degree of difficulty
	Other mechanisms including efflux	Performance could be difficult to predict
Regional permeability (colonic)	Poor absorption, $P_{app, CaCO2} < 10^{-6}$ cm s^{-1}, $k_a < 0.01$ min^{-1}	CR formulations with prolonged delivery duration may not be feasible. Likely will not be bioequivalent to IR
	Moderate absorption, $P_{app, CaCO2} = 10^{-6}–10^{-5}$ cm s^{-1}	CR development challenging but feasible. Might not be bioequivalent to IR
	Good absorption, $P_{app, CaCO2} > 10^{-5}$ cm s^{-1}, $k_a > 0.01$ min^{-1}	CR development should be feasible. Likely to be bioequivalent to IR
PK factors		
PK or PD half life	<1–2 h	Half life too short for CR development
	2–10 h	Acceptable half life
	>>10 h	Compound might not need CR for reducing dosing frequency
Metabolism and efflux	High presystemic or first-pass metabolism	Relative BA of CR formulation might be low
	Compound is P-gp or CYP3A4 substrate	CR performance difficult to predict (depends on dose and K_m, V_{max})

It should be understood that the ranges given in Table 7.2 are empirical simple rules of thumb and they can be different depending on the properties of the drug candidate and prior experience in a company. Ideally, each drug candidate should be analyzed on a case-by-case basis because many of the factors are interrelated

[a] Dose–solubility ratio calculated by dividing highest dose by lowest solubility in the pH range 1–7.5

Monolithic matrix systems are simple and cost-effective methods for fabrication of ER dosage forms. Most products employing the technology are presented as tablets. Manufacturing methods include blending, granulation, compression, and coating operations as used in conventional tablet manufacture. Specialized equipment is not usually required.

7.3 Release Mechanisms

7.3.1 Hydrophilic Matrices

Drug release from hydrophilic matrices has been extensively studied [22–31]. An in-depth discussion is beyond the scope of this chapter but some basic fundamentals are highlighted.

When a matrix tablet is exposed to GI fluids, the surface of the tablet is wetted and polymer hydrates to form a "gel layer," the surface layer polymer transitioning from a "glassy" solid to a "rubbery" gelatinous state. The tablet core remains essentially dry at this stage. Drug on the gel surface, if highly soluble and present at high concentration (high dose drug) is likely to pass into solution rapidly, providing an initial "pulse" of drug. The gel layer (rubbery state) grows progressively as fluid penetrates towards the core, increasing the thickness of the gel layer and diffusion path for drug release. As the outer layer becomes fully hydrated, polymer chain relaxation weakens the gel layer, with loss of integrity, disentanglement, and erosion from the matrix surface. Continuing fluid penetration through the gel layer ultimately leads to complete erosion and drug release.

Soluble drugs may be released by a combination of diffusion and erosion. Erosion is the predominant mechanism for insoluble drugs. Consistent and reliable release requires that polymer hydration and surface gel layer formation is rapid and consistent, to obviate compact disintegration and premature drug release. Consequently, polymers for hydrophilic matrices are usually supplied in smaller particle size ranges (such as METHOCEL™ CR grades).

Several mathematical models have been developed to describe release from hydrophilic matrices. The simplest and more widely used model is that derived by Korsmeyer et al. [32]:

$$\mathrm{M}t/\mathrm{M}a = kt^{n}, \tag{7.1}$$

where M*t*/M*a* is the fraction of drug released, *k* is the diffusion rate constant, *t* is the release time, and *n* is the release exponent indicative of the mechanism of release.

The equation was modified by Ford et al. [31] to account for any lag time (*l*) or initial burst release of the drug.

$$\mathrm{M}t/\mathrm{M}a = k(t-l)^{n}. \tag{7.2}$$

Both equations demonstrate that:

- When the exponent n has a value of 1.0, drug release rate is independent of time, viz., zero-order release kinetics apply (Case II transport). In such cases, polymer relaxation and erosion are rate-controlling steps.
- When $n=0.5$, Fickian diffusion is the rate-controlling step (Case I transport).
- Values of n between 0.5 and 1 indicate that both diffusion and polymer relaxation contribute to release kinetics (non-Fickian, anomalous or first-order release).

The extreme values of $n=0.5$ and 1 are only valid for slab geometry (square shape with flat surfaces). For cylindrical tablets, values range from $0.45<n<0.89$ for Fickian, anomalous, or Case II transport [26]. Use of an appropriate viscosity grade of polymer enables the design of matrices based on diffusion (n approaching ~0.45), diffusion and erosion (n between 0.45 and 0.9), or via erosion mechanisms (n approaching ~0.9).

Drugs with very low solubility (e.g., <0.01 mg/ml) may dissolve slowly and have slow diffusion through the gel layer of a hydrophilic matrix. Hence, release would primarily occur through erosion of the hydrated surface. In such cases, control over matrix erosion to achieve consistent release throughout the GI tract is critical. Hence low viscosity grades of polymer (e.g., low viscosity grades of hypromellose, METHOCEL K100LV CR and E50LV) are recommended to provide adequate erosion rates.

Drugs with very high water solubility can dissolve within the gel layer (even with small amounts of free water). It is important therefore to maintain gel layer integrity. A robust gel can be formed using high-viscosity polymers. It is also possible to use blends of polymers with differing viscosities to provide the requisite diffusivity and gel strength.

7.3.2 *Inert Matrices*

Drug release from inert matrices may involve several processes. These include fluid permeation of the matrix, dissolution and diffusion of the drug through and from the matrix, or erosion of the matrix material and consequent release of drug. Drug may also dissolve in the matrix material and be released by diffusion through the matrix or partition between the matrix and dissolution medium.

Higuchi derived the following relationship to describe drug release from inert matrices [1, 33]:

$$Q=\left[\left(\frac{D\epsilon Cs}{\tau}\right)(2A-\mathrm{Cs})t\right]^{1/2}, \tag{7.3}$$

where Q is the amount of drug released per unit surface area after time t, D is the diffusion coefficient of drug, ε is the tortuosity of the matrix, τ is the porosity of the matrix, Cs is the solubility of the drug in the dissolution medium, and A is the initial loading dose of the drug in the matrix.

Drug release is initiated by permeation of the dissolution medium into the matrix, dissolving drug, and creating channels through which diffusion takes place. The drug-depleted zone progressively extends towards the core of the matrix. High tortuosity means that the effective average diffusion path is large. The porosity term takes into account the space available for drug dissolution, with increased porosity resulting in increased release. Porosity and tortuosity are functions of drug load, the physicomechanical properties of the matrix, and the dispersion characteristics of the drug in the matrix.

If the drug is freely soluble in the dissolution medium, i.e., Cs>>A, such that the dissolution rate is rapid then the following equation applies:

$$Q = 2A\left(\frac{Dt}{\pi\tau}\right)^{1/2}. \tag{7.4}$$

Release rate is directly proportional to the amount of dispersed drug, A; it is proportional to $A^{1/2}$ for insoluble drugs if $2A = \mathrm{Cs}$. These relationships predict that plots of Q versus $t^{1/2}$ be linear.

7.4 Polymers for Modifying Release

7.4.1 Hydrophilic Matrices

This section describes the critical quality attributes of polymers used in hydrophilic matrix systems. These are listed in Table 7.3 along with FDA-recommended maximum use levels [34].

7.4.1.1 Hypromellose (Hydroxypropyl methylcellulose)

Hydroxypropyl methylcellulose (HPMC) is widely used in matrix applications. Key advantages include global regulatory acceptance, stability, nonionic nature (resulting in pH-independent release of drugs), and ease of processing by direct compression (DC) or granulation. Other advantages are versatility and suitability for various drugs and release profiles (different viscosity grades being available) and extensive history of use. It is a mixed alkyl hydroxyalkyl cellulose ether containing methoxyl and hydroxypropyl groups. Type and distribution of the substituent groups affect physicochemical properties such as rate and extent of hydration, surface activity, biodegradation, and mechanical plasticity. Matrices exhibit pH-independent drug release profiles while aqueous solutions are stable over a wide pH range (3–11) and are resistant to enzymatic degradation.

HPMC is available commercially from the Dow Chemical Company as METHOCEL™. Four grades are available (A, E, F, and K) having differing hydroxypropoxyl and methoxyl substitutions.

Table 7.3 FDA registered oral formulations containing commonly used hydrophilic polymers [34]

Polymer/material	No of hits on FDA Web page[a]	Maximum potency listed for oral formulations (mg)[b]
Hydrophilic polymers		
Cellulosics		
Methylcellulose	15	183.60
Hypromellose (hydroxypropyl methylcellulose, HPMC)	102	670.04
Hydroxypropylcellulose (HPC)	41	240.00
Hydroxyethyl cellulose (HEC)	11	150.00
Sodium carboxymethylcellulose (Na-CMC)	21	160.00
Noncellulosics: gums/polysaccharides		
Sodium alginate	9	350.00
Xanthan gum	22	109.52
Locust bean gum (Ceratonia)	2	74.25
Guar gum	9	40.00
Cross-linked high amylose starch		
Noncellulosics: others		
Polyethylene oxide (POLYOX™)	8	543.90
Homopolymers and copolymers of acrylic acid	14	195.00[c]

[a]Total number of listings on FDA web page for use in oral dosage forms

[b]The "maximum potency" specifies the maximum amount of inactive ingredient for oral route/oral dosage form containing that ingredient. Listed potency is for generic material; refer to FDA web page for specific grade listing. Also the maximum potency number may be higher if its status showed pending status at the time of writing this chapter

[c]Listing under poly(acrylic acid)

METHOCEL E (hypromellose 2910 USP) and K (hypromellose 2208, USP) are probably the most widely used grades in matrix formulations and are distributed worldwide by Colorcon Inc. The USP classification code is based on substitution. The first two digits represent the mean % methoxyl substitution and the last two the mean % hydroxypropyl substitution. HPMC is highly hydrophilic, hydrating rapidly in contact with water. Since the hydroxypropyl group is hydrophilic and the methoxyl group is hydrophobic, the ratio of hydroxypropyl to methoxyl content influences water mobility in a hydrated gel layer and therefore, drug release. METHOCEL grades for ER matrix formulations include E50LV, K100LV CR, K4M CR, K15M CR, K100M CR, E4M CR, and E10M CR. Viscosities of 2% aqueous solutions of these polymers range from 50 to 100,000 cPs at 20°C. Similar grades of HPMC are also available from suppliers such as Shin-Etsu Chemical Co., Ltd, Japan [35] and Ashland Aqualon Functional Ingredients [36].

HPMC exhibits glass transition temperatures, ranging from 160°C to 180°C depending on molecular weight and chemistry. It is classified as a nonthermoplastic material [37], thereby limiting its applicability in thermo-forming technologies such as hot melt extrusion or injection molding. Inclusion level can vary from 10 to 80%

of the total mass of the formulation, depending on the drug and desired release characteristics. A robust formulation with consistent performance and insensitivity to minor variations in materials or manufacturing processes may usually be obtained with a ≥30% (w/w) inclusion level [38–41].

Release rates from matrices depend on many interacting factors, such as polymer type and level, drug solubility and dose. Polymer:drug ratio, filler type and level, polymer:filler ratio, particle size of drug and polymer, and porosity and shape of the matrix are important [9, 42–57].

7.4.1.2 Hydroxypropyl Cellulose and Hydroxyethyl Cellulose

Hydroxypropyl cellulose (HPC) is a nonionic polymer, being a partially substituted poly (hydroxypropyl) ether of cellulose. It is available from Ashland Aqualon Functional Ingredients under the brand name of *Klucel* in different grades with differing solution viscosities. Molecular weight ranges from ~80,000 to 1,150,000 [36, 58]. High viscosity grades of HPC (e.g., Klucel HXF with fine particle size, 1,500–3,000 mPa of 1% solution) are generally used. Other high viscosity grades of HPC are also available from Nisso, namely, HPC-M and HPC-H (150–4,000 mPa of 2% solution) [59].

Inclusion levels can vary from 15 to 40%. Addition of an anionic surfactant (e.g., sodium lauryl sulfate) reportedly increases HPC viscosity and as a consequence reduces drug release rate [58, 60]. Combinations of HPC and other cellulosic polymers have been used to improve wet granulation and tableting characteristics and better control of drug release [58].

HPC is thermoplastic and its presence may enable processing of HPMC-containing formulations using hot melt extrusion or injection molding. It is not widely used because of its low swelling capacity and sensitivity to ionic strength of the dissolution media [61–65]. Gel strengths of HPC matrices decrease during dissolution, leading to less cohesive gel structures [61]. The lower tablet gel strength (Γ) of HPC matrices, compared to HPMC can cause poor in vitro/in vivo correlation [65].

Hydroxyethyl cellulose (HEC) is also a nonionic, partially substituted poly (hydroxyethyl) ether of cellulose. It is available in several grades from Ashland Aqualon Functional Ingredients under the brand name of Natrosol®. These vary in viscosity and degree of substitution [36]. High viscosity grades of HEC (1,500–5,500 mPa of 1% solution) are sometimes used in ER formulations. Typical inclusion levels are 15–40% of the total formulation mass. Swelling of HEC matrices has been reported to be considerably greater than HPC matrices. HEC matrices also exhibited relatively higher erosion rates, $t_{50\%}$ (time to 50% release) being shorter for HEC than for HPC matrices [66]. In contrast to its widespread use in pharmaceuticals, HEC is not currently approved for use in food products in Europe or the USA. This restriction is due to the high levels of ethylene glycol residues that are formed during its manufacture [67].

7.4.1.3 Sodium Carboxymethylcellulose

Sodium carboxymethylcellulose (Na CMC) is an anionic, water-soluble polymer, prepared by reacting cellulose with sodium monochloroacetate. Various viscosity grades are available, reflecting degree of substitution. Aqueous solutions are stable over the pH range 4–10, precipitation occurring below pH 2: solution viscosity decreases rapidly above pH 10. Generally, solutions exhibit maximum viscosity and stability at pH 7–9. At pH 4.5 and 6.8 matrices containing Na CMC exhibit the morphology of a swellable matrix. The macromolecular chains in the gel network are held together by weak bonds resulting in erosion-mediated drug release. At pH 1.0 in contrast, the gel layer is rigid, typical of a partially cross-linked hydrogel, resulting in diffusional release [68]. Such sensitivity to dissolution media pH is attributable to the ionic nature of the polymer. Hence, release mechanisms may be sensitive to media pH.

Na CMC has been used in blends with HPMC to prepare hydrophilic matrices [68–70]. Mixtures of Na CMC and HPMC in dilute solution exhibit higher-than-expected viscosities. This may be attributable to intermolecular cross-links between carboxyl and hydroxyl groups. Baveja et al. advocated combining HPMC with Na CMC to provide zero-order release profiles for propranolol, metoprolol, oxprenolol, and alprenolol [69]. They hypothesized that the polymer combination synergistically increased viscosity, allowing erosion at a rate determined by the transitioning of the front between glassy and the rubbery polymer states. It was later established that viscosity enhancement was not solely responsible for modulating drug release: Complex formation between the anionic polymer and cationic drug also played a role [71].

Aiman and coworkers showed that erosion-mediated release rate of dextromethorphan from matrix tablets containing Na CMC was significantly lower than from HPMC-containing matrices. Release was also pH sensitive. The slower release was attributable to drug/Na CMC complex formation [72]. Matrices comprising HPMC/Na CMC mixtures exhibited zero-order release profiles [69, 70]. Thus, pairing Na CMC with water-soluble basic drugs may have complex effects on drug release. For less soluble drugs, which are released principally by erosion, the above-reported effect appears to be reversed. A commercially available metformin hydrochloride ER tablet (Glucophage® XR, Bristol Myers Squibb) is reported to comprise a HPMC/Na CMC matrix to attain the desired release profile [73].

7.4.1.4 Sodium Alginate

Sodium alginate, a water-soluble salt of alginic acid is a natural linear unbranched polysaccharide extracted from marine brown algae. It consists of different proportions of α-D-mannuronic acid (M) and β-L-guluronic acid (G) units. The M and G units are $1 \rightarrow 4$ linked by glycosidic bonds, forming homopolymeric M- or G-blocks and heteropolymeric MG blocks [74]. Matrices incorporating either a single alginate salt or combinations of salts have been employed to sustain release in vitro and in vivo. Commercially, sodium alginate for ER applications is available from FMC under the brand name Keltone®. Grades LVCR, LKX, and HVCR are generally used [75].

The presence of carboxylate groups that can accept or release protons in response to pH changes makes sodium alginate pH sensitive. At pH values below the pK_a of the M (3.38) and G (3.65) monomers, the soluble sodium salt is converted to insoluble alginic acid. In a matrix tablet, sodium alginate pH sensitivity would affect the characteristics of the diffusion barrier and as a consequence drug release. Cryogenic electron microscopy reveals the hydrated surface layer formed by sodium alginate matrices in simulated gastric fluid to be particulate and porous, contrasting with the highly hydrated continuous gel layer formed in simulated intestinal fluid [76]. This difference in diffusion barrier properties affects hydration, swelling, and erosion kinetics leading to pH-dependent drug release. Cationic drugs (e.g., lidocaine) are released more slowly than anionic drugs (e.g., sodium salicylate), probably because of drug–polymer ionic interactions [77].

Matrices of sodium alginate are prone to lamination and crack formation at low pH (<3) which could result in "burst" release in the gastric environment [78, 79]. Crack formation does not occur in neutral environments. Such cracking can limit the use of sodium alginate in matrices because of the risk of dose dumping. However, its use may be facilitated by blending with HPMC, thereby obtaining a pH-independent release profile for basic drugs [80, 81].

Sodium alginate precipitates as alginic acid at low pH. The acid appears to confer a firm structure to the gel, reducing erosion. If drug solubility at this pH is high, diffusion through the gel matrix layer predominates as the release mechanism. At higher pH the alginate remains as the soluble salt, providing less resistance to erosion. This is likely to increase release rate. Such an effect can be beneficial for drugs whose solubility is lower at higher pH. Increased polymer erosion at such higher pH could compensate for the fall-off in driving force for diffusion/dissolution-mediated release as drug solubility decreases. An appropriate balance needs to be determined for each drug candidate. Verapamil hydrochloride ER matrices (Calan®SR, Pfizer) contain a combination of HPMC and sodium alginate to produce the desired drug release profile in vivo [82].

7.4.1.5 Xanthan Gum

Xanthan gum is an anionic high molecular weight polysaccharide produced by fermentation by the microorganism *Xanthomonas campestris*. Solutions exhibit weak gel-like properties at low shear rates. It does not form true gels at any concentration or temperature but can produce near zero-order drug release kinetics. Fickian diffusion was dominant during early dissolution of diclofenac minimatrices; erosion predominated during the later stages, suggesting that zero-order release was attainable. Rate of drug release is slowed by decreasing particle size of the polymer or increasing its concentration. Release is slightly faster in acidic media due to more rapid initial surface erosion. In tablets containing a large inclusion level of drug (50% theophylline), 20% xanthan gum proved to be an efficient release modifier but catastrophic failure occurred at 15% inclusion level [83].

Rheologically, xanthan gum exhibits rapid and marked shear thinning. In vitro drug release can depend somewhat on rate of agitation of the dissolution medium [83, 84]. Drug release was also found to be influenced by the ionic strength of the medium, particularly at ionic strengths similar to those in the GI tract. Thus, differences in GI fluid composition could affect in vivo performance. Commercially, xanthan gum is available from CP Kelco under the brand name of Xantural® [85]. The fine particle size grade of Xantural 75 (viscosity of 1,200–1,600 mPa) is generally recommended for use in ER formulations.

Combinations of xanthan gum with HPMC can retard drug release compared to single polymer systems. Such combination can also overcome the limitations of individual matrices. HPMC forms firm gels but does not hydrate as quickly as xanthan gum. However, xanthan gum does not form strong gels around a hydrating matrix and requires high concentrations to prevent rapid erosion. A combination of polymers may be more suitable for formulating ER matrices of high-solubility, high-dose drugs [86]. In such systems, the initial burst release, typical of high solubility drugs, is controlled by rapid gelation of the xanthan gum. Subsequent release and matrix integrity is maintained by the firm gel structure of HPMC.

The quick gelling property of xanthan gum has also been exploited in gas generating gastroretentive matrices of ciprofloxacin formulated with HPMC [87]. The instant viscolyzing behavior of xanthan gum enabled the entrapment of gas (carbon dioxide, formed by interaction of sodium bicarbonate in the matrix with the hydrochloric acid in the dissolution medium). Hydration and gelation of HPMC reduces the density of the formulation to provide early buoyancy for gastroretention. The subsequent hydration and firm gel layer formulation by HPMC further reduces bulk density, improving buoyancy and prolonging gastroretention.

7.4.1.6 Carbomers

Carbomers are synthetic high molecular weight polymers of acrylic acid that are cross-linked with either allyl sucrose or allyl ethers of pentaerythritol. They are commercially available from Lubrizol under the brand name of Carbopol® and are available in grades that vary in viscosity, polymer type, and polymerization solvent [88]. Being cross-linked, these polymers are not water soluble but are swellable and gel forming. Swelling and gel formation behaviors differ somewhat from other hydrophilic polymers like HPMC, where swelling follows polymer hydration, leading to relaxation of polymer chains and their subsequent entanglement (physical cross-linking) to form a viscous gel. With acrylic acid polymers, surface gel formation is not due to polymer chain entanglement (the polymers are already cross-linked) but to formation of discrete micro gels comprising many polymer particles [40].

Erosion, as occurs with linear polymers like HPMC does not occur because of the water insolubility. Instead, when the hydrogel is fully hydrated, osmotic pressure from within breaks up the structure, sloughing off discrete pieces of the hydrogel. The hydrogel remains intact and drug continues to diffuse uniformly through the gel layer.

In contrast to linear polymers, higher viscosity does not result in slower drug release. Lightly cross-linked polymers (lower viscosity) are generally more efficient in controlling release than highly cross-linked variants [89].

Release from carbomer matrices may depend on the pH of dissolution media, because of the anionic nature of the polymer (pK_a 6±0.5) [90]. Swelling and gel formation are pH dependent. At lower pH the polymer is not fully swollen and drug release is faster. As pH increases the polymer swells and rapidly forms a gel layer, prolonging drug release. Carbomers, being anionic may form complexes with cationic drugs depending on drug properties such as pK_a, solubility, amine group strength, steric orientation, molecular weight and size.

It has been reported that carbomer inclusion levels of about 30% produce comparable drug release profiles to HPMC in both water and 0.1 N HCl. Release was slower in pH 6.8 phosphate buffer. Carbomer matrices also exhibited significantly lower gel strengths compared to HPMC matrices in all three media. This has been postulated as the reason for their significantly faster drug release in vivo compared to HPMC matrices [65].

HPMC/carbomer matrices have been explored for controlling the release of various drugs [88, 91, 92]. Advantages include low inclusion levels, versatility in release modulation, and ability to extend release of some cationic drugs. Recently, the Research Group at Colorcon has shown that a mixed matrix incorporating HPMC/Carbomer/polyvinyl acetate phthalate (PVAP) provided slower release than single or binary systems. This was ascribed to a synergistic increase in viscosity/gel strength, possibly due to stronger hydrogen bonding between the hydroxy groups of HPMC and the carboxylic functions of the carbomer or PVAP. Stronger bonding provided a more rigid structure for drug diffusion [91, 93]. The influence of combination of carbomer, PVAP, and HPMC blend in a matrix formulation of Guaifenesin, a soluble drug is shown in Fig. 7.1 [94].

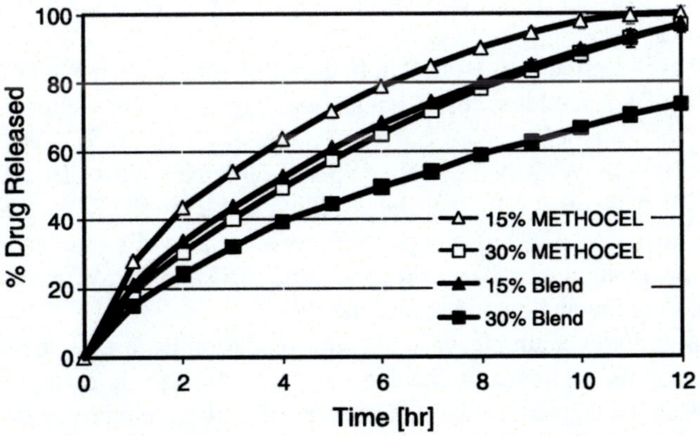

Fig. 7.1 Drug release profile of Guaifenesin from matrices containing 69% drug, 15% or 30% METHOCEL™ K4M CR or combination of METHOCEL™ K4M CR+carbomer and polyvinyl acetate phthalate, qs % Fast-flo lactose and 0.5%w/w each of Cab-O-Sil and magnesium stearate. Dissolution study was performed using USP apparatus II at 100 rpm and 900 ml of deionized water

A 15% inclusion level of the combined polymer provided a release profile similar to those exhibited with an inclusion level of 30% HPMC alone. Such a "polymer-sparing" effect can be beneficial where dose of drug is high, facilitating unit size reduction and cost savings.

The combined polymer matrix also engendered lower microenvironmental pH (3.5–4.5) within the gel layer (microenvironmental pH of HPMC alone matrix is 7.4–8.2). This may help improve solubility or possibly stability of some basic drugs. Moreover, the combination can produce matrices with higher gel strength that are less sensitivity to hydrodynamic conditions. Comparable dissolution profiles were evident at stirring rates of 50, 100, and 150 rpm in USP II Dissolution Apparatus [95].

7.4.1.7 Polyethylene Oxide

Polyethylene oxide (PEO) [POLYOX™] resins are water soluble, nonionic polymers manufactured by Dow Chemical Company and distributed worldwide by Colorcon [96]. They are free flowing white powders, soluble in water at temperatures up to 98°C and in certain organic solvents. Structures comprise the repeating sequence –$(CH_2CH_2O)_n$ where n represents the average number of oxyethylene groups. It is highly crystalline and available in molecular weight grades ranging from 1×10^5 to 7×10^6 Da. Their high molecular weights mean that the concentration of reactive end groups is very low. However, as their paired ether–oxygen electrons have a strong affinity for hydrogen bonding, they can form association complexes with a variety of monomeric and polymeric electron acceptors (e.g., gelatin, carbomer) as well as certain inorganic electrolytes, e.g., alkali halides [97].

PEO resins are among the fastest hydrating water soluble polymers, quickly forming hydrogels that initiate and regulate drug release. Systems using such resins are often superior in approaching zero-order release profiles. PEO can be used at 20–90% inclusion level depending on the drug and the desired release characteristics.

PEO behaves similarly to HPMC in hydrophilic matrix systems. With appropriate selection of a suitable viscosity grade, one may be able to achieve release profiles similar to hypromellose matrices [98]. Grades available are POLYOX WSR-205 NF, WSR-1105 NF, WSR N-12 K NF, WSR N-60 K NF, WSR-301 NF, WSR-303 NF, and WSR Coagulant NF. The high swelling capacity of PEO has been used in hydrophilic matrices to achieve expanded swelling, providing enhanced gastroretention. A formulation of gabapentin containing PEO and HPMC exhibited significant matrix swelling and gastric retention [99].

PEO can undergo chain cleavage via auto-oxidation leading to loss of viscosity in aqueous solution. Higher molecular weight grades are more sensitive to such oxidation. Rate of auto-oxidation can be minimized by including antioxidants and by controlling storage conditions. Commercially available PEO grades are supplied with the added antioxidant, usually butylated hydroxytoluene (BHT) at inclusion levels of 100–1,000 ppm, depending on the molecular weight of the polymer.

Inclusion of lactose, a reducing sugar, or mannitol, a reducible organic compound, as excipients in PEO matrices can cause instability of the PEO [100]. Destabilization has been attributed to the relative ease of aerobic auto-oxidation in the presence of these excipients, generating active oxygen species leading to heterolytic depolymerization of high molecular weight PEO, viscosity reduction, and faster drug release.

Poly (ethylene oxide) resins have melting points ranging from 63 to 67°C, becoming thermoplastic. Hence, they are suitable for hot melt extrusion, injection molding, or calendaring processes [98, 101–103] and discussed in a separate chapter in this book.

Several other materials can be useful gel matrix formers. They include methylcellulose [104], guar gum [105], chitosan [106], and cross-linked high amylose starch [107]. They are not widely used but, on occasion may be imminently suited for a specific drug, for delivery of a defined mode of release and absorption.

7.4.2 Inert Matrices

Table 7.4 lists water insoluble and lipidic materials commonly used for fabrication of inert matrices, along with their FDA listed maximum use level in designing oral formulations [34].

7.4.2.1 Ethylcellulose

Ethylcellulose is prepared by reacting alkali cellulose with ethyl chloride: It is characterized by degree of ethoxy substitution and associated molecular weight and is available in different molecular weights with varying viscosities in organic solvents. Multiple particle size grades are also available. It is manufactured by The Dow Chemical Company and commercially available as a dry powder (ETHOCEL™) or as an aqueous dispersion (Surelease®) from Colorcon.

Standard and fine particle size grades have been evaluated for manufacturing ER matrices by DC [108–110]. Utility as a sole matrix former in the dry state at high concentrations may, however, be limited by poor flow and static charge. Aqueous dispersions (Surelease) or in organic solution (ETHOCEL) have been used as wet granulation agents in ER inert matrix formulations [111]. Release profiles of theophylline from Surelease-granulated inert matrices are shown in Fig. 7.2a, b.

Tablets containing lactose as filler showed minimum influence of compression force on release profile. However, tablets containing microcrystalline cellulose had slower release rates at higher compression forces, to a threshold of 15–20 kN. The effects were attributed to the nature of these excipients, with mode and extent of deformation during compression affecting tablet porosity. Tablets containing lactose remained intact during dissolution while those containing microcrystalline

Table 7.4 FDA registered oral formulations containing commonly used water-insoluble polymers/materials [34]

Polymer/material	No of hits on FDA Web page[a]	Maximum potency listed for oral formulations (mg)[b]
Water-insoluble polymers		
Ethylcellulose	19	308.80
Cellulose acetate	10	47.49
Cellulose acetate phthalate (CELLACEFATE)	7	70.00
Methycrylic acid copolymers	50	430.80[c]
Poly(vinyl acetate)	2	46.00
Zein	4	135.00
Shellac	11	87.00
Fatty acids/alcohols/waxes		
Bees wax	8	16.80
Carnauba wax	22	300.00
Paraffin wax	5	150.20
Cetyl alcohol	5	59.00
Cetosterayl alcohol	2	70.00
Stearyl alcohol	4	244.00
Glyceryl behenate	10	50.60
Glyceryl monosterate	12	264.30
Hydrogenated vegetable oil	11	261.00
Hydrogenated cottonseed oil	7	402.00
Hydrogenated castor oil	11	410.82
Hydrogenated soybean oil	5	15.30

[a] Total number of listings on FDA web page for use in oral dosage forms

[b] The "maximum potency" specifies the maximum amount of inactive ingredient for oral route/oral dosage form containing that ingredient. Listed potency is for generic material; refer to FDA web page for specific grade listing. Also the maximum potency number may be higher if its status showed pending status at the time of writing this chapter

[c] Listing for enteric product 430.8

cellulose split. This effect was attributed to the capillary effect of microcrystalline cellulose facilitating migration of dissolution media into the inert matrix resulting in crack formation, splitting, and faster drug release. This example illustrates how type of filler may affect compact properties such as mechanical strength, porosity, and tortuosity and as a consequence drug release.

Ethylcellulose is thermoplastic, with a glass transition temperature of 120°C, making it suitable for melt extrusion as discussed in a separate chapter.

7.4.2.2 Polymethacrylates

Polymethacrylates (Eudragits®, Evonik) are synthetic cationic or anionic polymers of dimethylaminoethyl methacrylates, methacrylic acid, and methacrylic acid esters in varying ratios. Several types are commercially available as dry powders, aqueous

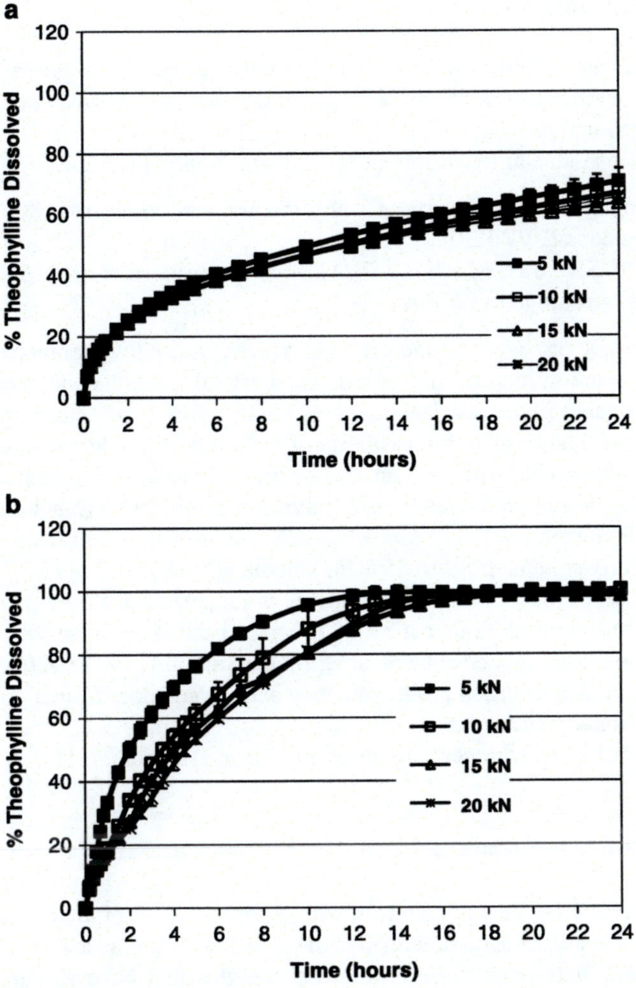

Fig. 7.2 Drug release profile of theophylline from Surelease granulated inert matrices. (**a**) Lactose as filler. (**b**) Microcrystalline cellulose as filler. Formulations consisted of 44% drug, 11% Surelease (on dry basis), 44% filler (lactose or microcrystalline cellulose), and 0.5% each of silicon dioxide and magnesium stearate. Studies utilized USP Apparatus II with sinkers, stirring at 100 rpm and 1,000 ml of DI water at 37±0.5°C

dispersions, or organic solutions. Polymethacrylates can be used as binders for both aqueous and organic solvent granulation, forming matrices with ER characteristics. In general, greater polymer inclusion levels (5–20%) are used to control release from matrices but inclusion levels of 10–50% may be required in direct-compression operations [112, 113]. Interactions between polymethacrylates and some drugs can occur depending on their ionic and physical properties. Drug release may also be affected by pH of the dissolution medium.

7.4.2.3 Polyvinyl Acetate

Polyvinyl acetate (PVAc) is a synthetic polymer, prepared by polymerization of vinyl acetate monomer. PVAc-based matrix formulations provide strong sustained release functionality.

PVAc is available in two different forms from BASF [114], viz.:

- Kollidon® SR, which consists of a physical mixture comprising 80% PVAc and 20% povidone (PVP K30).
- Kollicoat® SR 30 D, which is a PVAc aqueous dispersion stabilized with povidone and sodium lauryl sulfate.

Kollidon SR provides a coherent matrix even under low compression forces. When tablets are introduced into gastric or intestinal fluid, the water-soluble povidone component leaches out, leaving pores through which the active ingredient diffuses. The good flow and compressibility of Kollidon SR renders it suitable for DC tableting processes but matrices can also be prepared using wet granulation or melt extrusion: Suitability in the latter case is attributable to the thermoplastic nature of the PVAc. Inclusion level may depend on active ingredient solubility, varying from 15% for poorly soluble drugs to 55% for soluble actives [115].

Drug release from Kollidon SR matrices is independent of compression force. Aqueous dispersions of Kollicoat SR 30 D can be used as a release retarding binder in wet granulation processes. Depending on the solubility of the active ingredient and the required dissolution profile one may need to add pore former or channeling agent to modulate release rate.

PVAc is used in a commercially available ER matrix tablet [116].

7.4.2.4 Cellulose Acetate and Cellulose Acetate Butyrate

Cellulose acetate consists of cellulose with a portion or all of the hydroxyl groups acetylated. It is available commercially from Eastman Chemicals in a wide range of acetyl levels, chain lengths, and molecular weights that have varying properties [117]. They have been extensively used in the development of osmotic delivery systems, mainly as semipermeable coatings. ER inert matrix tablets can also be formulated with cellulose acetate as a directly compressible matrix former [118]. Release profile can be modified by changing the ratio of drug to cellulose acetate and incorporation of a plasticizer.

Cellulose acetate butyrate (CAB) has also been used as an inert matrix former. Drug release profiles from CAB matrices were reported to be slower than from CA matrices.

7.4.2.5 Fatty Acids, Alcohols, and Waxes

Fatty acids, alcohols, and waxes that do not melt at body temperature are used as insoluble matrix formers. Examples include hydrogenated castor oil, glyceryl behenate, glyceryl monostearate, stearic acid, cetyl alcohol, cetostearyl alcohol, and carnauba wax [55, 119, 120]. As these materials are insoluble in water, drug release from their matrices is mainly diffusional in nature. The presence of channeling (wetting) agents in the matrix can modulate drug release.

These hydrophobic materials are derived from natural products and tend to be complex multicomponent mixtures [119]. Hence, source or changes in isolation and purification may change composition in subtle ways, and consequently influence release profiles. Fatty acids and waxes exhibit complex solid state behavior, including polymorphism. Phase transition behavior may change cooling rate during processing and affect release rate. Waxes may also coat the drug particle during processing such that complete release is prevented. Moreover, low melting waxes (melting point <50°C) can cause sticking and picking during tableting [121, 122]. Digestion during GI transit could also affect release.

Matrices containing these materials can be formulated by DC or more commonly by fusion, drug and additive being mixed with the molten wax matrix at a temperature slightly above the melting point. The molten mass is then spray congealed, solidified, and milled to form granules for compression to tablets [1].

Some other materials may be useful matrix formers. These include zein [123], shellac [124], and cellulose acetate phthalate [125]. Enteric polymers such as HPMC acetate succinate have also been reported [126] but are not widely used.

In summary a dosage form needs to be designed to release drug in a mode or rate that delivers the requisite target drug plasma profile. In hydrophilic matrices, strategies to regulate gel strength, hydration rate, and pore formation in matrices can all be used, separately or in concert to achieve such delivery. Dissolution from gel forming or insoluble matrices generally is influenced by:

- Reducing gel strength to increase erosion rate and/or drug diffusivity in a hydrophilic matrix. This may be achieved by using lower viscosity polymers.
- Reducing hydration rate can mean less complete gel formation and faster release.
- Pore formation by incorporating additives with greater solubilities or by adding a greater percentage of pore forming materials.

7.5 Processing Characteristics

7.5.1 Hydrophilic Matrices

Hydrophilic matrix tablets can be manufactured by DC, wet granulation, dry granulation (roller compaction or slugging), or hot melt granulation/extrusion, depending on the drug, the formulation, and available equipment. HPMC polymers generally have very good compressibility, producing tablets with high mechanical strength [56]. High molecular weight grades may exhibit less plastic flow, requiring higher compaction pressures for deformation [54].

Wet (aqueous) granulation of hydrophilic matrices requires a spray system for application of the liquid binder to avoid forming a lumpy mass [127]. However, a binder may not be necessary with HPMC-containing matrices as this polymer has excellent binder properties. Overgranulation or use of high binder concentrations can adversely affect compressibility. If water is the granulating fluid, its uptake by the granulate can be slow as it causes surface hydration of the HPMC, with swelling and barrier formation resisting penetration. Hence modest quantities of granulating fluid are advised. Hydroalcoholic solutions have better penetration, reducing surface hydration, with faster and better solvent uptake [128]. Higher amounts of such granulation fluid are usually required but the resulting granules can be uniform, nonlumpy, and less friable and can provide superior tablet compacts.

A novel foam granulation approach has recently been introduced [129]. Air is incorporated into a solution of conventional water-soluble polymeric binder such as a low viscosity grade HPMC to generate foam. This improves distribution throughout the granulate and reduces the volume of granulating solvent required.

Hot melt granulation or extrusion can be used to prepare ER systems, although manufacturability is strongly dependent on the formulation (drug, polymer, etc). Numerous successful examples have been reported, particularly using PEO and HPC compositions [102, 130–132].

Techniques such as DC or wet granulation do not generally affect release from compacts, provided these have sufficient mechanical strength and contain optimized levels of polymer(s) [133]. Melt granulation or melt extrusion variables, however, may affect performance. Roth and coworkers compared a hot melt-extruded verapamil HCl formulation based on hypromellose/HPC with a matrix formulation manufactured using conventional technology comprising wet granulation, blending, and compression, to assess abuse deterrence propensity and dose dumping in the presence of ethanol [130]. Figure 7.3 illustrates the differing alcohol resistances of the melt extruded formulation and the conventional wet granulated product in in vitro dissolution tests. It was hypothesized that melt extrusion can lead to greater chain entanglement and a stronger gel layer.

PEO-based injection molding has been used to manufacture abuse-deterrent units that did not exhibit dose dumping in the presence of alcohol. Furthermore, there was no food effect (fasting vs. fed state) and release was consistent as well as prolonged, compared to units prepared by DC or wet granulation [134–137].

The mechanical attributes of hydrophilic matrices may be affected by manufacturing method. Melt granulation or use of hydroalcoholic solutions generally provide superior compacts to those incorporating granules formed by aqueous granulation, or to DC units [128]. Such differences in compact strength may reflect compact porosity. However, mechanical strength may have little influence on drug release when tablets are made with sufficient strength (to withstand handling) and contain optimized levels of polymer. A precompression step may ensure consistent porosity and avoid entrapment of air during compression.

Compression speed can affect tablet tensile strength [50–52]. Inclusion of small amounts of hydrophilic polymer as intergranular excipient (the remainder being added as extragranular component) may result in more robust compacts [45].

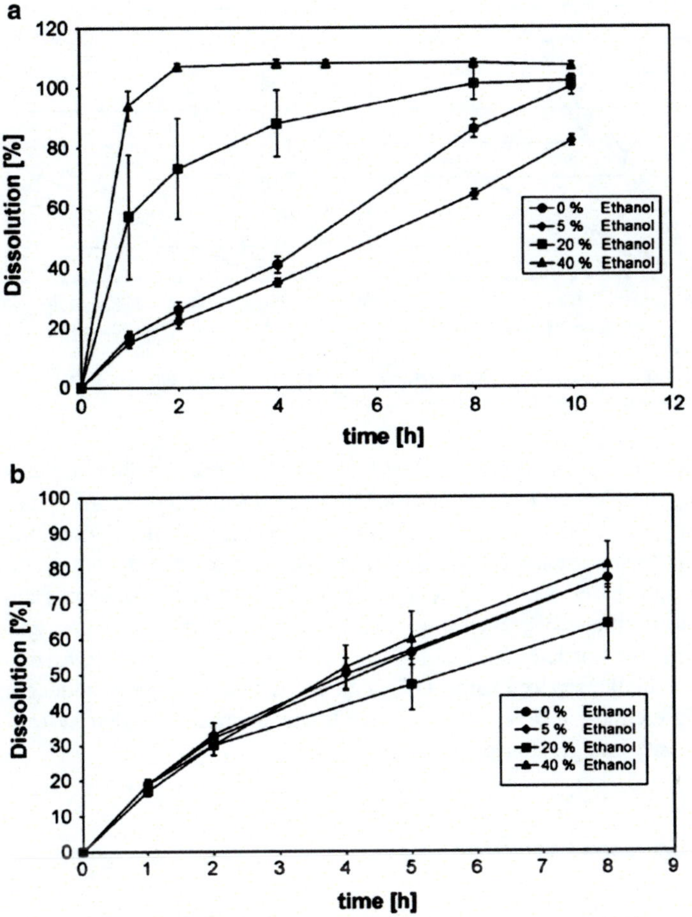

Fig. 7.3 Release profiles of Verapamil HCl hydrophilic matrix products prepared by (**a**) conventional technology of wet granulation, blending, and compression. (**b**) Melt extrusion in dissolution media with varying levels of ethanol (Reprinted with permission from [130])

7.5.2 Inert Matrix-Based Systems

Conventional manufacturing processes such as wet granulation [111] and DC [108–110] are appropriate for most inert-matrix systems. Melt extrusion also possesses advantages [108]. However, manufacturing variables and material attributes can affect release from inert matrices. Release can be influenced by porosity and tortuosity of the matrix. The type and level of pore former (water soluble or insoluble) can dramatically alter such porosity and tortuosity. Inclusion of water-insoluble excipients in inert matrices can reduce matrix wettability, reducing penetration of the dissolution medium and slowing drug release. Water-soluble excipients can enhance wetting or matrix porosity, providing faster drug release. Higher compaction forces during tableting generally leads to lower porosity and

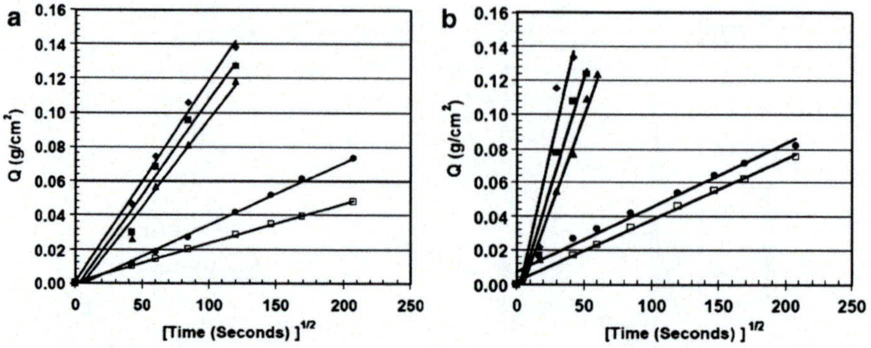

Fig. 7.4 Dissolution rate of ethylcellulose systems based on manufacturing process and ethylcellulose particle size (Reproduced with permission from [108])

slower release. Particle size of insoluble matrix formers can also influence release rate, larger particles producing matrices with more open structures and faster release.

Crowley et al. showed that the physical properties of ethylcellulose-based matrices reflected the matrix polymer properties and manufacturing process [108]. Ethylcellulose-based systems incorporating coarse particle size polymer (30–80 mesh) had significantly greater pore size, porosity, and tortuosity than systems prepared using fine-particle material (80–325 mesh). The difference was significantly greater in tablets prepared using DC. Melt extrusion, in contrast reduced particle-related differences. Dissolution rate was lower in melt extruded systems having lower porosity and greater tortuosity (Fig. 7.4).

7.5.3 Choosing the Matrix

Matrix gel formers and insoluble matrix systems each have unique features that need to be considered when considering the dosage form design strategy. Available manufacturing technology is also important. Cost of goods, target release, and plasma profile and food effects need to be considered as well as drug–excipient compatibility. Gel forming and insoluble matrices are both economic (low cost) operations and generally employ conventional processes (apart from melt extrusion equipment). Formulations based on hydrophilic matrix polymers are generally more robust in terms of sensitivity to minor variations in raw materials or manufacturing processes. Inert matrix systems generally tend to be more sensitive to such variations. However, formulations based on insoluble matrices generally tend to be more robust with respect to dissolution hydrodynamics, but be more sensitive to GI motility effects and "exhausted ghost" behaviors.

In some cases, the manufacturing process can increase polymer chain entanglement and gel strength, leading to improved hydrodynamic performance [130]. Novel techniques such as the multilayer Geomatrix® (SkyePharma) tablet may provide unique advantages for designing zero-order release profiles using conventional hydrogel-type matrices. Ultimately, careful regulation of gel strength and porosity by employing a combination of formulation additives and manufacturing techniques can facilitate the design of systems with the desired release characteristics.

It may be possible to evaluate or estimate performance a priori using mathematical models such as the sequential layer method [29]. Additionally, commercially available systems can be used to predict desired formulations based on molecular properties. The most notable of these systems is the HyperStart™ program [138], which utilizes properties of the compound, such as solubility, as starting points for formulation development.

Small-scale studies on compaction effects on matrix performance are also valuable. Making tablets at controlled pressure requires appropriate tablet instrumentation. Small (bench) scale apparatuses are now available for such a purpose, providing a wide range of compaction forces, minimal use of materials, and easy setup [139]. As these and other systems continue to be refined, their utility (and limitations) will become more evident.

7.6 Conclusions

ER oral delivery can improve drug efficacy, patient compliance, and as a consequence enhance quality of life of people with treatable clinical conditions. In recent years, new polymers, formulation techniques, and manufacturing processes have emerged, adding value to a long-standing platform. As the technology for oral extended release continues to evolve, possibilities will continue to emerge for improving efficacy, increasing patient compliance, reducing the risk of dose dumping, enhancing abuse deterance and supporting the development of personalized medications.

In an industrial context, it is important to be aware at the outset of equipment availability and constraints at the intended site for product manufacture so that the formulation and process are designed accordingly. A change in manufacturing strategy, e.g., moving manufacture to a different site may result in changes to release rate and an unsuitable process.

References

1. Lordi NG (1987) Sustained release dosage forms. In: Lachman L, Lieberman HA, Kanig JL (eds) The theory and practice of industrial pharmacy. Varghese, Bombay, pp 430–456
2. Chiao CSL, Robinson JR (1995) Sustained-release drug delivery systems. In: Gennaro AR (ed) Remington: the science and practice of pharmacy. Mack, Easton, Pennsylvania, pp 1660–1675
3. Kydonieus AF (1980) Fundamental concepts of controlled release. In: Kydonieus AF (ed) Controlled release technologies: methods, theory, and applications. CRC, Boca Raton, FL, pp 1–19

4. U.S. Food and Drug Administration, Center for Drug Evaluation and Research, Drug Approval Reports
5. GlaxoSmithKline. Our Histroy. See 1952. [Online]. http://www.gsk.com/about/histroy-noflash.htm#. Accessed 11 Aug 2006
6. Buri P, Doelker E (1980) Formulation of sustained-release tablets. II. Hydrophilic matrices. Pharm Acta Helv 55(7–8):189–197
7. Conte U et al (1979) Sustained release nitrofurantoin tablets by direct compression. Farmaco [Prat] 34(7):306–316
8. Conte U et al (1988) Swelling-activated drug delivery systems. Biomaterials 9(6):489–493
9. Lee B-J, Ryu S-G, Cui J-H (1999) Formulation and release characteristics of hydroxypropyl methylcellulose matrix tablet containing meatonin. Drug Dev Ind Pharm 25(4):493–501
10. Salomen JL, Doelker E (1980) Formulation of sustained release tablets. I. Inert matrices. Pharm Acta Helv 55(7):174–182
11. Wilding IR et al (1989) Gastrointestinal transit of Sinemet CR in healthy volunteers. Neurology 39(11 Suppl 2):53–58
12. Wilding IR et al (1991) Characterisation of the in vivo behaviour of a controlled-release formulation of levodopa (Sinemet CR). Clin Neuropharmacol 14(4):305–321
13. Zhu Y et al (2006) Controlled release of a poorly water-soluble drug from hot-melt extrudates containing acrylic polymers. Drug Dev Ind Pharm 32(5):569–583
14. Patel N, Madan P, Lin S (2011) Development and evaluation of controlled release ibuprofen matrix tablets by direct compression technique. Pharm Dev Technol 16(1):1–11
15. Badshah A, Subhan F, Rauf K (2010) Controlled release matrix tablets of olanzapine: influence of polymers on the in vitro release and bioavailability. AAPS PharmSciTech 11(3): 1397–1404
16. Chansanroj K, Betz G (2010) Sucrose esters with various hydrophilic-lipophilic properties: novel controlled release agents for oral drug delivery matrix tablets prepared by direct compaction. Acta Biomater 6(8):3101–3109
17. Pavli M, Vrecer F, Baumgartner S (2010) Matrix tablets based on carrageenans with dual controlled release of doxazosin mesylate. Int J Pharm 400(1–2):15–23
18. Tadros MI (2010) Controlled-release effervescent floating matrix tablets of ciprofloxacin hydrochloride: development, optimization and in vitro-in vivo evaluation in healthy human volunteers. Eur J Pharm Biopharm 74(2):332–339
19. Al-Zoubi NM, Alkhatib HS, Obeidat WM (2011) Evaluation of hydrophilic matrix tablets based on Carbopol(R) 971P and low-viscosity sodium alginate for pH-independent controlled drug release. Drug Dev Ind Pharm 37(7):798–808
20. Thombre AG (2005) Assessment of the feasibility of oral controlled release in an exploratory development setting. Drug Discov Today 10(17):1159–1166
21. Frutos P et al (2001) In vitro release of metoclopramide from hydrophobic matrix tablets. Influence of hydrodynamic conditions on kinetic release parameters.. Chem Pharm Bull 49(10):1267–1271
22. Siepmann J et al (1999) HPMC-matrices for controlled drug delivery: a new model combining diffusion, swelling, and dissolution mechanisms and predicting the release kinetics. Pharm Res 16(11):1748–1756
23. Siepmann J et al (2000) Calculation of the required size and shape of hydroxypropyl methylcellulose matrices to achieve desired drug release profiles. Int J Pharm 201(2):151–164
24. Siepmann J, Lecomte F, Bodmeier R (1999) Diffusion-controlled drug delivery systems: calculation of the required composition to achieve desired release profiles. J Control Release 60(2–3):379–389
25. Siepmann J, Peppas NA (2000) Hydrophilic matrices for controlled drug delivery: an improved mathematical model to predict the resulting drug release kinetics (the "sequential layer" model). Pharm Res 17(10):1290–1298
26. Siepmann J, Peppas NA (2001) Modeling of drug release from delivery systems based on hydroxypropyl methylcellulose (HPMC). Adv Drug Deliv Rev 48(2–3):139–157
27. Siepmann J, Peppas NA (2001) Mathematical modeling of controlled drug delivery. Adv Drug Deliv Rev 48(2–3):137–138

28. Siepmann J et al (1999) A new model describing the swelling and drug release kinetics from hydroxypropyl methylcellulose tablets. J Pharm Sci 88(1):65–72
29. Siepmann J, Streubel A, Peppas NA (2002) Understanding and predicting drug delivery from hydrophilic matrix tablets using the "sequential layer" model. Pharm Res 19(3): 306–314
30. Caraballo I (2010) Factors affecting drug release from hydroxypropyl methylcellulose matrix systems in the light of classical and percolation theories. Expert Opin Drug Deliv 7(11):1291–1301
31. Ford JL et al (1991) Mathematical modelling of drug release from hydroxypropylmethylcellulose matrices: effect of temperature. Int J Pharm 71(1–2):95–104
32. Korsmeyer RW et al (1983) Mechanisms of potassium chloride release from compressed, hydrophilic, polymeric matrices: effect of entrapped air. J Pharm Sci 72(10):1189–1191
33. Higuchi T (1963) Pharm Sci 52:1145. Direct Link: Abstract PDF (413 K) References
34. Inactive ingredient search for approved drug products. http://www.accessdata.fda.gov/scripts/cder/iig/index.cfm. Accessed 29 Sept 2010
35. Metolose®. http://www.metolose.jp/e/pharmaceutical/metolose.shtml. Accessed 30 Mar 2011
36. Ashland functional ingredients. http://www.ashland.com/solutions/products-services/functional-ingredients. Accessed 30 Mar 2011
37. Repka MA et al (2005) Characterization of cellulosic hot-melt extruded films containing lidocaine. Eur J Pharm Biopharm 59(1):189–196
38. Ghimire M et al (2010) In-vitro and in-vivo erosion profiles of hydroxypropylmethylcellulose (HPMC) matrix tablets. J Control Release 147(1):70–75
39. Levina M, Gothoskar A, Rajabi-Siahboomi A (2006) Application of a modelling system in the formulation of extended release hydrophilic matrices. Pharm Technol Eur 18(7):20–26
40. Tiwari SB, Rajabi-Siahboomi AR (2008) Extended-release oral drug delivery technologies: monolithic matrix systems. Methods Mol Biol 437:217–243
41. Tajarobi F et al (2009) The impact of dose and solubility of additives on the release from HPMC matrix tablets – identifying critical conditions. Pharm Res 26(6):1496–1503
42. Gao P et al (1996) Swelling of hydroxypropyl methylcellulose matrix tablets. 2. Mechanistic study of the influence of formulation variables on matrix performance and drug release. J Pharm Sci 85(7):732–740
43. Hardy IJ, Cook WG, Melia CD (2006) Compression and compaction properties of plasticised high molecular weight hydroxypropylmethylcellulose (HPMC) as a hydrophilic matrix carrier. Int J Pharm 311(1–2):26–32
44. Hogan JE (1989) Hydroxypropylmethylcellulose sustained release technology. Drug Dev Ind Pharm 15(6 & 7):975–999
45. Huang Y et al (2003) Effects of manufacturing process variables on in vitro dissolution characteristics of extended-release tablets formulated with hydroxypropyl methylcellulose. Drug Dev Ind Pharm 29(1):79–88
46. Levina M, Rajabi-Siahboomi A (2004) The influence of excipients on drug release from hydroxypropyl methylcellulose matrices. J Pharm Sci 93(11):2746–2754
47. Mitchell K et al (1993) The influence of drugs on the properties of gels and swelling characteristics of matrices containing methylcellulose or hydroxypropylmethylcellulose. Int J Pharm 100:165–173
48. Mitchell K et al (1990) The influence of additives on the cloud point, disintegration and dissolution of hydroxypropyl methylcelluose gels and matrix tablets. Int J Pharm 66:233–242
49. Mitchell SA, Balwinski KM (2008) A framework to investigate drug release variability arising from hypromellose viscosity specifications in controlled release matrix tablets. J Pharm Sci 97(6):2277–2285
50. Nokhodchi A et al (1996) The effects of compression rate and force on the compaction properties of different viscosity grades of hydroxypropylmethylcellulose 2208. Int J Pharm 129:21–31
51. Nokhodchi A et al (1996) The effect of moisture on the heckel and energy analysis of hydroxypropylmethylcellulose 2208 (HPMC K4M). J Pharm Pharmacol 48(11):1122–1127

52. Nokhodchi A et al (1996) The influence of moisture content on the consolidation properties of hydroxypropylmethylcellulose K4M (HPMC 2208). J Pharm Pharmacol 48(11): 1116–1121
53. Nokhodchi A et al (2002) The effect of various surfactants on the release rate of propranolol hydrochloride from hydroxypropylmethylcellulose (HPMC)-Eudragit matrices. Eur J Pharm Biopharm 54(3):349–356
54. Nokhodchi A, Rubinstein MH (2001) An overview of the effects of material and process variables on the compaction and compression properties of hydroxypropylmethyl cellulose and ethyl cellulose. STP Pharm Sci 11:195–202
55. Tiwari SB et al (2003) Controlled release formulation of tramadol hydrochloride using hydrophilic and hydrophobic matrix system. AAPS PharmSciTech 4(3):18–23
56. Rajabi-Siahboomi AR, Nokhodchi A, Rubinstein MH (1998) Compaction behaviour of hydrophilic cellulose ether polymers. Pharmaceutical technology tableting and granulation yearbook. Advanstar Communications, Santa Monica, CA, pp 32–40
57. Hardy IJ et al (2007) Modulation of drug release kinetics from hydroxypropyl methyl cellulose matrix tablets using polyvinyl pyrrolidone. Int J Pharm 337(1–2):246–253
58. Kabir MA, Reo JP (2009) Hydroxypropyl cellulose. In: Rowe RC, Sheskey PJ, Quinn ME (eds) Handbook of pharmaceutical excipients. Pharmaceutical Press and American Pharmacists Association, London, Chicago, pp 317–322
59. Nisso HPC (hydroxypropyl cellulose). http://www.nissoamerica.com/hpc/. Accessed 30 March 2011
60. Application note: controlled drug release using HPC in a matrix tablet. http://www.nissoamerica.com/hpc/CR02.pdf. Accessed 30 Mar 2011
61. Roshdy MN, Schnaare RL, Schwartz JB (2001) The effect of formulation composition and dissolution parameters on the gel strength of controlled release hydrogel tablets. Pharm Dev Technol 6(4):583–593
62. Lee C-Y et al (2006) Physical characterization and drug release profiling for hard capsules prepared with hydroxypropylcellulose or polyethylene oxide. Chin Pharm J 58:75–84
63. Vueba ML et al (2004) Influence of cellulose ether polymers on ketoprofen release from hydrophilic matrix tablets. Eur J Pharm Biopharm 58(1):51–59
64. Vueba ML et al (2006) Influence of cellulose ether mixtures on ibuprofen release: MC25, HPC and HPMC K100M. Pharm Dev Technol 11(2):213–228
65. Roshdy MN et al (2002) The effect of controlled release tablet performance and hydrogel strength on in vitro/in vivo correlation. Pharm Dev Technol 7(2):155–168
66. Sinha Roy D, Rohera BD (2002) Comparative evaluation of rate of hydration and matrix erosion of HEC and HPC and study of drug release from their matrices. Eur J Pharm Sci 16(3): 193–199
67. Hapgood KP (2009) Hydroxyethyl cellulose. In: Rowe RC, Sheshey PJ, Quinn ME (eds) Handbook of pharmaceutical excipients. Pharmaceutical Press and American Pharmacists Association, London, Chicago, pp 311–314
68. Conti S et al (2007) Matrices containing NaCMC and HPMC 2. Swelling and release mechanism study. Int J Pharm 333(1–2):143–151
69. Baveja SK, Ranga Rao KV, Padmalata Devi K (1987) Zero order release hyrophilic matrix tablets of b-adrenergic blockers. Int J Pharm 39:39–45
70. Conti S et al (2007) Matrices containing NaCMC and HPMC 1. Dissolution performance characterization. Int J Pharm 333(1–2):136–142
71. Dabbagh MA et al (1999) Release of propranolol hydrochloride from matrix tablets containing sodium carboxymethylcellulose and hydroxypropylmethylcellulose. Pharm Dev Technol 4(3):313–324
72. Obaidat AA, Rashdan LA, Najib NM (2002) Release of dextromethorphan hydrobromide from matrix tablets containing sodium carboxymethylcellulose and hydroxypropylmethylcellulose. Acta Pharm Turcica 44(2):97–104
73. Glucophage (metformin hydroxhloride) Tablets. http://packageinserts.bms.com/pi/pi_glucophage.pdf. Accessed 30 Sept 2010

74. Giunchedi P et al (2000) Evaluation of alginate compressed matrices as prolonged drug delivery systems. AAPS PharmSciTech 1(3):E19
75. FMC Biopolymer, sustained release, matrix tablets. http://www.fmcbiopolymer.com/Pharmaceutical/Applications/SustainedRelease/MatrixTablets.aspx. Accessed 30 Mar 2011
76. Hodsdon AC et al (1995) *Structure and behaviour in hydrophilic matrix sustained release dosage forms: 3.* The influence of pH on the sustained-release performance and internal gel structure of sodium alginate matrices. J Control Release 33(1):143–152
77. Park HY et al (1998) Effect of pH on drug release from polysaccharide tablets. Drug Deliv 5(1):13–18
78. Ching AL et al (2008) Modifying matrix micro-environmental pH to achieve sustained drug release from highly laminating alginate matrices. Eur J Pharm Sci 33(4–5):361–370
79. Qui Y, Zhang G (2005) Research and development aspects of oral controlled release dosage forms. In: Wise DL (ed) Handbook of pharmaceutical controlled release technology. Marcel Dekker, New York, pp 465–503
80. Howard JR, Timinis P (1988) Controlled release formulation. US Patent 4,792,452
81. Timmins P, Delargy AM, Howard JR (1997) Optimization and characterization of a pH independant extended release hydrophilic matrix tablet. Pharm Dev Technol 2(1):25–31
82. CALAN® SR (verapamil hydrochloride) sustained-release oral caplets. http://media.pfizer.com/files/products/uspi_calan_sr.pdf. Accessed 30 Mar 2011
83. Melia CD (1991) Hydrophilic matrix sustained release systems based on polysaccharide carriers. Crit Rev Ther Drug Carrier Syst 8(4):395–421
84. Fitzpatrick JP (1995) Xanthan gum in hydrophilic matrix drug delivery systems. In: Karsa DR, Stephenson RA (eds) Excipients and delivery systems for pharmaceutical formulations (Spl Publ S. No. 161). Royal Society of Chemistry, London
85. CP Kelco, Pharmaceutical. http://www.cpkelco.com/market_personalcare/app-pharm.html. Accessed 30 Mar 2011
86. Gohel MC et al (2009) Fabrication of modified release tablet formulation of metoprolol succinate using hydroxypropyl methylcellulose and xanthan gum. AAPS PharmSciTech 10(1):62–68
87. Talwar N, Sen H, Staniforth JH (2003) Orally administered controlled drug delivery system providing temporal and spatial control. EP Patent 1,107,741
88. Carbopol® polymers, product overview and recommendations. http://www.lubrizol.com/Pharmaceutical/Products/CarbopolPolymers.html. Accessed 30 Mar 2011
89. Draganoiu E, Rajabi-Siahboomi A, Tiwari S (2009) Carbomer. In: Rowe RC, Sheskey PJ, Quinn ME (eds) Handbook of pharmaceutical excipients. Pharmaceutical Press and American Pharmaceutical Association London, Chicago, pp 110–114
90. Oral solid dosage introduction. http://www.lubrizol.com. Accessed 30 Mar 2011
91. Samani SM, Montaseri H, Kazemi A (2003) The effect of polymer blends on release profiles of diclofenac sodium from matrices. Eur J Pharm Biopharm 55:351–355
92. Tiwari SB, Rajabi-Siahboomi AR (2009) Applications of complementary polymers in HPMC hydrophilic extended release matrices. Drug Deliv Tech 9(7):20–27
93. Khamanga SM, Walker RB (2005) Comparison of method of manufacture on in vitro performance of verapamil matrix tablets. Controlled Release Society 32nd Annual Meeting and Exposition, Abstract #718, Miami, FL, 18–22 July 2005
94. Tiwari SB, Rajabi-Siahboomi AR (2008) Modulation of drug release from hydrophilic matrices. PharmTech.com 1 Sept 2008
95. Tiwari S, Martin L, Rajabi-Siahbhoomi A (2008) The influence of hydrodynamic conditions on Verapamil hydrochloride release from a hydrophilic matrices using ionic and non-ionic polymers. In: AAPS Annual Meeting, Atlanta, 15–20 Dec 2008
96. POLYOX™ Water-soluble polymers for pharmaceutical applications. https://www.colorcon.com/products/core-excipients/extended-controlled-release/polyox. Accessed 30 Mar 2011
97. Braun DB (1980) Poly(ethylene oxide). In: Davidson RL (ed) Handbook of water-soluble gums and resins. McGraw-Hill, New York
98. Maggi L, Bruni R, Conte U (2000) High molecular weight polyethylene oxides (PEOs) as an alternative to HPMC in controlled release dosage forms. Int J Pharm 195(1–2):229–238

99. Berner B et al (2007) Gastric retentive gabapentin dosage forms and methods for using same. WO Patent WO/2007/079,195
100. L'Hote-Gaston J, Wallick D (2011) Effect of filler type on the stability of polyethylene oxide in a hydrophilic matrix tablet. www.colorcon.com/literature/…/polyethylene_oxide_matrix_tab.pdf. Accessed 30 Mar 2011
101. Crowley MM et al (2002) Stability of polyethylene oxide in matrix tablets prepared by hot-melt extrusion. Biomaterials 23(21):4241–4248
102. Li L, AbuBaker O, Shao ZJ (2006) Characterization of poly (ethylene oxide) as a drug carrier in hot-melt extrusion. Drug Dev Ind Pharm 32(8):991–1002
103. Quinten T et al (2009) Evaluation of injection moulding as a pharmaceutical technology to produce matrix tablets. Eur J Pharm Biopharm 71(1):145–154
104. Sanghavi N, Kamath P, Amin D (1990) Sustained release tablets of theophylline. Drug Dev Ind Pharm 16(11):1843–1848
105. Khullar P, Khar RK, Agarwal SP (1998) Evaluation of guar gum in the preparation of sustained-release matrix tablets. Drug Dev Ind Pharm 24(11):1095–1099
106. Ilium L (1998) Chitosan and its use as a pharmaceutical excipient. Pharm Res 15(9): 1326–1331
107. Lenaerts V et al (1998) Cross-linked high amylose starch for controlled release of drugs: recent advances. J Control Release 53(1–3):225–234
108. Crowley MM et al (2004) Physicochemical properties and mechanism of drug release from ethyl cellulose matrix tablets prepared by direct compression and hot-melt extrusion. Int J Pharm 269(2):509–522
109. Katikaneni PR et al (1995) Ethylcellulose matrix controlled release tablets of a water-soluble drug. Int J Pharm 123(1):119–125
110. Upadrashta SM et al (1993) Direct compression controlled release tablets using ethylcellulose matrices. Drug Dev Ind Pharm 19(4):449–460
111. Application of ethylcellulose in preparation of extended release theophylline inert matrix tablets by wet granulation. https://www.colorcon.com/literature/marketing/mr/Extended%20Release/Surelease/English/ads_surelease_appl_theo_er_v1_07.21.2009kaf%20DFB-HD%2008.20.pdf. Accessed 28 Mar 2011
112. Chang RK et al (2009) Polymethacrylates. In: Rowe RC, Sheskey PJ, Quinn ME (eds) Handbook of pharmaceutical excipients. Pharmaceutical Press and American Pharmacists Association, London, Chicago, pp 525–533
113. Gallardo D, Skalsky B, Kleinebudde P (2008) Controlled release solid dosage forms using combinations of (meth) acrylate copolymers. Pharm Dev Technol 13(5):413–423
114. Controlled release. http://www.pharma-ingredients.basf.com/Kollidon/TheKollidonProductFamily/ControlledRelease.aspx. Accessed 31 Mar 2011
115. Kollidon® SR. http://www.pharma-ingredients.basf.com/Products.aspx?PRD=30071321. Accessed 31 Mar 2011
116. RYZOLT™ (tramadol hydrochloride) extended-release tablets. http://www.rxlist.com/ryzolt-drug.htm. Accessed 28 Mar 2011
117. Daugherity PD, Nause RG (2009) Cellulose acetate. In: Rowe RC, Sheskey PJ, Quinn ME (eds) Handbook of pharmaceutical excipients. Pharmaceutical Press and Amerian Pharmacists Association, Chicago, London, pp 141–143
118. Yum J, Stephen W (2000) A feasibility study using cellulose acetate and cellulose acetate butyrate. Pharm Technol 24(10):92–106
119. Ozyazici M, Gokce EH, Ertan G (2006) Release and diffusional modeling of metronidazole lipid matrices. Eur J Pharm Biopharm 63(3):331–339
120. Gokce EH, Ozyazici M, Ertan G (2009) The effect of geometric shape on the release properties of metronidazole from lipid matrix tablets. J Biomed Nanotechnol 5(4):421–427
121. Boles M, Deasy P, Donnellan M (1993) Design and evaluation of a sustained-release aminophylline tablet. Drug Dev Ind Pharm 19(3):349–370
122. Bodmeier R et al (1990) Formation of sustained release wax matrices within hard gelatin capsules in a fluidized bed. Drug Dev Ind Pharm 16(9):1505–1519

123. Katayama H, Kanke M (1992) Drug release from directly compressed tablets containing zein. Drug Dev Ind Pharm 18(20):2173–2184
124. Pearnchob N, Siepmann J, Bodmeier R (2003) Pharmaceutical applications of shellac: moisture-protective and taste-masking coatings and extended-release matrix tablets. Drug Dev Ind Pharm 29(8):925–938
125. Lin SY, Kawashima Y (1987) Drug release from tablets containing cellulose acetate phthalate as an additive or enteric-coating material. Pharm Res 4(1):70–74
126. Hilton AK, Deasy PB (1993) Use of hydroxypropyl methylcellulose acetate succinate in an enteric polymer matrix to design controlled release tablets of amoxicillin trihydrate. J Pharm Sci 82(7):737–743
127. Liu C-H et al (1993) Properties of hydroxypropylmethylcellulose granules produced by water spraying. Int J Pharm 100:241–248
128. Shah N et al (1996) Effects of processing techniques in controlling the release rate and mechanical strength of hydroxypropyl methylcellulose based hydrogel matrices. Eur J Pharm Biopharm 42:183–187
129. Foamed binder technology. www.dow.com/scripts/litorder.asp?filepath=/198-02124.pdf. Accessed 30 Mar 2011
130. Roth W et al (2009) Ethanol effects on drug release from Verapamil Meltrex®, an innovative melt extruded formulation. Int J Pharm 368(1–2):72–75
131. Repka MA et al (2007) Pharmaceutical applications of hot-melt extrusion: Part II. Drug Dev Ind Pharm 33(10):1043–1057
132. Crowley MM et al (2007) Pharmaceutical applications of hot-melt extrusion: part I. Drug Dev Ind Pharm 33(9):909–926
133. Velasco MV et al (1999) Influence of drug:hydroxypropylmethylcellulose ratio, drug and polymer particle size and compression force on the release of diclofenac sodium from HPMC tablets. J Control Release 57(1):75–85
134. Haahr A et al (2007) Drug abuse resistant, controlled release, using Egalet® dosage units
135. Tygesen PH et al (2010) Pharmaceutical compositions resistant to abuse. Google Patents
136. Corporate presentation: Egalet. http://www.egalet.com/multimedia/Egalet_Corp_Presentation_200913.pdf. Accessed 30 Mar 2011
137. Washington N, Wilson CG (2006) The future for controlled & sustained drug delivery. Drug Deliv Technol 6(9):71–74
138. HyperStart® formulation service. https://www.colorcon.com/formulation/formulation-tools. Accessed 28 Mar 2011
139. Gamlen tablet press. http://www.compressibility.com/. Accessed 12 Mar 2011

Chapter 8
Ion-Exchange Approaches to Controlling Drug Release

Der-Yang Lee, Timothy Kutch, and Rick S. Chan

Abstract Most pharmaceutically active compounds are prepared as salts and as such possess at least one ionizable functional group. Hence they may be incorporated into resinates using appropriate ion-exchange resins. In the context of controlling drug release, resinates are unique in that formulation in a liquid presentation may be feasible. Resinate formation may also mask a bitter taste in liquid formulations or those that are formulated for buccal delivery, the drug being bound ionically to the resin and displaced only at gastric pH. This chapter reviews the principles of ion-exchange resin formation, and of drug release from such resinates. Product examples are also presented and discussed.

8.1 Introduction

An ion-exchange resin (IER) comprises an insoluble, commonly synthetic matrix possessing ionizable groups capable of exchanging ions with those in bulk solution with which it is in contact. Thus, under appropriate conditions it can deliver to or sequester chemical species from an aqueous environment. The process is reversible, exchange capability being regenerated by washing the resin with an excess of the originally bound ions. The technology is utilized in many industrial applications such as chemical and biosynthesis, food processing and agriculture. In the pharmaceutical industry it is used to separate and purify proteins, nucleotides and amino acids. Use in dosage form design may improve bioavailability of poorly soluble drugs, mask bitter taste and control drug release, either to enhance effectiveness or possibly inhibit narcotic abuse. Resins are also used as therapeutic agents for lowering cholesterol, potassium reduction and in chronic renal failure.

D.-Y. Lee (✉)
Johnson & Johnson, McNeil Consumer Healthcare, Fort Washington, PA, USA
e-mail: DLee1@its.jnj.com

C.G. Wilson and P.J. Crowley (eds.), *Controlled Release in Oral Drug Delivery*, Advances in Delivery Science and Technology, DOI 10.1007/978-1-4614-1004-1_8,

Synthetic resins were introduced in 1935 [1] and were first suggested in 1956 for sustaining drug release from dosage forms [2]. The FDA approved the first IER product in 1958 viz Kayexalate®, for treating hyperkalemia. Resins have since been widely investigated for modifying oral drug delivery in particular but also for nasal [3], transdermal [4] and iontophoretically assisted transdermal [5] drug delivery, as evidenced by the plethora of patent applications covering the topic [6–8]. The high incidence of tobacco addiction in US stimulated the development of an IER nicotine chewing gum (Nicorette®) in 1984 which was a landmark IER therapy. Applications in oral controlled-release medications are discussed in this chapter.

8.2 Chemical Nature of Resins

IERs comprise a polymer matrix structure containing ionizable functional groups as shown in Fig. 8.1. The matrix is insoluble across the entire pH range, providing an inert platform for ion exchange [10]. The functional components are ionizable groups bonded to the matrix backbone. These have the capability to exchange their native ions with identically charged counter-ions in solution.

IERs exhibit Donnan exclusion, i.e., they are insoluble but the exchangeable ions can interchange with counter-ions in the surrounding liquid. They typically comprise copolymers, cross-linked for stabilization. Greater cross-linking provides a dense internal structure (gel resins); conversely, less cross-linking produces multichanneled macroporous resins.

Resins are classified by the charge on the exchangeable ion (cationic or anionic) and the binding affinity of the functional group (strong or weak) viz:

1. Strong cation IERs possess sulphonic acid functional groups
2. Weak cation IERs possess carboxylic acid functional groups

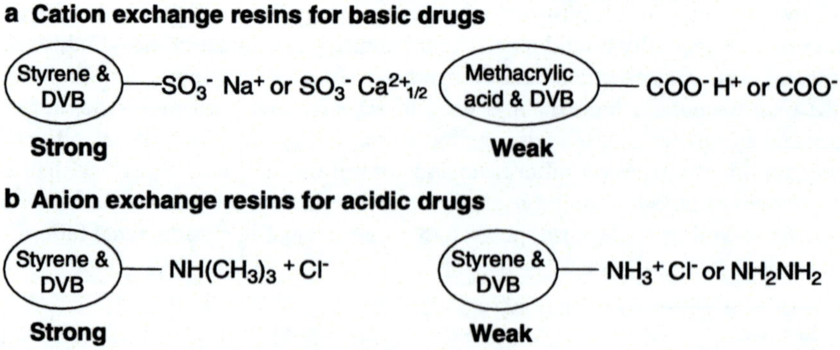

Fig. 8.1 Common structures of IER backbone and functional group [9]

3. Strong anion IERs possess trimethyl ammonium chloride
4. Weak anion IERs possess ammonium chloride or primary amine functional groups

The exchange capacity of resins is defined as the number of chemical equivalents available for exchange per unit. This can be expressed in milliequivalents per dry gram or milliequivalents per wet milliliter of resin. Structure, polymer backbones and exchange capacities of selected products are outlined in Table 8.1 for resins available as products in their own right or for forming resinates with medicinal compounds.

Polysaccharide matrices are also used in IERs but use at this time is limited to nontherapeutic areas such as separation chromatography.

Examples of pharmaceutical products, formulated as ion-exchange resinates are presented in Table 8.2.

8.3 Preparation of Resinates

The process for drug–resin complex manufacture is relatively simple. An ionizable active pharmaceutical ingredient (API) is loaded on to the resin by exchanging with the resin functional group's native ion, forming a drug–resin complex or "resinate" (also referred to as Polistirex in some literature). Drug structure (and biological activity) is not affected but bound drug is essentially inert and not "available" to the environment. Binding energy is determined by the ionic attractions between the resin functional group and the exchanging ions.

Drug may also be adsorbed on the resin surface in accordance with surface energy requirements [11]. Equation (8.1) describes the reaction of sodium sulphonate resin, R, being loading with a basic drug, X, of hydrochloride salt. Figure 8.2 illustrates the loading process and resulting resinate, IER-X.

$$R-SO_3^{-}Na^{+}+X\bullet H^{+}Cl^{-}\Rightarrow R-SO_3^{-}H^{+}\bullet X+Na^{+}Cl^{-} \tag{8.1}$$

The chemistry governing drug loading adheres to the Law of Mass Action requiring that drug be in concentrated solution to exchange onto the resin. The equilibrium reaction is expressed in (8.2) where, R is a resin and, D, is a drug, while X and Y are counter-ions.

$$aRX+b\mathrm{DY}\overset{eq}{\Leftrightarrow}c\mathrm{RD}+d\mathrm{XY} \quad \text{and} \quad K_{eq}=\frac{[\mathrm{RD}]^{c}[\mathrm{XY}]^{d}}{[\mathrm{RX}]^{a}[\mathrm{DY}]^{b}}. \tag{8.2}$$

Batch or column techniques can be employed for loading [12]. The batch technique in simple terms involves stirring a solution of the drug and suspended resin until equilibrium is reached. Column loading involves passing drug in solution through a column packed with resin until eluent and eluate concentrations are equal. The batch process is preferred for fine particles, whereas column operations are better suited for larger resin particles. Drug loaded resin is washed with deionized water to ensure that the native functional group ions and the drug's salt counter-ions are removed. Resinate is then dried to remove residual moisture.

Table 8.1 Commercially available ion exchange resins

Commercial name (compendial name)	Structure	Polymers	Ionizable group	Exchange capacity	Pharmaceutical uses
Amberlite™ IRP69 (sodium polystyrene sulphonate USP)	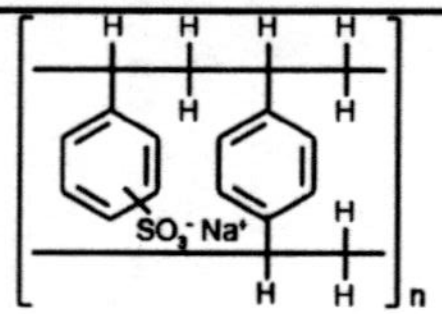	Styrene/Divinylbenzene	Strong acid $-SO_3^- Na^+$	5 meq/g	Reduce serum potassium
Purolite® C100CaMR (calcium polystyrene sulphonate BP/JP)	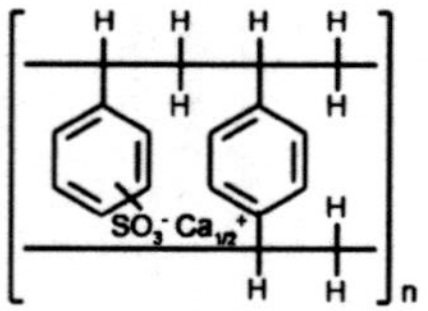	Styrene/Divinylbenzene	Strong acid $-SO_3^- Ca_{1/2}^{2+}$	1.3–2 meq/g	Reduce serum potassium
Amberlite™ IRP64 (polacrilex resin)	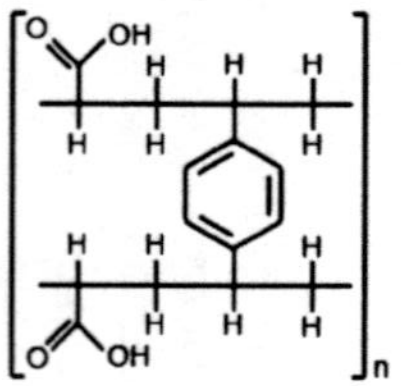	Methacrylic acid/ Divinylbenzene	Weak acid $-COO^- H^+$	10 meq/g	Vitamin B12 and nicotine stabilization

Amberlite™ IRP88, Purolite® C115KMR (polacrilin potassium NF)	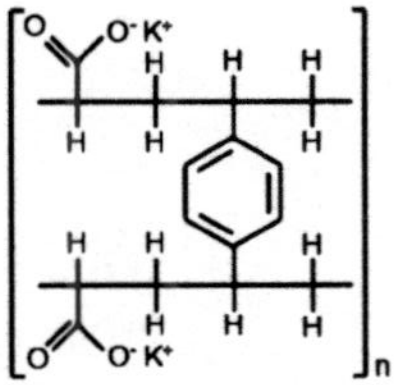	Methacrylic acid/ divinylbenzene	Weak acid –$COO^- K^+$	10 meq/kg	Taste masking, β-lactam antibiotics
Duolite™ AP143 (cholestyramine resin USP)	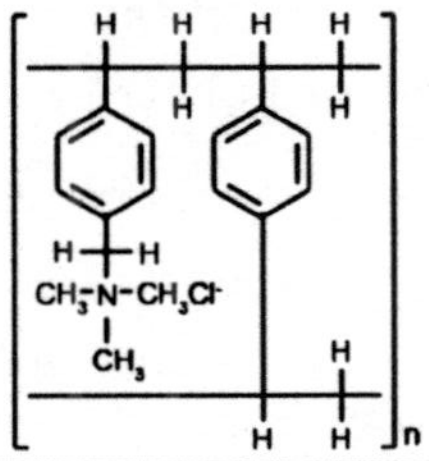	Styrene/Divinylbenzene	Strong base –$N^+(R)_3\ Cl^-$	1.8–2.2 g/g	Reduce serum cholesterol

Table 8.2 Ion exchange resin-containing pharmaceutical products

Trade name	Active Ingredient	Polymers	Indication	Dose	FDA approval
Renagel®	Selevamer	Allylamine and epichlorohydrin	Hyperphosphataemia	400 mg or 800 mg	1998
Paxil® oral Suspension	Paroxetine	Methacrylic acid and divinylbenzene	Antidepressant	10 mg	1992
Betoptic S® ionic Suspension	Betaxolol	Styrene and divinylbenzene	Ocular hypertension	2.8 mg/ml	1989
Tussionex® 12-hour	Hydrocodone and chlorpheniramine	Styrene and divinylbenzene	Antitussive and decongestant	10 mg and 8 mg	1987
Nicorette® Gum	Nicotine	Methacrylic acid and divinylbenzene	Smoking cessation	2 mg or 4 mg	1984
Delsym® 12-hour	Dextromethorphan	Styrene and divinylbenzene	Antitussive	30 mg	1982
Colestid®	Colestipol	Diethylenetriamine and epichlorohydrin	Hypercholesterolemia	5 g	1977

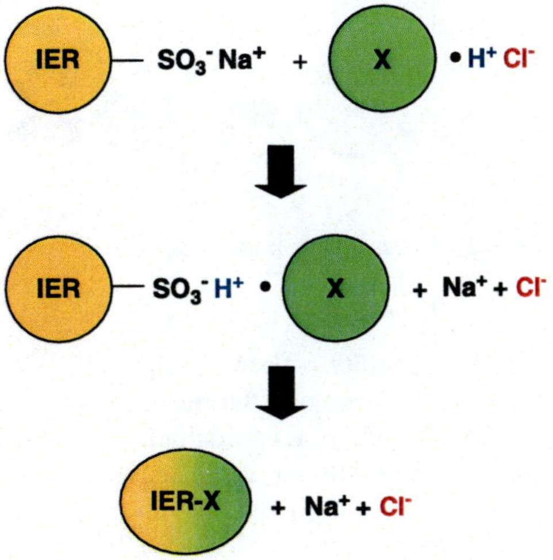

Fig. 8.2 Loading of basic drug onto cation exchange resin

Time, temperature and pH can affect drug loading. Maximum loading occurs when equilibrium is reached so longer loading times will generally promote greater drug loading. Increasing temperature usually improves drug solubility, leading to increased ionization of drug and possibly help loading of poorly ionized drug. Higher temperatures also reduce the activation energy and ease drug attachment. For weak cation exchange resins, neutral pH retards ionization and results in slower release rates [13]. The Henderson–Hasselbach relationship (8.3) predicts that, for a weakly acidic drug more than 50% of the drug species are ionized when solution pH is adjusted to greater than the drug's pK_a.

$$\text{For an acidic drug, pH} = \text{p}K_a + \log\frac{[\text{A}^-]}{[\text{HA}]}. \tag{8.3}$$

Conversely, (8.4) predicts that, for a weakly basic drug more than 50% is ionized when solution pH is less than the drug's pK_a. If the solution pH is below (acidic drug) or above (basic drug) the drug's pK_a the excess hydrogen and hydroxide ions will compete for binding sites on the resin and reduce loading efficiency.

$$\text{For a basic drug, pH} = \text{p}K_a + \log\frac{[\text{B}]}{[\text{BH}^+]} \tag{8.4}$$

8.4 Drug Release Mechanisms

Drug release from resinates can be influenced by numerous variables. Drug on the resin surface passes directly to the external solution but drug exchanged with the counter-ion must diffuse through the resin matrix to enter the bulk medium. Factors such as matrix

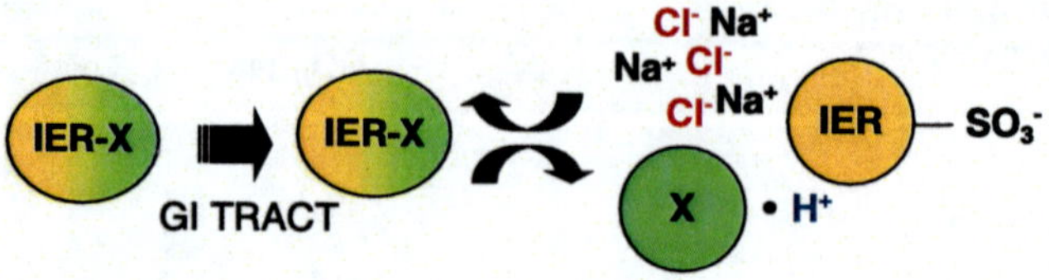

Fig. 8.3 Release of drug from resinate

porosity, tortuosity, molecular weight and diffusion path length can determine diffusion rate and as a consequent the rate of drug release.

Figure 8.3 illustrates cationic electrolytes exchanging a basic drug when exposed to a counter-ion environment (e.g., gastrointestinal tract).

A drug, ionically bound within a resin core can be considered as being released in three stages viz:

- The counter-ion diffuses into the matrix.
- The counter-ion exchanges with bound drug at a rate determined by chemical equilibrium.
- Free drug diffuses through the matrix for release at the resin/medium interface.

Therefore, drug release is influenced by the resin's intrinsic properties and the external environment.

8.4.1 Intrinsic Properties of the Resin

8.4.1.1 Physical Structure

The degree of cross-linking induced at resin manufacture can affect pore diameter. Inorganic electrolytes permeating the matrix are sufficiently small such that porosity should not affect their diffusion rates. In contrast, drug molecules are usually many times larger which can influence diffusion through the matrix. Cross-linking also affects matrix swellability which in turn can affect diffusion. Swellability can also be affected by the polymerization process at resin manufacture [14]. Drug release rate can be affected by particle size of the resinate. Release can be faster from smaller particles due to their greater surface area. Larger particles may be utilized to provide slower release.

8.4.1.2 Chemical Structure

Drug release is usually rapid from resins that possess weak cationic or anionic (carboxylic acid or ammonium/primary amine) functional groups. Gradual release is provided by resins with strong cationic or anionic capacities (sulphonic acid or

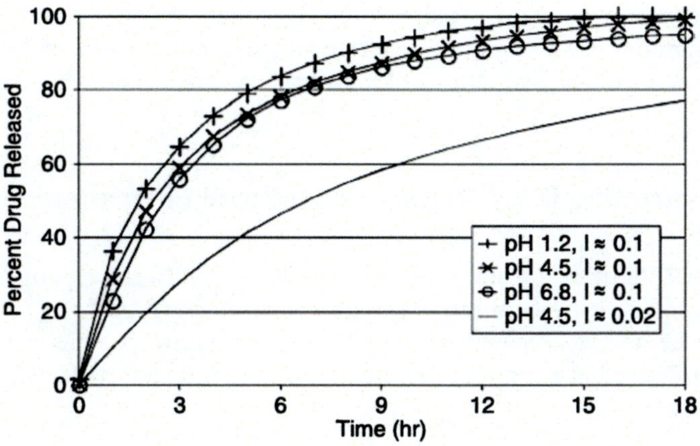

Fig. 8.4 Effect of pH and ionic strength on Drug Release from resinate (unpublished information)

tertiary ammonium functional groups). The pK_a value of the resin functional group can also significantly affect release rate in acidic media. Such differences can enable selection of an appropriate resin matrix to influence drug release.

8.4.2 External Environment

When a resinate is introduced to the GI tract endogenous electrolytes act as counter-ions to drive drug release. Equation (8.5) describes the chemical reaction of a basic drug, X, exchanging with sodium to liberate the drug from the resin, R.

$$R-SO_3^-H^+-X+Na^+Cl^- \Rightarrow R-SO_3^-Na^+ + X-H^+Cl^-. \qquad (8.5)$$

The ionic strength of the surrounding medium affects the rate of ion exchange. Higher ionic strength provides more counter-ions for exchange and can increase drug release. Ionic strength, I, can be determined as in (8.6) where, c is the molar concentration and z is the charge number.

$$I = \frac{1}{2}\sum_{i=1}^{n} c_i z_i^2. \qquad (8.6)$$

In the case of strong cation exchange resins pH does not significantly affect the rate of drug release [13]. However, greater ionic strength of the release medium results in faster and more complete drug unloading as shown in Fig. 8.4. Such behavior can be utilized for taste masking a bitter drug by resinate formation. Release in the oral cavity is constrained due to low ionic strength of saliva but drug is readily released for absorption in the more ion-rich gastric region.

Release from a resinate during gastrointestinal transit can only occur if counter-ions for exchange are present. Thus, ion exchange can be influenced by GI tract location, as well as the nature of the ion-exchange group (strong or weak exchanger). Such possibilities need to be considered when designing a controlled delivery system. Factors affecting gastrointestinal transit are considered in a separate chapter.

8.5 Controlling Drug Release by Resinate Formation

Preparation of drug as a resinate can increase its solubility and rate of dissolution which in turn may enhance absorption as a unit transits the GI tract. This may extend or enlarge the "absorption window" by providing more drugs in solution for absorption. Such absorption extension/enhancement could prolong therapeutic concentrations of drug in plasma.

8.5.1 Solubility Enhancement

IER complexation can help overcome solubility issues because each individual drug molecule is bound to a functional group. Dissolution does not require overcoming crystal lattice energy. The hydrophilic nature of some IER also reduces problems of "wetting-out" encountered with hydrophobic drugs. This effect has been demonstrated with indomethacin, whose solubility is about 6 ppm. Stirring excess indomethacin in simulated gastric fluid for 3 days achieved a concentration of only 1 ppm. However, when indomethacin resinate was exposed to the same conditions a saturated solution was obtained in 30 min [10].

Tawakkul et al. [15] observed faster dissolution of risperidone from a drug–Amberlite IRP 64 complex than from a physical mixture of the drug and ion-exchange resin. They attributed this to possible wetting, improved solubility of the drug in the resinate complex and other factors such as modification of the microenvironment pH. This has signification applications in the formulation of poorly soluble drugs to improve bioavailability (Fig. 8.5).

If rapid release of drug in the GI tract is desired resinates with weak ion-exchange properties are probably most suitable. Such enhancement may be achieved with drug bound to resin with carboxylic acid functional groups. Low cross-linking, small particle size and high drug loading on the resin are also desirable. Anionic and strong cationic exchange resins in contrast are likely to limit ion exchange in gastric conditions and delay release.

8.5.2 Bioavailability Enhancement

Thairs et al. [16] examined the effect of dose, food and coating of an ion-exchange resin on its gastric residence time and distribution after ingestion. Using scintigraphy, they showed that approximately 20% of the resin persisted in the stomach for the

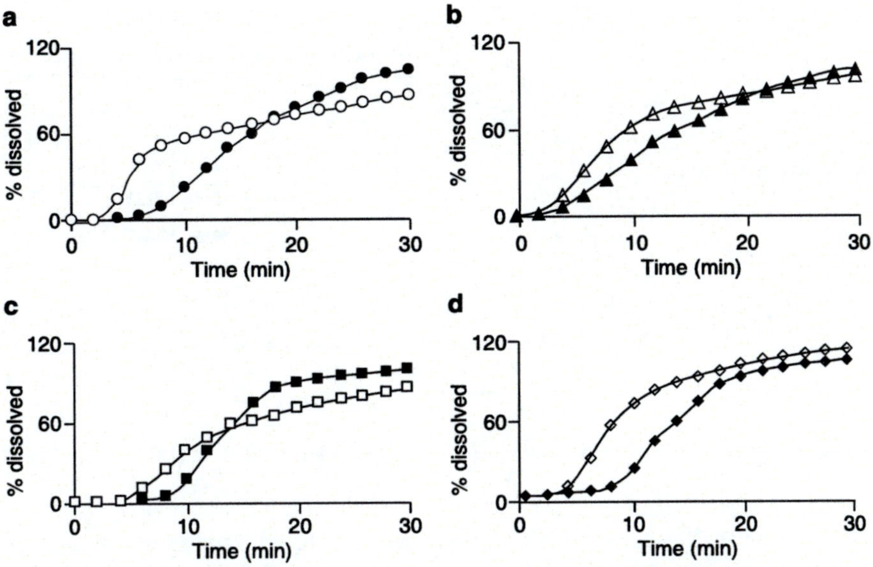

Drug release of risperidone from [a] Physical mixture 1:1 (●) & complex 1:1 (○),[b]Physical mixture 1:2 (▲) & complex 1:2 (△), [c] Physical mixture 1:4 (■), & complex 1:4 (□)and [d] Physical mixture 1:6 (◆) & complex 1:6 (◇)

Fig. 8.5 Drug release of risperidone: effect of resinate formation. Reprinted from [15]

entire 6 h of study and in all cases the resin was distributed evenly throughout the fundus, body and antrum of the stomach. Gastric retention was also longer with resin particles without polymer coating. Dose and food had insignificant effects on gastric residence time.

Cuna et al. [17] prepared amoxicillin-loaded ion-exchange resinate, encapsulated in mucoadhesive polymers (polycarbophil and Carbopol 934) to achieve increased efficacy of amoxicillin in peptic ulcer treatment by targeted delivery to the gastric mucosa and prolonged drug release at the site. Fluorescence microscopy indicated that gastric residence time and distribution of particles on the mucosa was better for nonpolymer-coated particles. While there was no ready explanation for the mechanism for increased gastric residence time and/or mucoadherence, the findings suggest that there may be opportunities for formulating drug molecules having "absorption windows" by increasing the gastric retention time with resinates, to provide better or more complete absorption. Another possibility would be to target drug delivery to treat gastric disease, for example gastric reflux caused by *H. pylori*.

8.5.3 *Prolonging or Extending Drug Release*

Sustaining drug release may be best affected using resins with strong exchange propensity, high cross linkage and having larger particle size. Reduced load of API is

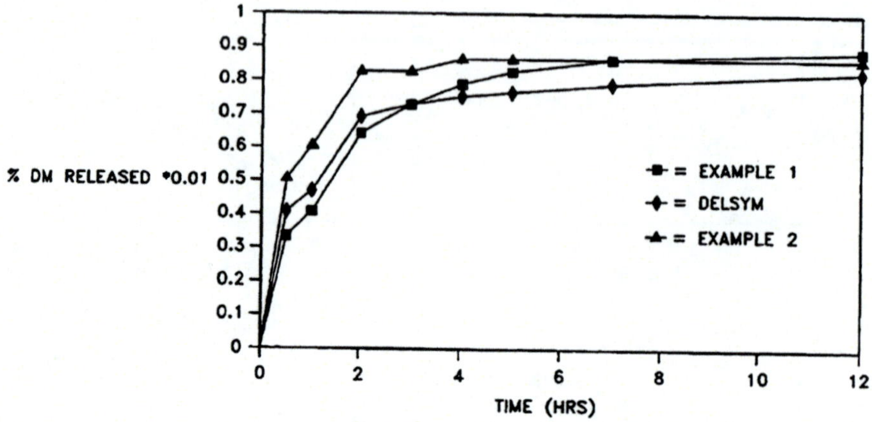

Fig. 8.6 Release profile of dextromethorphan (DM) from Delsym® 12-Hour [18]

also beneficial. Release can also be sustained or delayed by coating the resinate particles with a semipermeable membrane, providing a rate limiting barrier to diffusion to the surrounding medium. Membrane structure and thickness can be designed to complement the release properties of the resinate and synchronize delivery with therapeutic requirements. The nature of release-controlling polymers and their properties such as permeability and pH sensitivity are discussed elsewhere in this book.

An immediate-release drug layer can also be applied on top of a controlled-release membrane, or a combination of immediate-release and controlled-release particles can be included in the dose unit, as happens with other controlled-release techniques. This can provide early plasma levels (and associated onset of action) while the slow release from the coated resin prolongs the therapeutic effect with certain drugs. Release profiles from a formulation of the antitussive agent, dextromethorphan hydrobromide, comprising a mixture of immediate-release and controlled-release particles to provide 12-h cover are shown in Fig. 8.6.

Resinates usually swell on hydration and such expansion can crack a controlled-release membrane, destroying its release-modifying properties. The Pennwalt Corporation developed a process to inhibit such fracture by pretreating resinate with an impregnating agent to keep resinate hydrated and swollen during coating [19]. Glycerin and polyethylene glycol are suitable impregnating agents, as are propylene glycol, mannitol, lactose and methylcellulose [20]. Such treatment did not affect drug release [21]. However, the possibility that residual moisture acquired during coating might destabilize moisture-sensitive drugs needs to be borne in mind.

8.6 Pharmaceutical Applications

In addition to modifying a drug's dissolution and release resinates have other unique features that could contribute to optimal dosage form design. Features include:

- Their insolubility and nonabsorbable nature renders resins safe for ingestion
- Masking of bitter tasting or irritant drugs
- Enhance chemical stability of a labile drugs

These features are exemplified in the following examples.

8.6.1 *Controlled-Release Oral Liquids*

While ion-exchange technology is extensively used to control drug release from solid dosage forms (tablets and capsules), resins are unique in offering the possibility to design controlled-release oral liquids. Most if not all other technologies release drug when exposed to liquid, or have coating/matrix polymers ruptured or otherwise compromised by mastication or "chewing" if formulated as liquids. Ion-exchange complexation keeps the drug bound to the resin, prevents its leaching and avoids contact with counter-ions in solution. Ions in the gastric milieu then initiate release from the resinate. If the release profile needs to be further adjusted, prolonged polymeric coatings can be applied to the drug-resinate to modify release. A practical example concerns the resin, Dowex® 50WX4, a sulphonic acid cation exchanger. This was loaded with terbutaline and coated with Eudragit® RS/RL in a ratio of 70:30. Coated resinate particles were suspended in 0.75% w/w HPMC solution. Drug release had not changed after 6 months and the particle size used (mean 200 μm) did not evince a gritty mouth feel [22].

Tussionex® Extended Release Pennkinetic Suspension utilized ion-exchange technology in which both hydrocodone and chlorpheniramine were separately adsorbed on sulphonic acid ion-exchange resins. Hydrocodone-resinate was then coated with a semipermeable membrane to control its release, whereas no release modifier was applied to the chlorpheniramine-resinate. Following multiple dosing, a peak plasma concentration for hydrocodone of 22.8 ng/ml, occurred at 3.4 h. Chlorpheniramine mean plasma concentration was 58.4 ng/ml after 6.3 h, contrasting with T_{max} values of 1.5 h and 2.8 h respectively obtained with hydrocodone and chlorpheniramine formulated as an "immediate release" suspension [23]. Use of ion-resinates successfully slowed the release of both drugs.

Theophylline is formulated as a controlled-release liquid. It is a weak acid, and while binding onto an ion-exchange resin is possible, it is difficult to control drug release by the equilibrium-driven ion-exchange process. Motycka et al. [24] succeeded in modifying its release by coating the (anion exchange) resin with ethylcellulose and hard paraffin.

8.6.2 *Taste Masking and Stability Enhancement*

Ion-exchange resins have been used as taste masking agents since the 1970s. Bitter drugs are adsorbed on the resin, which can then be formulated to deliver drug in immediate or controlled-release mode. Nicotine, a volatile oil and an irritant, is adsorbed to a weakly acidic methacrylic and divinylbenzene resin e.g., Amberlite™ IRP64, which is then incorporated into a flavored chewing gum base. Chewing and contact with salivary fluid results in the nicotine being gradually released from the resinate and absorbed through the buccal mucosa [25]. The resinate allows nicotine, which otherwise is unstable and unpalatable, to be formulated as a pharmaceutical product for treating nicotine addiction.

Borodkin used a sulphonic acid ion-exchange resin to absorb iron (ferrous state) and formulate a controlled-release iron suspension supplement. About 25% of the dose (as ferrous ion) is gradually released from the resinate in the stomach, preventing irritation typically associated with a bolus dose. In addition, the resinate was better-tasting and reduced potential tooth staining [26].

8.6.3 Chewing Gum

Chewing gums have been developed as delivery systems in nicotine replacement therapy (NRT) for smoking cessation. Nicotine is complexed with a weakly acidic methacrylic and divinylbenzene resin (e.g., Amberlite™ IRP64) [25]. The resinate is then incorporated in a gum base, which is chewed gently, then held close to the cheek to facilitate saliva access to the resinate particles. Nicotine is slowly released, producing a tingling sensation. When the sensation ends the gum can be chewed again to expose other resinate particles, the process being repeated until all the nicotine is released. Buccal absorption of nicotine from a resinate formulation is slower and delivers lower plasma levels than those obtained following cigarette smoking (pulmonary absorption).

8.6.4 Other Applications

Hughes proposed the concept of resinates to obviate or minimize abuse potential of controlled substances such as narcotics [27]. An aversive agent (for example a bitter agent like denatonium chloride or an irritant like capsaicin), either alone or in combination with a controlled substance, is mixed with an ion-exchange resin to form a resinate. The resinate is then compounded into suitable dosage forms. Attempts to extract the controlled substance from the dosage form results in an extract that is unpalatable (when the aversive agent is released from the ion-exchange resin), or in minimal extraction of the substance of abuse due to resin binding.

8.7 Future Perspectives

The increasing prominence of novel molecular structures as therapeutic agents presents formidable delivery and formulation challenges to noninvasive delivery. Protein therapeutic agents, monoclonals and other such macromolecules do not ostensibly seem suited for combining with resin-type materials. However, polymeric materials that may possess an electric charge (charged polymers) have some "resin-like" features. Hence, some of the principles and strategies associated with resinate formation might also be germane to formulation of biologicals for oral delivery.

Cationic polymers have the potential to be effective nonviral DNA vehicles [28]. The surface of polyvinyl alcohol (PVA) microspheres (50–160 μm) was aminated, and loaded with negatively charged DNA. Microcapsules were then prepared by treating with cellulose acetate butyrate and poly(*N*-isopropylacrylamide-co-methyl methacrylate-co-methacrylic acid). This enteric coating is insoluble in gastric fluid (pH 1.3) but easily solubilized in the small intestine. No DNA degradation occurred during the loading and encapsulation processes.

Polymer micelles may have the potential to deliver nanoscale therapeutics [29]. Block polymers of polyethylene oxide and methacrylic acid (PEO-b-PMA) can be self-assembled into micelles around a metal cation core followed by cross-linking. Chelating the metal ion confers a negative charge to the micelle and allows loading with the cationic chemotherapeutic agent, cisplatin. Controlled release is initiated using a chloride-containing constitution vehicle; 40% drug being released in over 20 h.

Grafted starch-g-poly (acrylic acid) copolymers and starch/poly (acrylic acid) mixtures inactivated the digestive enzymes trypsin and α-chymotrypsin [30]. The polymer binding affinity for Ca^{2+} and Zn^{2+} deprived the proteases of essential cofactors and produced an inhibition factor up to 3.80 ± 0.40. Sequestering Ca^{2+} helped to open epithelial tight junctions and can improve peptide absorption [31]. Such concepts are at relatively early stages of evaluation but offer hope that oral delivery of peptides and proteins may ultimately become feasible.

8.8 Summary

IER technology offers the potential to enhance or otherwise control oral drug delivery of small molecule therapeutic agents. There is also some promise that they might facilitate oral delivery of biologics such as proteins and DNA. Further work in this area is warranted to turn promise to reality, in the light of the exciting possibilities that biopharmaceuticals offer as novel therapeutic agents.

References

1. Adams B, Holmes E (1935) Adsorptive properties of synthetic resins. J Soc Chem Ind 54:1–6
2. Chaudhary N, Sauders L (1956) Sustained release of drugs from ion-exchange resins. J Pharm Pharmacol 8:975–986
3. Cheng Y-H, Watts P, Hincliffe M, Hotchkiss R, Nankervis R, Faraj N, Smith A, Davis S, Illum L (2002) Development of a novel nasal nicotine formulation comprising an optimal pulsatile and sustained plasma nicotine profile for smoking cessation. J Control Release 79:243–254
4. Banerjee P (1987) Compositions and method of controlling penetration of topical and systemic agents. US Patent 4,692,462

5. Banga A (1998) Electrically assisted transdermal and topical drug delivery. Taylor & Francis, Bristol, PA
6. Keating J (1961) Pharmaceutical preparations comprising cation-exchange resin adsorption compounds and treatment therewith. US Patent 2,990,332
7. Hays E (1961) Pharmaceutical compositions containing a resin-narcotic compound and a resin-antihistamine compound. US Patent 3,035,979
8. Keating J (1964) Pharmaceutical preparations comprising phosphorous containing cation exchange resins having a basic drug adsorbed thereon; and treatment therewith. US Patent 3,143,465
9. Sigma-Aldrich (2010) Ion exchange resins: classification and properties. http://www.sigmaaldrich.com/etc/medialib/docs/Aldrich/Instructions/ion_exchange_resins.Par.0001.File.tmp/ion_exchange_resins.pdf
10. Hughes L (2010) Ion exchange resins: unique solutions to formulation problems. http://www.rohmhaas.com/ionexchange/pharmaceuticals/Formulations_doc/unique_solutions.pdf
11. Anand V, Kandarapu R, Garg S (2001) Ion-exchange resins: carrying drug delivery forward, Drug Discov Today 6(17)
12. Jeong S, Park K (2008) Drug loading and release properties of ion-exchange resin complexes as a drug delivery matrix. Int J Pharm 361:26–32
13. Sriwongjanya M, Bodmeier R (1998) Effect of ion exchange resins on the drug release from matrix tablets. Eur J Pharm Biopharm 46:321–327
14. Guo X, Chang R, Hussain M (2009) Ion-exchange resins as drug carriers. J Pharm Sci 98(11):3886–3902
15. Tawakkul M, Shah R, Zidan A, Sayeed V, Khan M (2009) Complexation of risperidone with a taste-masking resin: novel application of near infra-red and chemical imaging to evaluate complexes. Pharm Dev Technol 14(4):409–421
16. Thairs S, Ruck S, Jackson SJ, Steele RJC, Feely LC, Washington C, Washington N (1998) Effect of dose size, food and surface coating on the gastric residence and distribution of an ion-exchange resin. Int J Pharm 17:47–53
17. Cuna M, Alonso MJ, Torres D (2001) Preparation and in vivo evaluation of Mucoadhesive Microparticles. Preparation and in-vivo evaluation of mucoadhesive microspheres containing amoxicillin-resin complexes for drug delivery to the gastric mucosa. Eur J Pharm Biopharm 51:199–200
18. Wen B, Ramsay M, Scheurer H, Dokuzovic V, Lam V (1999) Antitussive drugs delivered by ion exchange resins. US Patent 6,001,392
19. Raghunathan Y (1980) Prolonged release pharmaceutical preparations. US Patent 4,221,778
20. Raghunathan Y (1989) Controlled release pharmaceutical preparations. US Patent 4,847,077
21. Breunig T, Hall H, Madsen J (2005) Drug release and spectroscopic examination of an ion exchange resin. http://www.encap.com/articles.htm
22. Cuña M, Jato V, Torres D (2000) Controlled-release liquid suspensions based on ion-exchange particles entrapped within acrylic microcapsules. Intl J Pharm 199:151–158
23. Hollenbeck G, McDermott JJ, Attkisson CC (2010) Sustained release drug delivery compositions and methods. US Patent 2010/0092562A1 (15 Apr 2010)
24. Motycka S, Newth JL, Nairn JG (1985) Preparation and evaluation of microencapsulated and coated ion-exchange resin beads containing theophylline. J Pharm Sci 74(6):643–646
25. Lichtneckert S, Lundgren C, Ferno O (1975) Chewable smoking substitute composition. US Patent 3,901,248
26. Borodkin S (1976) Iron-resin adsorbate. US Patent 3,947,572
27. Hughes L (2010) Pharmaceutical formulation including a resinate and an aversive agent. US Patent 7,655,256
28. Fundueanu G, Constantin M, Bortolotti F, Cortesi R, Ascenzi P, Menegatti E (2007) Cellulose acetate butyrate-pH/thermosensitive polymer microcapsules containing aminated poly(vinyl alcohol) microspheres for oral administration of DNA. Eur J Pharm Biopharm 66:11–20
29. Bontha S, Kabanov V, Bronich T (2006) Polymer micelles with cross-linked ionic cores for delivery of anticancer drugs. J Control Release 114:163–174

30. Ameye D, Voorspoels J, Foreman P, Tsai J, Richardson P, Geresh S, Remon J (2001) Trypsin inhibition, calcium and zinc ion binding of starch–g-poly(acrylic acid) copolymers and starch/poly(acrylic acid) mixtures of peroral peptide drug delivery. J Control Release 75:357–364
31. Borchard G, Lueβen A, de Boer J, Verhoef C, Junginger HE (1996) The potential of mucoadhesive polymers in enhancing intestinal peptide drug absorption. III: Effects of chitosan-glutamate and carbomer on epithelial tight junctions in vitro. J Control Release 39:131–138

Chapter 9
Pulsatile Delivery for Controlling Drug Release

Sumalee Thitinan and Jason T. McConville

Abstract The mechanisms and modes of drug action are such that systems providing constant and persistent plasma levels may not be optimally effective with some drugs and clinical conditions. At times, drug may need to be delivered at a specific time, for a certain period, to a specific location. "Delivery on demand" may even be optimal in response to the presence of a specific "biomarker-type" mediator. There is in consequence much interest in systems that can deliver drug from a unit in a time, location, or stimulus-driven mode.

This chapter discusses the potential, possibilities, and limitations for pulsatile delivery so that therapy can meet the above requirements.

9.1 Introduction

The opening chapter in this book outlines how drug delivery and attendant dosage form design has evolved. In essence, earlier forms of medication comprised a unit containing an accurate dose that was administered orally (or possibly parenterally or topically) the intent being that sufficient (drug) found its way to the site of action to alleviate the condition, with the rest not causing too much damage *en route*. "Controlled" delivery sought to modify a drug's effect by dosage form design and was largely facilitated by the development of analytical techniques to monitor drug in biological compartments and tissues, linking to efficacy or side effects.

Many early "controlled release" systems were based on the paradigm that plasma presence and duration of action were essentially synonymous. "Sustained," "prolonged," or "extended" release formulations were designed accordingly, as were

J.T. McConville (✉)
College of Pharmacy, The University of Texas at Austin, Austin, TX, USA
e-mail: jtmcconville@mail.utexas.edu

C.G. Wilson and P.J. Crowley (eds.), *Controlled Release in Oral Drug Delivery*, Advances in Delivery Science and Technology, DOI 10.1007/978-1-4614-1004-1_9,

"delayed release" systems to provide therapy coinciding with onset of clinical symptoms (e.g. during sleeping). Many good medications currently available are based on such considerations. However, as insights on molecular biology, drug–receptor interactions, and physiological/pathological process have been accrued it has become abundantly clear that, in many cases simply "soaking" a receptor with a solution of drug does may not optimize response. There is usually little information in such a "constant signal." Time and timing are essential components in most biological processes, along with associated variables such as frequency and amplitude of the signal. Such insights dictate that controlling drug release requires more sophisticated systems than simply delaying delivery or sustaining drug presence in the biosystem. Some such concepts have already been reduced to practice, e.g. with the advent of "chronotherapeutic" drug delivery.

In light of the above considerations the capability to deliver "pulses" of drug, where the condition or mode of treatment warrants is gaining much interest. This chapter reviews various clinical conditions, their potential to be treated with pulsatile drug delivery (PDD), and the state of the art of principles, materials, and experiences in pulsatile delivery.

Pulsatile delivery can have other functions in addition to aligning activity with the biology of a condition. Enteric coating, to safeguard drug from gastric (low pH) degradation is well established and may require rapid release of drug (i.e. a pulse) on reaching the favored region of the GI tract. There are also possibilities for "pulsed" dosing to be employed to enhance drug absorption and reduce dosing frequency.

9.2 Definition of Pulsatile Dosing

PDD may concern temporal targeting, related to the circadian rhythms of a clinical state, and/or site-specific drug delivery for localized therapy. Chapter 1 offers a useful definition of pulsatile release viz "the release of more than one dose of drug from a given system." Pulses are invariably separated by a time interval. Doses or pulses need not be limited to delivery of the same drug. Different medications could well be delivered in such sequential mode.

9.3 Dosage Form Design Considerations

The design of a pulse-release system must incorporate features that are calculated to deliver the requisite target plasma profile of the drug in question. Such a profile must in turn be optimal for efficacy and safety of the medication. If design concerns more than one "pulse," the modes (kinetics) of such sequential release can differ, e.g. an "immediate" release fraction could be followed by zero order or other forms of slower release, if this is optimum for the particular drug or clinical condition. The "immediate release" component is usually the less complex aspect of dosage form design for pulsing. A unit that rapidly disintegrates on ingestion usually provides the "first" pulse. Hence, techniques discussed in this chapter mostly concern later pulsing.

9.4 Benefits of Pulsed Release

9.4.1 Reduced Dose Frequency

Many therapies require relatively constant levels of drug to be present in plasma or at the site of drug action [1]. Hence, effective treatment can involve multiple administrations of a single dose over time, if the timecourse (kinetics) of the drug and associated activity in the biosystem is short. Frequent dosing usually leads to poor compliance, with the medication being less effective in consequence. Hence, a dosage form where, for instance, twice daily dosage can be replaced by dosing a unit, dosed once daily that releases drug in a pulsed fashion such that plasma levels are comparable to twice daily administration could be more patient-friendly and hence more effective. Lundberg and Sjöblom [2] developed systems for delivering at least two consecutive pulses of Proton Pump Inhibitors (PPIs) that are separated in time, by using suitable coated units that release drug in the small intestine to simulate twice daily administration. Commercial products of a number of PPIs utilize this approach.

9.4.2 Absorption Enhancement

Drugs that are substrates for intestinal or hepatic Cytochrome P450 may be difficult to formulate in classic "prolonged release" systems. The relatively low levels of drug encountering these enzymes while *en route* to the systemic compartments, being consequent to lower rates of input to the GI tract from a gradually releasing system, make them more prone to metabolism, because of higher enzyme/drug ratios than, if higher levels of drug were to be encountered at the site(s) of enzyme activity. Amidon and Leesman [3] proposed that prolonging the action of such drugs could be best effected by "pulse" delivery. Higher levels of drug, consequent to delivery as a "pulse are less likely to be extensively metabolized by Cytochrome P450 in the intestinal wall."

9.4.3 Time-specific Drug Delivery

It is now well established that maintaining a relatively constant plasma drug level throughout the dosage interval is not optimum in many conditions. Relationships between drug presence, duration of action, and safety may be influenced by, among other factors circadian rhythms. Varying drug concentrations in the biosystem may be more effective if coinciding with and being capable of managing peak manifestations of the clinical condition. This is the case with anti-inflammatory conditions and antiasthma therapy among others [4]. Pulse, rather than persistent delivery may also alleviate or eliminate side effects [5]. Targeting a specific rhythm of a disease could reduce dosage, thereby reducing drug exposure and unwanted effects. Targeting rhythms may also prevent drug interactions, providing wider treatment options for patients suffering from multiple ailments.

Over 100 drugs display temporal "activity" behaviors [6]. Examples include:

- Theophylline [7] effect is enhanced when dosed in the evening corresponding to the peak effect of asthma during nighttime or early morning.
- Propranolol displays more activity at the times of greatest stress from cardiovascular disease [8].
- Statins have greater cholesterol-lowering capability when dosed in the evening.
- Survival in ovarian cancer is quadrupled if doxorubicin is dosed in the morning along with evening dosing of cisplatin.

Many more "time-associated" effects are likely to emanate from increasing understanding of the molecular basis of clinical conditions and greater insights on how novel and existing medication perform. Such understanding is likely to greatly expand the opportunities for drug delivery systems that align receptor presence with optimum effects. Diseases such as Type 1 diabetes can currently be managed reasonably well by a combination of blood glucose monitoring and a strict dosing regimen involving parenteral delivery of insulin [9]. However, it is far from convenient, particularly for overnight cover and requires heroic patient compliance [10]. An orally administered dose would be far more desirable and probably more effective as a consequence.

9.5 Circadian Rhythms and Diseases

Cardiovascular, pulmonary, hepatic, and renal functionalities vary throughout the day, being controlled within a periodicity of 24 h. They are naturally synchronized with internal body clocks and the sleep-wake cycle. This can mean that some functionalities have time-related peaks and troughs. Disease states can affect such functionality and, as a consequence can also exhibit peak times of activity within a circadian rhythm.

Clinical research has been conducted with varying numbers of study subjects to explore circadian rhythms associated with disease states. Peak time of disease activity is shown in Table 9.1. Several potential oral delivery methods have been described for targeting circadian rhythms, as described below. Such awareness is useful for diagnosis and treatment [16]. Osteoarthritis may be distinguished

Table 9.1 Commonly occurring disease states and their circadian rhythms

Disease	Peak symptoms (24-h clock)	Reference
Allergic rhinitis	Morning	[7]
Arthritis		
Osteoarthritis	21.00	[11]
Rheumatoid	06.00–09.00	
Nocturnal Asthma	19.00–07.00	[7]
Cardiovascular disease		
Angina pectoris	06.00–12.00	[12]
Acute myocardial infarction	08.00–11.00	[13]
Diabetes	06.00	[14]
Peptic ulcer disease	21.00	[15]

from the rheumatoid form by the time of day at which it is most active. Cancer therapies may also be timed for a more effective outcome [17]. Other examples described below.

9.5.1 Allergic rhinitis

Allergic rhinitis is very common in the USA, with more than 50 million estimated sufferers [18]. Inhaled allergens (e.g. pollen, mold, animal dander) cause release of histamine and subsequent symptoms such as itching, excessive mucus production, and swelling. Symptoms are most severe in the morning [19]. Antihistamine drugs are used in therapy and can be taken orally to control symptoms such as sneezing, rhinorrhea, itching, and conjunctivitis. Several studies have been conducted by Reinberg to determine the efficacy of antihistamine medications delivered at specific times [20]. They indicated that for selected first- and second-generation H1-receptor antagonists duration of action was prolonged when administered in the morning rather than evening.

9.5.2 Arthritis

Pain from osteoarthritis becomes progressively worse throughout the day [19]. Rheumatoid arthritis in contrast is worst in early morning, following sleep [19]. A pulsed release delivery system containing NSAIDs should be timed to ensure that the highest plasma levels of the drug coincide with peak pain of either osteoarthritis or rheumatoid arthritis. Meloxicam has been formulated as a multiparticulate floating-PDD system for administering at night but releasing drug in early morning for treating rheumatoid arthritis [21]. A floating alginate bead system was developed to investigate various formulations. The lag time preceding pulsatile release could be controlled within the range 1.9–7.8 h by varying bead density and concentration of the hydrophobic drug. Hollow calcium pectinate beads were developed by Badve et al. [22]. The pectinate beads performed in a similar fashion to the alginate beads described in the previous study, and a scintigraphy study in rabbits demonstrated gastro retention of up to 5 h before pulsed release of the diclofenac sodium model drug.

9.5.3 Nocturnal Asthma

Symptoms of asthma occur more frequently during the night than during the day [7]. Furthermore, some circadian changes seem to be associated with worsening nocturnal asthmatic symptoms. Cortisol levels are lowest in the middle of the night, increasing and peaking in the early morning [23] while higher levels of histamine (a mediator of bronchoconstriction) coincide with greater bronchoconstriction during sleep time [24].

Since bronchoconstriction and exacerbation of nocturnal asthmatic symptoms vary in a circadian fashion, chronotherapies have been studied for asthma, employing theophylline, β_2-agonists, anticholinergic drugs, and corticosteroids. Sustained release formulations of theophylline are administered in the evening to alleviate airways obstruction and nocturnal asthma [25]. Oral administration of corticosteroids at 8:00 a.m. and 3:00 p.m. was more effective in controlling nocturnal asthma than the same doses administered at 3:00 p.m. and 8:00 p.m. [26].

9.5.4 Cardiovascular Disease

Several functions (e.g. heart rate, blood pressure, stroke volume, cardiac output, blood flow, peripheral resistance) of the cardiovascular system exhibit circadian rhythms [8]. There is greater risk of myocardial infarction and associated sudden death in the early morning [27], with capillary resistance and vascular reactivity being greater at this time [4]. Consequently, medications with dosing schedule established have been formulated in an attempt to provide therapeutic concentration of a drug at the target site when the drug is most needed. Antihypertensive products such as Verelan PM and Covera HS provide chronotherapy for hypertension, releasing drug during the vulnerable period of 6 a.m. to noon after night-time administration [28].

9.5.5 Diabetes

Release of endogenous insulin is reputedly pulsatile [29], being aligned with peak plasma glucose levels for optimum effect. This natural rhythm changes to inadequate secretion in the case of Type 1 diabetes or altering the pulsatile mode in the Type 2 condition [23]. Thus, the goal with insulin therapy is to mimic the normal-patterned secretion of the endogenous mediator. A glucose-responsive hydrogel, comprising copolymers of acrylamide and allyl glucose was developed by Obaidat and Park [30]. They suggested that glucose-sensitive phase-reversible hydrogels might be able to deliver requisite amounts of insulin when exposed to varying glucose concentrations. Later Zhang et al. reported a simplified method for the production of D-glucose responsive hydrogels to deliver pulses of proteins such as insulin [31]. Use of CM-dextran, in the hydrogel would seem to enable delivery in a variety of pH and ionic strength environments.

It may not yet be possible to deliver insulin or other proteinaceous medications orally but the above-reported approaches illustrate attempts to provide "delivery-on-demand" for the requisite therapeutic agents.

9.5.6 *Peptic Ulcer Disease*

Peptic ulcer pain is reportedly most frequent at night [4], being attributable to higher levels of gastric acid production in the afternoon and evening [32]. Disease management might accordingly take the form of evening dosing of an H_2-receptor antagonist or morning administration of a single dose of a PPI [4, 19]. The H_2-receptor antagonist famotidine has a longer duration of action than cimetidine (10–12 h cf. 4–8 h) [33]. It is accordingly considered more suited for ulcer therapy by decreasing gastric acid secretion for longer periods during night time [19].

Dexlansoprazole, a PPI is presented as a biphasic delivery system (Dexilant Capsules) [34]. The first population of granules starts releasing a pulse of drug within 1–2 h after administration. A second pulse follows within 4–5 h, prolonging the therapeutic effect [35].

The gastro-intestinal tract is designed to progressively disintegrate, digest, and absorb nutrients and finally excrete waste or other unwanted materials. Structure, composition, and movements are broadly designed to reflect such functions but variability is great, with respect to conditions that can affect drug dissolution and absorption. These are discussed in detail in Chap. 2. GI tract-related variables that can have a major effect on drug dissolution/absorption-related factors include, but are not limited to:

- Environmental pH.
- Fluid volume in the various regions.
- Transit rate of the dosage form.
- Highly localized sections for absorption of some drugs e.g. amoxicillin, vitamin B12.

Clinical condition and coadministered medication can also play a role. Such variables are important considerations in pulsed dosage form design and illustrate how knowledge of the GI tract as well as the clinical condition, pharmacokinetic and physicochemical properties of the drug are vital for intelligent dosage form design. The reader is strongly encouraged to become familiar with such diversity, as discussed elsewhere in this book.

9.6 Modulation of Oral Drug Delivery

External or self-modulation may be employed for the temporal targeting of oral dosage forms.

- External modulation devices have been conceptualized to control drug release in response to an externally generated signal such as an electronic input to alter rate of delivery from a mechanical pump, an oscillating magnetic field, ultrasound, temperature or an electronic signal [36].
- Self-modulated devices use approaches such as rate-controlled mechanisms: pH-sensitive polymers, enzyme substrate reaction, pH-sensitive drug solubility, competitive binding, and metal concentration-dependant hydrolysis [37].

These modulations may be employed for site-specific and/or time-delayed drug delivery. It may also be desirable for certain therapies to be delivered to specific sites in the GI tract. For example, peptic ulcer disease may concern gastric or duodenal ulceration [38] and specific targeting might improve efficacy. Conditions such as irritable bowel disease (IBD) viz Crohn's Disease or ulcerative colitis, and bowel cancer also offer the potential for more localized drug delivery [39]. Conventional therapies for the treatment of IBD rely on daily administration of high doses of drug. Crohn's disease is often treated with high-dose anti-inflammatory agents [40], and in the case of corticosteroid treatment there is a risk of dependency, relapse, and the development of irreversible adverse effects [41].

Localized delivery might also reduce dose and side effects, particularly if drugs can vary, in terms of absorption from different regions of the GI tract. Leuprolide absorption is variable throughout the GI tract in rats, with maximum absorption in the colon [42]. Region-specific absorption has also been demonstrated with levetiracetam [43]. Dosage form design to liberate the drug from its carrier at such local sites may be achieved by pulse-delivery.

9.6.1 Modes of Self-Modulation

Self-modulation for delivery to the large intestine focuses on protective coatings and biodegradable polymers. These polymers may be used to target specific enzyme activity or dissolve at specific sites in the GI tract when a pH change occurs (Table 9.2).

Polymers, such as pectin, xylan, guar gum, and azo bond-containing hydrogels (examples of which include copolymers of 2-hydroxyethyl methacrylate and methyl

Table 9.2 Manufacturing options available for self-modulated site-specific drug delivery

Substance	Comments	References
Azo bond containing hydrogels	Specific enzyme activity: azo-reductase in the colon due to resident bacteria	[44, 45]
	Swelling controlled by the localized pH of the colon	
Enteric coatings	Commercially available Eudragit® (Evonik) dissolves at specific elevated pH	[46–48]
Glycosidic bond containing hydrogels	Specific enzyme activity: glycosidases in the colon Used as a film coating onto the drug-containing component	[49, 50]
Guar gum, Pectin, Xylan	Enzymatic degradation: The result of resident bacteria in the distal GI tract	[51–56]
Enzyme containing substrates	Enzymatic degradation of pectin by pectinase irrespective of external environment	[57]

methacrylate) can be degraded by colon-resident bacteria [57]. Glycosidic bonds are also susceptible to hydrolysis by enzymatic activity in the colon [58, 59]. Such self-modulated systems require appropriate pH and presence of the appropriate microbes. However, intersubject pH differences and other as yet undefined variables could affect functionality. Further research is warranted to establish the potential and limitations of such systems for pulsatile colonic delivery.

9.7 Pulsatile Drug Delivery Formulations

Pulsatile dosage was defined at the start of the chapter as providing the release of one or more "doses" from a unit (separated in time). If more than one dose (of the same drug) is required, the delivery system must utilize either external or self-modulation to produce a "lag-time" to separate the pulses. Strategies can capitalize on the functions, and differing environments in the GI tract. Examples are:

- Gastro retention, where one segment of the dose remains in the stomach due to its low density, e.g. floating on the stomach contents.
- Swelling mechanisms such that release of a drug segment is delayed or otherwise constrained by having to diffuse through a gel layer.
- Erosion/degradation of a component or protective agent/coating before a drug segment is released.

9.7.1 Floating Doses

Gastro-retentive principles and mechanisms are considered in another chapter. Floating dosage forms comprise low-density materials or those that rapidly lower

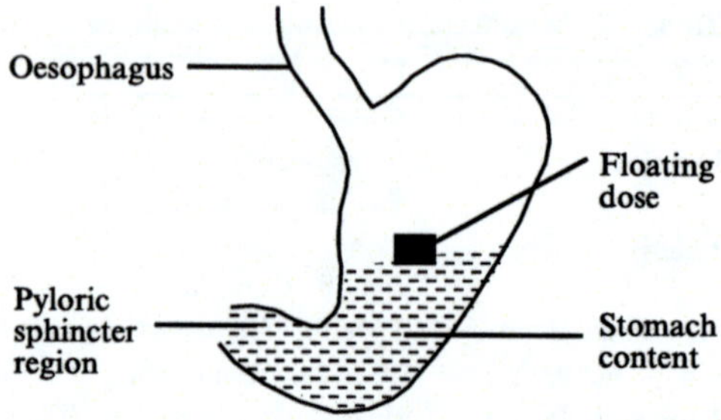

Fig. 9.1 Schematic of the stomach showing retention of a floating dosage form

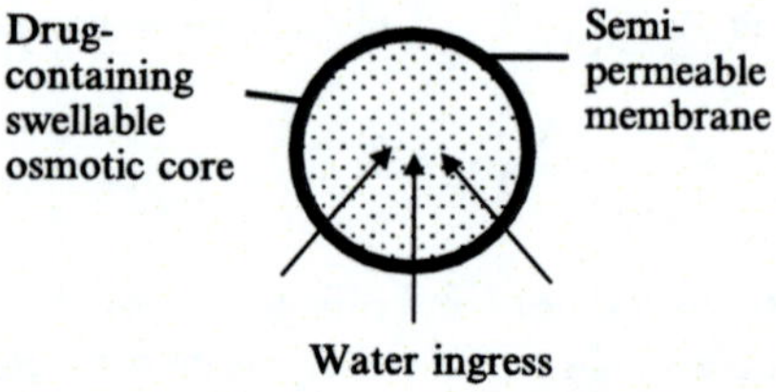

Fig. 9.2 Osmotic core pulsed release pellets

their density to enable buoyancy in the stomach contents and increase gastric residence (Fig. 9.1). Residence can be terminated by increasing the density of the floating device or by complete emptying of the stomach contents. Low-density polyvinyl pyrrolidone [60], low-density hollow polycarbonate microspheres [61], and rapidly swelling porous hydrogel composites [62], have all exhibited flotation. However, none have been effectively developed to a PDD system. Sigmoidal release profiles have been achieved using low-density calcium alginate beads [63, 64]. These are insoluble in the stomach [65], and dissolve as the pH rises in the small intestine.

A true PDD has been achieved by using an effervescent core to generate carbon dioxide to produce a low-density polymeric coated tablet [66]. The device remained intact until a semipermeable film ruptured after a certain lag time followed by a pulse release of drug.

9.7.2 Pellets or Granules

Swelling mechanisms are normally associated with the incorporation of a drug-containing osmotic core and a semipermeable outer layer (Fig. 9.2). Water ingress into these devices causes them to swell, rupturing the external coat and releasing the drug in pulsatile mode (Table 9.3).

Granules and pellets have a disadvantage relative to tablets or capsules due to their smaller size compromising their osmotic capabilities. Film rupture occurs following very little water ingress, with too-early pulse release and inadequate lag, or time to reach specific release regions. Consequently they may need to be incorporated into a coated capsule or tablet.

Table 9.3 Summary of pellet-based formulations showing pulsed release

Osmotic system	Comments	Reference
Particulate delivery system	Permeability reducing agent and soluble component are control rate of water ingress. Swellable core ruptures outer coat	[67]
Multiparticulate delivery system	Diffusion of soluble core in a series of pulses[a] can provide multiple doses and lag-times	[68]
Multiparticulate	Sigmoidal release.[b] Eudragit® RS coating with succinic acid incorporated to vary permeability	[69]
Time-controlled explosion system (TES)	A drug coated inert core coated with an insoluble swellable layer is in turn coated with an insoluble semipermeable outer layer	[70–72]
Membrane coated pellet	Lipophilic plasticizers alter permeability of the coat to control water ingress	[73, 74]
Particulate pulse release system	Permeability of the outer coat is controlled by photo-initiation. Length of exposure to a UV source affects permeability	[75]
Colon targeted	Eudragit® outer coat to dissolve at pH 6.8 then allow diffusion of drug through a diffusive coat	[47]

[a] Outer coat remains intact
[b] Sustained release following a lag-time

Table 9.4 Summary of tablet formulations showing pulsed release

Device	Comments	Reference
Osmotic tablet (developed as OROS®)	Zero order delivery of drug followed by a pulsed delivery. OROS® incorporates a hydrophilic barrier beneath a semipermeable tablet coating	[76]
Coated laminated tablet	Polymer layers erode in sequence with only one uncoated face	[77]
Time Clock®	A soluble drug-containing core coated with a variable thickness hydrophilic/hydrophobic polymer layer	[78]
Erosion tablet	Enteric outer coat dissolves in SI (> pH 5), then HPMC coat erodes to expose drug after a lag-time	[46]
Laminated tablet device	Laminated pH buffered layers to protect proteins erode and deliver in a series of pulses	[79]
Disintegration tablet	Drug-containing core press coated with insoluble ethyl cellulose	[80]
Press-coated tablets	A highly soluble drug-containing core tablet is press coated with a disintegrating outer layer of L-HPC. Lag-time is controlled by varying the thickness of the outer layer	[81]

9.7.3 Tablets

Tablets developed for pulsatile release may incorporate an osmotically active core, as with pellets (Table 9.3). Others utilize erodible polymers to slowly expose single layer or multiple layers of drug, thereby providing multiple pulses. These are summarized in Table 9.4.

Pulsed release from osmotically activated devices tends to deliver a sigmoidal release profile. However, an exception is the OROS® device which provides a sustained release profile following a lag-time [76]. The release of a drug is initiated from within the still-intact tablet by water penetrating the core through an outer coat, resulting in swelling of the hydrophilic polymer which "pushes" drug through the laser-drilled orifices.

9.7.4 Capsule-Based Devices

A range of lag-times can be selected, to enable an increased range of temporal targeting from a single capsule device.

Capsules can be coated with a semipermeable or water impervious layer. A semipermeable coat provides the basis for osmotically driven release. The pressure caused by internal swelling forces the capsule open or ruptures the body to deliver drug. A water-impervious, low-density capsule might also float on the gastric contents, prolonging residence time in the GI tract before providing a pulsed release of a drug after a lag-time. A nonfloating device requires a barrier to external fluids for preventing premature release. The capsule cap may be removed by a swelling osmotic action, or a seal may be inserted in the end of the capsule body, being dislodged after a desired lag-time. Examples of capsule devices capable of delivering pulses are indicated below.

9.7.4.1 Pulsincap

Pulsincap (Fig. 9.3) has an outer impermeable capsule body Fig. 9.3 (v), which houses the drug formulation (Fig. 9.3 [iii]). This can be separated from or be mixed with an expulsion excipient layer (Fig. 9.3 [iv]). Sodium bicarbonate/citric acid effervescing mixtures may be used as expulsion agents [82–84].

The Pulsincap device incorporates expanding low-substitute hydroxypropyl cellulose (L-HPC). In a study evaluating the regional GI targeting of dofetilide, the excipient layer contained sucrose as a soluble excipient [85]. Release was controlled by the length of the hydrogel plug (Fig. 9.3 [ii]) [86]. Figure 9.3 illustrates how the plug is removed by expansion on uptake of water during GI tract transit.

The challenging design of the swellable hydrogel plug and complex insertion process led to the exploration of alternative versions of Pulsincap™. A hydroxypropylmethyl cellulose (HPMC)/lactose-based erodible tablet was subsequently developed as studies had indicated that lag-time could be more readily controlled by erosion mechanisms, HPMC being readily erodible in GI tract fluids [87].

9.7.4.2 Egalet

This technology comprises an impermeable shell (Fig. 9.4 [iii]) containing a drug core (Fig. 9.4 [ii]) and two erodible outer layers (Fig. 9.4 [i]) at each open end.

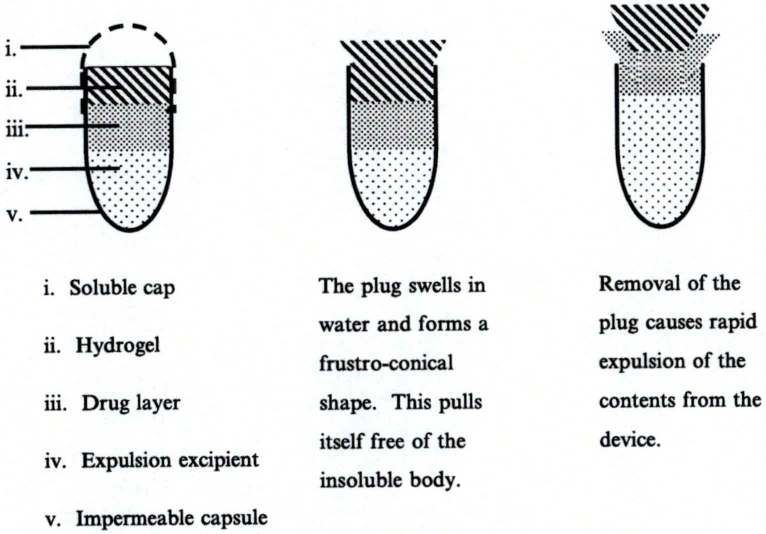

Fig. 9.3 The Pulsincap™ capsule device

Fig. 9.4 The Egalet™ device

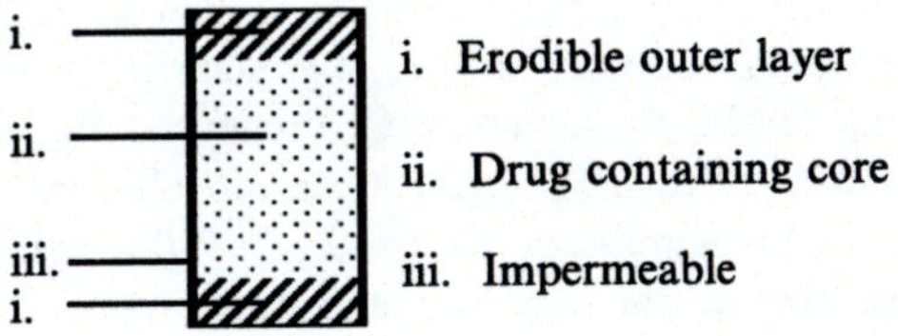

On erosion of the outer layer the drug in the core dissolves on exposure to GI tract contents.

In another variant it is possible to include another drug in the erodible outer layer for sustained release followed by pulsed release of a different drug from the inner core.

The Egalet device is potentially susceptible to inconsistent release due to reliance on uniform erosion at its terminal ends. Krögel and Bodmeier evaluated a similarly constructed device and highlighted the issue of asymmetrical erosion [88].

9.7.4.3 Chronset

Wong et al. describe the Chronset technology [89]. This system can deliver drugs in a pulsatile fashion using osmotic pressure generated inside the semipermeable membrane (Fig. 9.5).

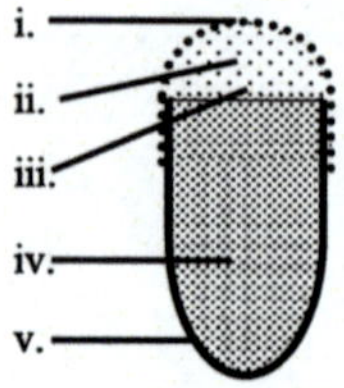

Fig. 9.5 The Chronset® device

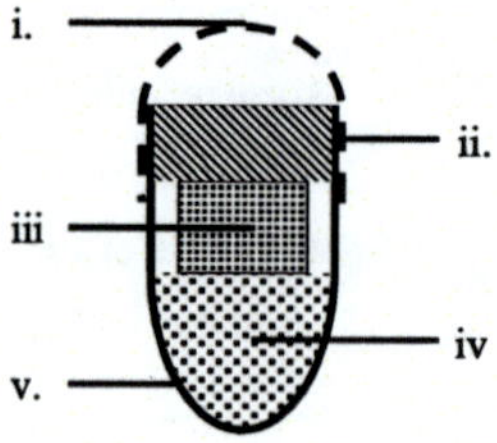

Fig. 9.6 The time delayed capsule (TDC)

The capsule shell is water-insoluble (Fig. 9.5 [v]) and houses the therapeutic agent and soluble filler(s) (e.g. sucrose, lactose) (Fig. 9.5 [iv]). A rigid barrier on top of the drug-containing contents (Fig. 9.5 [iii]), transfers the driving from osmotic expansion against the capsule shell. Water from the GI tract is absorbed into the osmotic core (Fig. 9.5 [ii]) through the semipermeable cap (Fig. 9.5 [i]), causing expansion of the osmotic core. This subsequent expansion "pushes" against the rigid barrier, causing separation. Continuing expansion of the osmotic core causes the shell and cap to become separated, exposing the drug-containing compartment to the GI fluids and enabling delivery of the contained drug as a "pulse."

9.7.4.4 Time-Delayed Capsule

A time-delayed capsule (TDC) (Fig. 9.6) comprises a capsule body, coated to render insoluble in water and containing a swelling expulsion excipient, a drug-containing tablet and an erodible tablet (ET).

Gastrointestinal (GI) fluids erode the ET (Fig. 9.6 [ii]), while the capsule body contents are protected by the ethyl cellulose coat (Fig. 9.6 [v]). After a lag-time, determined by erosion time of the ET, water enters the capsule, causing rapid swelling of the expulsion excipient (Fig. 9.6 [iv]). This pushes the undissolved drug tablet (Fig. 9.6 [iii]) free of the capsule body followed by disintegration and rapid drug

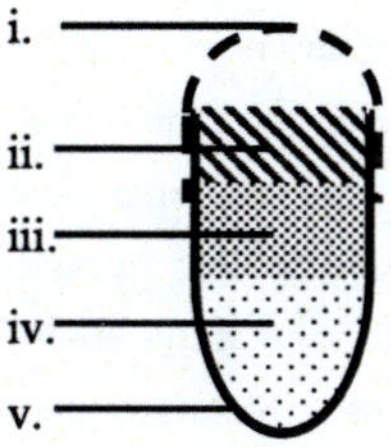

Fig. 9.7 The PORT System®

release as a "pulse." For esthetic purposes and ease of swallowing a soluble gelatin cap is placed on top of the device (Fig. 9.6 [i]).

9.7.4.5 Programmable Oral Release Time System®

The programmable oral release time (PORT) device (Fig. 9.7) utilizes osmotic pressure to drive drug release in a similar manner to the Chronset technology. However, the body of the capsule is semipermeable to aqueous fluid, contrasting with the impermeable nature of the TDC (Fig. 9.7 [v]). It contains a swellable osmotic component (Fig. 9.7 [iv]) and drug-containing layer (Fig. 9.7 [iii]).

An insoluble wax plug at the top of the capsule body seals the contents (Fig. 9.7 [ii]). As the contents swell, on ingress of water the waxy plug is dislodged and the drug-containing layer (Fig. 9.7 [iii]) is made available for release/dissolution. A soluble cap seals the device (Fig. 9.7 [i]). Several variables can be manipulated to control release lag-time. e.g. capsule body wall thickness and composition, concentration of the osmotic contents, and the length/composition of the waxy plug.

9.7.4.6 Pressure-Controlled Colon Delivery Capsule

The pressure-controlled colon delivery capsule (PCDC) capsule design (Fig. 9.8) utilizes two coatings and capitalizes on the varying physiological conditions along the GI tract. Configuration is as follows:

- An insoluble coating is present on the inside wall of a standard capsule (Fig. 9.8 [v]).
- The drug-containing core may comprise an oily liquid base (Fig. 9.8 [iv]).
- An insoluble waxy plug seals the top of the insoluble coat (Fig. 9.8 [iii]).
- The capsule containing the oily coated core with the insoluble waxy plug is sealed with a standard cap (Fig. 9.8 [ii]).
- The entire capsule is then enteric coated (Fig. 9.8 [i]) [90].

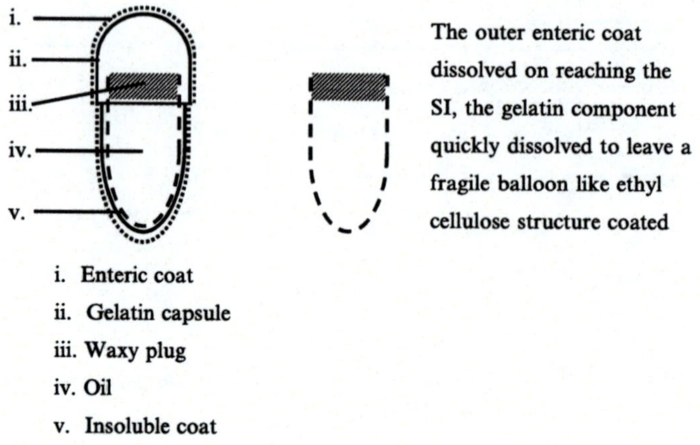

Fig. 9.8 Structure and function of the PCDC

Fig. 9.9 Structure and function of the CTDC

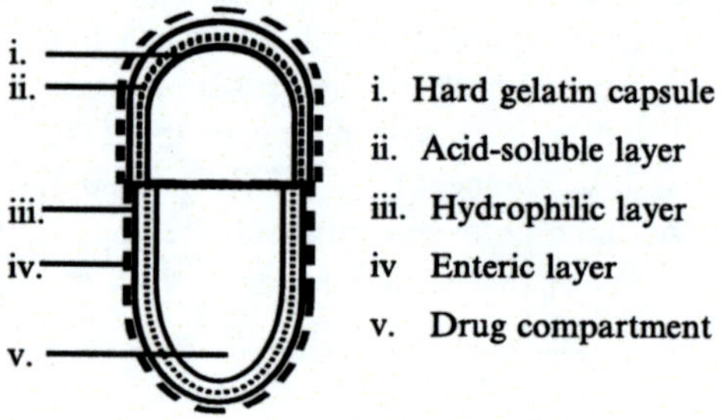

Once ingested the capsule remains intact at gastric pH. The enteric coat and gelatin dissolve in the intestine, leaving a relatively fragile balloon-like structure that is ruptured by the elevated pressure at the illeo-cecal junction. Drug is accordingly released at colonic entry.

9.7.4.7 Colon-Targeted Delivery Capsule

A schematic of the colon-targeted delivery capsule (CTDC) is shown in Fig. 9.9. It comprises a capsule (Fig. 9.9 [i]) coated with three layers.

The innermost coat comprises an acid soluble layer (Fig. 9.9 [iv]), which in turn is coated with a water soluble layer (Fig. 9.9 [iii]). The outermost coat comprises an enteric polymer (Fig. 9.9 [ii]) that does not dissolve until the capsule device has emptied from the stomach.

The enteric and water-soluble coats dissolve in the small intestine to expose an acid-soluble layer that dissolves in the colon, liberating the contained drug [91].

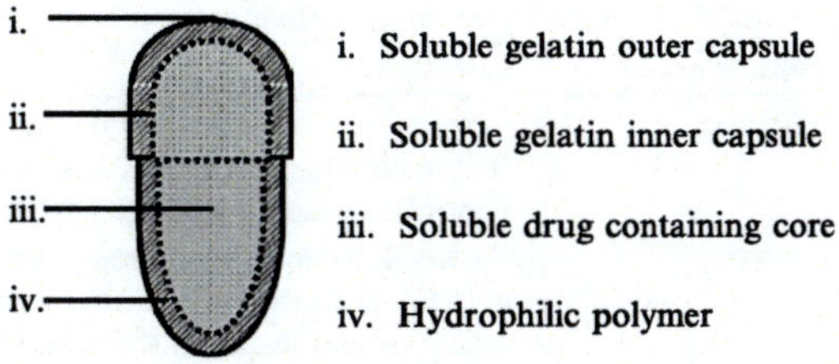

Fig. 9.10 The HS capsule

9.7.4.8 Hydrophilic Sandwich Capsule

The hydrophilic sandwich (HS) concept (Fig. 9.10) comprises:

- Two different-sized capsules, one contained within the other [i].
- The void space contains various concentrations of a hydrophilic polymer such as HPMC, creating a hydrophilic "sandwich" [iv].
- A drug-containing core [iii] is housed within the inner capsule [ii].

The outer capsule rapidly dissolves on exposure to an aqueous environment. The hydrophilic "sandwich" then forms a gel barrier, protecting the inner drug layer for a predetermined lag time depending on gel layer thickness and concentration/type of hydrophilic polymer [92].

The capsule-based systems described above have a number of potential disadvantages with respect to consistency of performance. The Chronset, PORT System, and Pulsincap devices all rely on swelling mechanisms that are vulnerable to frictional forces associated with plug removal or other release-determining sequences, e.g.:

- Rugosity of the internal surface of the Pulsincap capsule is not uniform. Thus, frictional forces vary as the hydrogel plug swells, giving variable lag-times.
- Frictional forces on the external surface of the Chronoset capsule during cap removal can cause variable release times. The rigid barrier layer, which acts as a swelling block, can also restrict drug release from the core.
- The PORT System is subject to the influence of frictional forces as the osmotic core swells to expel the waxy plug. Additional studies have highlighted the importance of a tight fitting seal by using hot-melt wax plugs [93]. These prevented premature drug release.
- The formation of an ethyl cellulose pressure-sensitive balloon structure in the PCDC can also cause variability. Premature bursting may occur due to intersubject GI pressure variations. The flexible balloon structure formed after dissolution of the outer coat may also compromise plug fit.

Capsule-based devices are summarized in Table 9.5.

Table 9.5 Capsule-based pulsed delivery devices

Capsule device	Comments	References
Pulsincap™	A water impermeable capsule body consisting with hydrogel plug. Plug length and insertion depth controls lag-time control (Fig. 9.3)	[94]
Egalet™	An insoluble tube[a] with erodible plugs inserted at either end. Plugs are comprised of selected M_w PEG, waxy materials and surfactants. Composition controls lag-time (Fig. 9.4)	[95]
Chronset®	An osmotically coated active compartment within a semipermeable cap This swells to pushes against a rigid barrier layer and removes the cap. Lag-time is controlled by the osmotic potential (Fig. 9.5)	[89]
Time-delayed capsule	Water impermeable coat on gelatin capsule with an erodible tablet. Erosion of the tablet allows water to enter the capsule, swelling of an expulsion excipient causes expulsion of the drug (Fig. 9.6)	[96–100]
Programmable Oral Release Time (PORT) System®	A water-permeable coated gelatin capsule with a swellable osmotic core and sealed with an insoluble wax plug. The contents swell to remove the plug. The wall thickness and composition, concentration of the osmotic contents and the length of the hydrogel plug control lag-time (Fig. 9.7)	[101–103]
Pressure-controlled colon delivery capsule (PCDC)	Internally coated capsule. Drug is filled into the PCDC in an oily liquid base, and sealed with an insoluble waxy plug. Elevated pressure exerted by the ileocaecal junction, ruptures the device to release drug into the colon (Fig. 9.8)	[90, 104, 105]
Colon-targeted delivery capsule (CTDC)	Enteric outer coat dissolved in the SI to expose an acid soluble layer, which dissolves in the colon (Fig. 9.9)	[91]
Hydrophilic sandwich capsule	HPMC layer sandwiched between a large outer gelatin capsule and a smaller inner gelatin capsule-containing drug. Erosion of the HPMC layer provides a lag-time, controlled by grade and/or thickness (Fig. 9.10)	[91]

[a] Capsule shaped

9.8 Conclusions and Future Perspectives

Pulsed delivery is a relatively new "arrival" for designing better delivery systems for oral medications. Additional insights that are likely to emanate from the increasing use of biomarkers is likely to present even more challenges for optimizing delivery, but also opportunities for approaches such as "pulse-mode" delivery.

It is also conceivable that dosage form and device combinations could be utilized in concert to deliver medication. Improvements in other areas of controlled delivery, such as better gastro-retention could well lead to systems with prolonged residence in the GI tract. It is then possible to conceive of external stimuli that can activate a "pulse" of drug, in response to an externally measured biological efficacy parameter.

As a new area in controlled oral release there is much potential for the growth and maturing of pulsatile delivery.

References

1. Khan MA, Sastry SV, Vaithiyalingam SR, Agarwal V, Nazzal S, Reddy IK (2000) Captopril gastrointestinal therapeutic system coated with cellulose acetate pseudolatex: evaluation of main effects of several formulation variables. Int J Pharm 193:147–156
2. Lundberg JP, Sjöblom B (1999) Oral pharmaceutical pulsed release dosage form. WO/1999/032093, July 1999
3. Amidon GL, Leesman GD (1993) Pulsatile drug delivery system. US Patent 5,229,131 WO/1999/032093, July 1993
4. Mandal AS, Biswas N, Karim KM, Guha A, Chatterjee S, Behera M, Kuotsu K (2010) Drug delivery system based on chronobiology – a review. J Control Release 147:314–325
5. Yano H, Hirayama F, Kamada M, Arima H, Uekama K (2002) Colon-specific delivery of prednisolone-appended alpha-cyclodextrin conjugate: alleviation of systemic side effect after oral administration. J Control Release 79:103–112
6. Lemmer B (1989) Temporal aspects of the effects of cardiovascular active drugs in humans. In: Lemmer B (ed) Chronopharmacology: cellular and biochemical interactions. Marcel Dekker, New York, pp 525–541
7. Smolensky MH, Lemmer B, Reinberg AE (2007) Chronobiology and chronotherapy of allergic rhinitis and bronchial asthma. Adv Drug Deliv Rev 59:852–882
8. Lemmer B (2007) Chronopharmacology of cardiovascular medications. Biol Rhythm Res 38:247–258
9. Brod M, Thomsen TL, Christensen T (2010) Evaluating the choice of pen versus syringe: implications for long-term diabetes management. Value Health 13:A529
10. Markman M, Kennedy A, Webster K, Peterson G, Kulp B, Belinson J (1999) An effective and more convenient drug regimen for prophylaxis against paclitaxel-associated hypersensitivity reactions. J Cancer Res Clin Oncol 125:427–429
11. Labrecque G, Reinberg AE (1989) Chronopharmacology of nonsteroid anti-inflammatory drugs. In: Lemmer B (ed) Chronopharmacology: cellular and biochemical interactions. Marcel Dekker, New York, pp 525–541
12. Gassler JP, Bisognano JD (2009) Commentary: revisiting the early morning blood pressure surge and myocardial ischemia. J Clin Hypertens 11:348–350
13. Bogaty P, Waters DD (1989) Possible mechanisms accounting for the temporal distribution of anginal attacks. In: Lemmer B (ed) Chronopharmacology: cellular and biochemical interactions. Marcel Dekker, New York, pp 525–541
14. Jarret RJ (1989) Temporal aspects of blood glucose regulation: implications for diabetic patients. In: Lemmer B (ed) Chronopharmacology: cellular and biochemical interactions. Marcel Dekker, New York, pp 525–541
15. Sanders SW, Moore JG (1992) Gastrointestinal chronopharmacology – physiology, pharmacology and therapeutic implications. Pharmacol Ther 54:1–15
16. Elliott WJ (2001) Timing treatment to the rhythm of disease – a short course in chronotherapeutics. Postgrad Med 110:119–122
17. Levi F, Okyar A, Dulong S, Innominato PF, Clairambault J (2010) Circadian timing in cancer treatments. Annu Rev Pharmacol Toxicol 50:377–421
18. Kilgore D, Najm W (2010) Common respiratory diseases. Prim Care 37:297–324
19. Ohdo S (2010) Chronotherapeutic strategy: rhythm monitoring, manipulation and disruption. Adv Drug Deliv Rev 62:859–875

20. Reinberg AE (1989) Chronopharmacology of H1-Antihistamines. In: Lemmer B (ed) Chronopharmacology: cellular and biochemical interactions. Marcel Dekker, New York, pp 525–541
21. Sharma S, Pawar A (2006) Low density multiparticulate system for pulsatile release of meloxicam. Int J Pharm 313:150–158
22. Badve SS, Sher P, Korde A, Pawar AP (2007) Development of hollow/porous calcium pectinate beads for floating-pulsatile drug delivery. Eur J Pharm Biopharm 65:85–93
23. Haus E (2007) Chronobiology in the endocrine system. Adv Drug Deliv Rev 59:985–1014
24. Sajan J, Cinu TA, Chacko AJ, Litty J, Jaseeda T (2009) Chronotherapeutics and chronotherapeutic drug delivery systems. Trop J Pharm Res 8:467–475
25. Reinberg A, Pauchet F, Ruff F, Gervais A, Smolensky MH, Levi F, Gervais P, Chaouat D, Abella ML, Zidani R (1987) Comparison of once-daily evening versus morning sustained-release theophylline dosing for nocturnal asthma. Chronobiol Int 4:409–419
26. Martin RJ, Banks-Schlegel S (1998) Chronobiology of asthma. Am J Respir Crit Care Med 158:1002–1007
27. Portaluppi F, Hermida RC (2007) Circadian rhythms in cardiac arrhythmias and opportunities for their chronotherapy. Adv Drug Deliv Rev 59:940–951
28. Smith DHG (2001) Pharmacology of cardiovascular chronotherapeutic agents. Am J Hypertens 14:2965–3015
29. Waldhausl W (1989) Circadian-rhythms of insulin needs and actions. Diabet Res Clin Pract 6:S17–S24
30. Obaidat AA, Park K (1997) Characterization of protein release through glucose-sensitive hydrogel membranes. Biomaterials 18:801–806
31. Zhang R, Tang M, Bowyer A, Eisenthal R, Hubble J (2006) Synthesis and characterization of a D-glucose sensitive hydrogel based on CM-dextran and concanavalin A. React Funct Polym 66:757–767
32. Gandhi BR, Mundada AS, Gandhi PP (2011) Chronopharmaceutics: as a clinically relevant drug delivery system. Drug Deliv 18:1–18
33. Youan B-BC (2004) Chronopharmaceutics: gimmick or clinically relevant approach to drug delivery? J Control Release 98:337–353
34. Howden CW, Larsen LM, Perez MC, Palmer R, Atkinson SN (2009) Clinical trial: efficacy and safety of dexlansoprazole MR 60 and 90 mg in healed erosive oesophagitis – maintenance of healing and symptom relief. Aliment Pharmacol Ther 30:895–907
35. Abel C, Desilets AR, Willett K (2010) Dexlansoprazole in the treatment of esophagitis and gastroesophageal reflux disease. Ann Pharmacother 44:871–877
36. Paolino D, Fresta M, Sinha P, Ferrari M (2006) Drug delivery systems. In: Webster JG (ed) Encyclopedia of medical devices and instrumentation, 2nd edn. Wiley, New York, pp 437–495
37. Goldbart R, Kost J (1999) Calcium responsive bioerodible drug delivery system. Pharm Res 16:1483–1486
38. Li ZS, Zou DW, Ma XQ, Chen J, Shi XG, Gong YF, Man XH, Gao L, Zhao YF, Wang R, Yan XY, Dent J, Sung JJ, Wernersson B, Johansson S, Liu WB, He J (2010) Epidemiology of peptic ulcer disease: endoscopic results of the systematic investigation of gastrointestinal disease in China. Am J Gastroenterol 105:2570–2577
39. Patel M, Shah T, Amin A (2007) Therapeutic opportunities in colon-specific drug-delivery systems. Crit Rev Ther Drug Carrier Syst 24:147–202
40. van Hogezand RA, Witte AMC, Veenendaal RA, Wagtmans MJ, Lamers C (2001) Proximal Crohn's disease: review of the clinicopathologic features and therapy. Inflamm Bowel Dis 7:328–337
41. Rutgeerts PJ (2001) Review article: the limitations of corticosteroid therapy in Crohn's disease. Aliment Pharmacol Ther 15:1515–1525
42. Zheng YQ, Qiu YH, Lu MYF, Hoffman D, Reiland TL (1999) Permeability and absorption of leuprolide from various intestinal regions in rabbits and rats. Int J Pharm 185:83–92

43. Stockis A, Sargentini-Maier ML, Otoul C, Connor A, Wilding I, Wray H (2010) Assessment of levetiracetam bioavailability from targeted sites in the human intestine using remotely activated capsules and gamma scintigraphy: open-label, single-dose, randomized, four-way crossover study in healthy male volunteers. Clin Ther 32:1813–1821
44. Brøndsted H, Kopecek J (1991) Hydrogels for site-specific oral-drug delivery – synthesis and characterization. Biomaterials 12:584–592
45. Van den Mooter G, Samyn C, Kinget R (1994) The relation between swelling properties and enzymatic degradation of azo polymers designed for colon-specific drug-delivery. Pharm Res 11:1737–1741
46. Gazzaniga A, Busetti C, Moro L, Sangalli ME, Giordano F (1995) Time-dependent oral delivery systems for colon targeting. Stp Pharma Sci 5:83–88
47. Gupta VK, Beckert TE, Price JC (2001) A novel pH- and time-based multi-unit potential colonic drug delivery system. I. Development. Int J Pharm 213:83–91
48. Kao CC, Chen SC, Sheu MT (1997) Lag time method to delay drug release to various sites in the gastrointestinal tract. J Control Release 44:263–270
49. Brondsted H, Hovgaard L, Simonsen L (1995) Dextran hydrogels for colon-specific drug-delivery. 4. Comparative release study of hydrocortisone and prednisolone sodium-phosphate. Stp Pharma Sci 5:65–69
50. Wakerly Z, Fell JT, Attwood D, Parkins D (1996) Pectin/ethylcellulose film coating formulations for colonic drug delivery. Pharm Res 13:1210–1212
51. Adkin DA, Kenyon CJ, Lerner EI, Landau I, Strauss E, Caron D, Penhasi A, Rubinstein A, Wilding IR (1997) The use of scintigraphy to provide "proof of concept" for novel polysaccharide preparations designed for colonic drug delivery. Pharm Res 14:103–107
52. Ahrabi SF, Madsen G, Dyrstad K, Sande SA, Graffner C (2000) Development of pectin matrix tablets for colonic delivery of model drug ropivacaine. Eur J Pharm Sci 10:43–52
53. Gliko-Kabir I, Yagen B, Baluom M, Rubinstein A (2000) Phosphated crosslinked guar for colon-specific drug delivery II. In vitro and in vivo evaluation in the rat. J Control Release 63:129–134
54. Krishnaiah YSR, Satyanarayana S, Prasad YVR, Rao SN (1998) Gamma scintigraphic studies on guar gum matrix tablets for colonic drug delivery in healthy human volunteers. J Control Release 55:245–252
55. Macleod GS, Fell JT, Collett JH, Sharma HL, Smith AM (1999) Selective drug delivery to the colon using pectin: chitosan: hydroxypropyl methylcellulose film coated tablets. Int J Pharm 187:251–257
56. Sinha VR, Kumria R (2001) Polysaccharides in colon-specific drug delivery. Int J Pharm 224:19–38
57. Krogel I, Bodmeier R (1999) Evaluation of an enzyme-containing capsular shaped pulsatile drug delivery system. Pharm Res 16:1424–1429
58. Gloster TM, Turkenburg JP, Potts JR, Henrissat B, Davies GJ (2008) Divergence of catalytic mechanism within a glycosidase family provides insight into evolution of carbohydrate metabolism by human gut flora. Chem Biol 15:1058–1067
59. Chourasia MK, Jain SK (2003) Pharmaceutical approaches to colon targeted drug delivery systems. J Pharm Pharm Sci 6:33–66
60. Iannuccelli V, Coppi G, Leo E, Fontana F, Bernabei MT (2000) PVP solid dispersions for the controlled release of furosemide from a floating multiple-unit system. Drug Dev Ind Pharm 26:595–603
61. Joseph NJ, Lakshmi S, Jayakrishnan A (2002) A floating-type oral dosage form for piroxicam based on hollow polycarbonate microspheres: in vitro and in vivo evaluation in rabbits. J Control Release 79:71–79
62. Chen J, Blevins WE, Park H, Park K (2000) Gastric retention properties of superporous hydrogel composites. J Control Release 64:39–51
63. Lee BJ, Ryu SG, Cui JH (1999) Controlled release of dual drug-loaded hydroxypropyl methylcellulose matrix tablet using drug-containing polymeric coatings. Int J Pharm 188:71–80

64. Lee BJ, Cui JH, Kim TW, Heo MY, Kim CK (1998) Biphasic release characteristics of dual drug-loaded alginate beads. Arch Pharm Res 21:645–650
65. Durig T, Venkatesh GM, Fassihi R (1999) An investigation into the erosion behaviour of a high drug-load (85%) particulate system designed for an extended-release matrix tablet. Analysis of erosion kinetics in conjunction with variations in lubrication, porosity and compaction rate. J Pharm Pharmacol 51:1085–1092
66. Krogel I, Bodmeier R (1999) Floating or pulsatile drug delivery systems based on coated effervescent cores. Int J Pharm 187:175–184
67. Chen C-M (1995) Pulsatile particles drug delivery system. US Patent 5,472,708, Dec 1995
68. Chen C-M (1993) Multiparticulate pulsatile drug delivery system. US Patent 5,260,068, Nov 1993
69. Narisawa S, Fukui E, Yoshino H, Hirakawa Y, Noda K (1994) Porosity-controlled ethylcellulose film coating. 5. Mechanism of drug-release from beads coated with porous ethylcellulose film. Chem Pharm Bull 42:2131–2134
70. Ueda S, Hata T, Asakura S, Yamaguchi H, Kotani M, Ueda Y (1994) development of a novel drug-release system, time-controlled explosion system (TES). 1. Concept and design. J Drug Target 2:35–44
71. Ueda S, Ibuki R, Kimura S, Murata S, Takahashi T, Tokunaga Y, Hata T (1994) Development of a novel drug-release system, time-controlled explosion system (Tes). 3. Relation between lag time and membrane thickness. Chem Pharm Bull 42:364–367
72. Ueda S, Yamaguchi H, Kotani M, Kimura S, Tokunaga Y, Kagayama A, Hata T (1994) Development of a novel drug-release system, time-controlled explosion system (Tes). 2. Design of multiparticulate Tes and in-vitro drug-release properties. Chem Pharm Bull 42:359–363
73. Schultz P, Kleinebudde P (1997) A new multiparticulate delayed release system. 1. Dissolution properties and release mechanism. J Control Release 47:181–189
74. Schultz P, Tho I, Kleinebudde P (1997) A new multiparticulate delayed release system. 2. Coating formulation and properties of free films. J Control Release 47:191–199
75. Kok P, Vonk P, Kossen NWF (2000) A particulate pulse-release system and mathematical description with the Maxwell-Stefan theory. J Control Release 66:293–306
76. Magruder PR, Barclay B, Wong PSL, Theeuwes F (1989) Composition comprising a therapeutic agent and a modulating agent. US Patent 4,851,229, July 1989
77. Fujioka K, Sato S, Takada Y, Sasaki Y, Tamura N (1991) Sustained pulsewise release pharmaceutical preparation. US Patent 5,011,692, April 1991
78. Pozzi F, Furlani P, Gazzaniga A, Davis SS, Wilding IR (1994) The time clock system – a new oral dosage form for fast and complete release of drug after a predetermined lag time. J Control Release 31:99–108
79. Jiang HL, Zhu KJ (2000) Pulsatile protein release from a laminated device comprising of polyanhydrides and pH-sensitive complexes. Int J Pharm 194:51–60
80. Lin SY, Lin KH, Li MJ (2001) Micronized ethylcellulose used for designing a directly compressed time-controlled disintegration tablet. J Control Release 70:321–328
81. Ghimire M, McInnes FJ, Watson DG, Mullen AB, Stevens HNE (2007) In-vitro/in-vivo correlation of pulsatile drug release from press-coated tablet formulations: a pharmacoscintigraphic study in the beagle dog. Eur J Pharm Biopharm 67:515–523
82. Gohel MC, Sumitra GM (2002) Modulation of active pharmaceutical material release from a novel 'tablet in capsule system' containing an effervescent blend. J Control Release 79:157–164
83. Kroegel I, Bodmeier R (1997) Development of floating or pulsatile DDS based on effervescent cores. Proc Int Symp Control Release Bioact Mater 24:273–274
84. Krogel I, Bodmeier R (1999) Floating or pulsatile drug delivery systems based on coated effervescent cores. Int J Pharm 187:175–184
85. Stevens HNE, Wilson CG, Welling PG, Bakhshaee M, Binns JS, Perkins AC, Frier M, Blackshaw EP, Frame MW, Nichols DJ, Humphrey MJ, Wicks SR (2002) Evaluation of Pulsincap (TM) to provide regional delivery of dofetilide to the human GI tract. Int J Pharm 236:27–34

86. Binns J, Stevens HNE, McEwen J, Pritchard G, Brewer FM, Clarke A, Johnson ES, McMillan I (1996) The tolerability of multiple oral doses of Pulsincap(TM) capsules in healthy volunteers. J Control Release 38:151–158
87. Makino K, Idenuma R, Ohshima H (1996) A model for erosion kinetics of a hydrogel matrix. Colloids Surf B Biointerfaces 8:93–100
88. Krogel I, Bodmeier R (1999) Development of a multifunctional matrix drug delivery system surrounded by an impermeable cylinder. J Control Release 61:43–50
89. Wong PSL, Theeuwes F, Larsen SD, Dong LC (1995) Osmotic device for delayed delivery of agent. US Patent 5,443,459, Aug 1995
90. Jeong YI, Ohno T, Hu ZP, Yoshikawa Y, Shibata N, Nagata S, Takada K (2001) Evaluation of an intestinal pressure-controlled colon delivery capsules prepared by a dipping method. J Control Release 71:175–182
91. Ishibashi T, Ikegami K, Kubo H, Kobayashi M, Mizobe M, Yoshino H (1999) Evaluation of colonic absorbability of drugs in dogs using a novel colon-targeted delivery capsule (CTDC). J Control Release 59:361–376
92. Stevens HNE (2003) Pulsincap and hydrophilic sandwich (HS) capsules: innovative time-delayed oral drug delivery technologies. Drugs Pharm Sci 126:257–262
93. Yu LX, Lipka E, Crison JR, Amidon GL (1996) Transport approaches to the biopharmaceutical design of oral drug delivery systems: prediction of intestinal absorption. Adv Drug Deliv Rev 19:359–376
94. McNeill ME, Rashid A, Stevens HNE (1994) Dispensing device. British Technology Group, London, UK
95. Bar-Shalom D, Slot L, Lee WW, Wilson CG (2003) Development of the Egalet technology. Drugs Pharm Sci 126:263–271
96. Stevens HNE, Rashid A, Bakhshaee M (1995) Dispensing device. British Technology Group, London, UK
97. Ross AC, MacRae RJ, Walther M, Stevens HNE (2000) Chronopharmaceutical drug delivery from a pulsatile capsule device based on programmable erosion. J Pharm Pharmacol 52:903–909
98. McConville JT, Hodges LA, Jones T, Band JP, O'Mahony B, Lindsay B, Ross AC, Florence AJ, Stanley AJ, Humphrey MJ, Wilson CG, Stevens HNE (2009) A pharmacoscintigraphic study of three time-delayed capsule formulations in healthy male volunteers. J Pharm Sci 98:4251–4263
99. McConville JT, Ross AC, Chambers AR, Smith G, Florence AJ, Stevens HNE (2004) The effect of wet granulation on the erosion behaviour of an HPMC-lactose tablet, used as a rate-controlling component in a pulsatile drug delivery capsule formulation. Eur J Pharm Biopharm 57:541–549
100. McConville JT, Ross AC, Florence AJ, Stevens HNE (2005) Erosion characteristics of an erodible tablet incorporated in a time-delayed capsule device. Drug Dev Ind Pharm 1:77–87
101. Amidon GL, Leesman GD, Sherman LB (1995) Multi stage drug delivery system. US Patent 5,387,421, Feb 1995
102. Crison JR, Siersma PR, Amidon GL (1996) A novel programmable oral release technology for delivering drugs: human feasibility testing using gamma scintigraphy. Proc Int Symp Control Release Bioact Mater 23:51–52
103. Crison JR, Siersma PR, Taylor MD, Amidon GL (1995) Programmable oral release technology. PORT Systems: a novel dosage form for time and site specific oral drug delivery. Proc Int Symp Control Release Bioact Mater 22:278–279
104. Takaya T, Ikeda C, Imagawa N, Niwa K, Takada K (1995) Development of a colon delivery capsule and the pharmacological activity of recombinant human granulocyte colony-stimulating factor (rhG-CSF) in beagle dogs. J Pharm Pharmacol 47:474–478
105. Shibata N, Ohno T, Shimokawa T, Hu ZP, Yoshikawa Y, Koga K, Murakami M, Takada K (2001) Application of pressure-controlled colon delivery capsule to oral administration of glycyrrhizin in dogs. J Pharm Pharmacol 53:441–447

Chapter 10
Ordered Mesoporous Silica for the Delivery of Poorly Soluble Drugs

Michiel Van Speybroeck, Randy Mellaerts, Johan Adriaan Martens, Pieter Annaert, Guy Van den Mooter, and Patrick Augustijns

Abstract Ordered mesoporous silica (OMS) materials offer much promise as carriers for poorly soluble drugs because of their high porosity, large specific surface area, and uniform pore shape and dimensions. Liquid as well as solid type phases of drugs, confined and stabilized in the pores of OMS, can exhibit special physicochemical properties and enhanced dissolution rates compared to crystalline forms. The ability to design mesopore size precisely provides the formulation scientist with the potential to readily attain and closely control drug release. Absorption enhancement may require stable supersaturation of released drug. If this can be effected (viz. drug precipitation attenuated by suitable formulation adjuvants), systemic absorption can be enhanced.

In vivo proof of concept of OMS as a dissolution-enhancing technology has been demonstrated in various animal species. The findings are promising and suggest that adsorption on OMS can successfully enhance and control absorption of poorly soluble drugs.

10.1 Historical Overview

Contemporary approaches to drug discovery often lead to drug candidates with high lipophilicity and low aqueous solubility [1]. This has increased interest in technologies that overcome poor solubility/dissolution rate. Products incorporating several such technologies, including complexation with cyclodextrins, particle size reduction (micronization and nanonization), high-energy solids and lipid/surfactant-based

---

P. Augustijns (✉)
Laboratory for Pharmacotechnology and Biopharmacy, Katholieke Universiteit Leuven, Leuven, Belgium
e-mail: Patrick.Augustijns@pharm.kuleuven.be

C.G. Wilson and P.J. Crowley (eds.), *Controlled Release in Oral Drug Delivery*, Advances in Delivery Science and Technology, DOI 10.1007/978-1-4614-1004-1_10,

systems, have been commercialized. An additional and emerging strategy to enhance drug dissolution concerns drug adsorption on ordered mesoporous silica (OMS) materials. These exhibit an array of uniform mesopores (according to the IUPAC nomenclature, pores with a diameter between 2 and 50 nm are termed mesopores), very high surface area, and large pore volume. All of these features render them suitable for enhancing drug dissolution as will be outlined in this chapter.

The first documented attempts to disperse poorly soluble drugs on high surface area carriers occurred in the early 1970s. Pyrogenic silica (silicon dioxide), produced by a flame process, sometimes called fumed silica and known commercially as (e.g.) Aerosil® was used as a carrier. This material does not exhibit porosity but has a high specific surface area due to its small particle size. Monkhouse and Lach loaded poorly soluble drugs on such supports and observed enhanced dissolution rates compared to crystalline drugs [2]. The drug molecules were shown to interact with the pyrogenic silica surface via hydrogen bonding and van der Waals interactions [3]. These bonds were predicated to break easily on contact with aqueous medium, resulting in enhanced dissolution rate compared to bulk crystalline material. Several papers were subsequently published on enhancement of drug dissolution using these carriers but the technology has not been used in commercial drug products, probably because of inadequate in vivo performance.

Vallet-Regi et al. were the first to explore an OMS material for controlling drug delivery [4], reporting on controlled release of ibuprofen from OMS material MCM-41. Early papers focused primarily on controlling drug release [5, 6]. In 2003, Tozuka et al. filled the pores of an OMS material (coded FSM-16) with the poorly soluble drug salicylamide and reported that the amorphous drug phase constrained in the FSM-16 mesopores exhibited enhanced dissolution compared to crystalline drug [7]. Similar findings were reported by Charnay et al. [8]. Finally, in 2007, it was demonstrated that the adsorption of isolated molecules of the poorly soluble drug itraconazole on the walls of an OMS material was a promising approach to obtain a high release rate [9]. A subsequent in vivo study demonstrated that the OMS-based itraconazole was bioequivalent to the commercial polymer-based solid dispersion product Sporanox® [10].

OMS were initially developed as catalysts and adsorbents, but applications and opportunities are now expanding rapidly in the pharmaceutical and biomedical sciences. Today, the research on OMS for pharmaceutical/biomedical applications has become multidisciplinary, with material sciences, physical chemistry and pharmaceutical sciences intersecting. OMS-based materials are investigated for applications in the pharmaceutical and biomedical sciences such as oral drug delivery, systemic delivery, stimuli-responsive nanodevices, bone tissue engineering, gene transfection, and cell tracking. This chapter provides context for their use in oral delivery, thereby mainly focusing on their potential to enhance drug dissolution rate.

10.2 Specific Benefits

10.2.1 Physical Stabilization of Amorphous and Molecularly Dispersed Drugs

The large numbers of poorly soluble compounds emanating from drug discovery pipelines has stimulated growing interest in formulation approaches to overcome solubility/dissolution-limited oral absorption. In such a context, amorphous drug forms have been widely investigated. The high internal energy of the amorphous state, relative to the crystalline state can provide increased apparent solubility and dissolution rate, which may translate to increased bioavailability. However, the amorphous state is physically unstable and may convert to the energetically favored crystalline state during processing or storage. To enhance physical stability, amorphous drugs are usually formulated as molecular or near-molecular dispersions in a physiologically inert carrier matrix, typically a polymer [11]. Although this approach, the so-called solid dispersion approach, has evinced widespread scientific interest, the number of commercial products based on solid dispersions remains limited. In addition to issues related to the method of preparation and dosage form development, metastability has been reported [12].

When organic molecules are introduced in a porous OMS material, two scenarios are possible:

- Molecules are attracted to the pore wall, i.e. are adsorbed. The adsorption equilibrium may be such that the molecules are spread over the surface, with no intermolecular interactions.
- Intermolecular interactions take place within the pore; the contained compound then possesses the properties of a confined liquid or solid phase. Such confinement can significantly affect the properties of a contained drug.

Numerous studies have shown that reducing the dimensionality of the confined space to approach the range of intermolecular forces can lead to significant alterations of the behavior of a confined phase, most notably: freezing point depression [13, 14]. The change in freezing temperature T_f can be related to the pore size D on the basis of the Gibbs–Thomson equation

$$\Delta T_f = T_{f,pore} - T_{f,bulk} = -\frac{2(\gamma_{ws} - \gamma_{wf})\nu}{D\lambda_{f,bulk}},$$

where

ν is the crystal molar volume,

γ_{ws} and γ_{wf} are the wall-solid and wall-fluid surface tensions, respectively, and

$\lambda_{f,bulk}$ is the bulk heat of melting.

In this equation, the sign of the shift in freezing temperature is determined by the difference of the surface tensions $\gamma_{ws} - \gamma_{wf}$. Thus, freezing temperature is decreased (or increased) with respect to the bulk value if the pore wall prefers the liquid phase to the solid phase (or prefers the solid phase over the liquid phase). Although increases in freezing temperature have been observed experimentally in some exceptional cases [15], a decrease in freezing temperature upon confinement is much more prevalent. According to the Gibbs–Thomson equation, ΔT_f is inversely correlated to the pore size. This has been demonstrated experimentally [14].

However, below a certain pore diameter, freezing does not occur. The physics underlying this effect are not fully understood, mainly due to the numerous complications that can occur in experimental systems, as reviewed in detail elsewhere [14]. At present, a substantial body of knowledge of the physical nature of confined systems has been obtained from molecular simulation studies. In one of these studies, it was demonstrated that, for pore diameters smaller than 12 times the diameter of the confined molecules, the confined phase was amorphous throughout the pores whereas its bulk was a rigid solid [16]. These results were confirmed in a subsequent experimental study [17]. At this time, it is not well understood whether confined phases represent long-lived metastable states or true thermodynamic equilibrium.

Most experimental studies on confined systems have used nontherapeutic model compounds, e.g. cyclohexane, benzene, or nitrobenzene. Studies using more complex organic molecules such as drugs have been limited. One of the most comprehensive studies was conducted by Azaïs et al. [18]. Using NMR spectroscopy, they clearly demonstrated that ibuprofen confined in OMS material MCM-41 exhibited liquid-like properties at ambient temperatures. At low temperature (–50°C), ibuprofen crystallized in MCM-41 material having a pore size of 11.6 nm, whereas vitrification occurred in material with 3.5 nm pore size. In the latter material, no crystallization was observed after at least 1 year when stored at 5°C.

Using a combination of differential scanning calorimetry and X-ray photoelectron spectroscopy, Mellaerts et al. demonstrated that, at a drug load of 20%, the poorly soluble drug itraconazole was molecularly dispersed on the internal surface of the OMS material SBA-15 [19]. This monomolecular state was retained for 1 year, when stored at 25°C/52% relative humidity. More recently, ten physicochemically diverse poorly soluble drugs were successfully dispersed on SBA-15 using a generic solvent impregnation procedure and a drug load of 20% [20]. There was no evidence for the presence of amorphous or crystalline drug phases. Such dispersions remained stable for at least 6 months storage at 25°C/52% relative humidity.

As a dissolution-enhancing approach, adsorption onto OMS is most closely related to polymeric solid dispersion techniques. Both can provide a stable, amorphous, molecularly dispersed drug on a physiologically inert carrier. However, mechanisms of stabilization differ substantially. In the polymeric solid dispersions, the drug is kinetically *frozen* in the carrier matrix. If the mobility of the drug–polymer dispersion is sufficiently low, crystallization can be prevented for pharmaceutically relevant time scales. However, the development of polymeric systems that truly stabilize the amorphous/molecularly dispersed drug during manufacturing and storage is cumbersome. This may account for the limited number of commercial products based on solid dispersion technology [21]. OMS materials, in contrast, offer new possibilities for amorphous drug forms, in that their capability to stabilize amorphous forms is a function of dimensions, rather than mobility. If the ratio of pore diameter to molecular diameter is sufficiently low, intermolecular interactions leading to crystallization of the confined molecules can be prevented [16].

In addition to physical stability, drug–carrier miscibility can influence formulation as a solid dispersion in that miscibility determines maximum drug load [21]. For OMS, the strength of the interaction between drug and the silica surface appears less important for maximum drug load capacity [20]. For instance, ibuprofen can be loaded in the OMS material MCM-41 such that the carrier pores are virtually completely filled [18]. In contrast, the interaction between ibuprofen and the silica surface is very weak [22]. Furthermore, the results of a recent study demonstrate that the poorly soluble drug fenofibrate – a drug that is hard to formulate as a solid dispersion due to its poor miscibility with most polymeric carriers – was successfully processed into a dissolution-enhancing formulation using OMS [23].

10.2.2 Pore Diameter and Controlled Release of Poorly Soluble Drugs

A major advantage of OMS concerns flexibility of synthesis. Judicious selection of the template and conditions can provide an end product that is engineered to meet user requirements. Although other porous silicates have been used for drug delivery, such materials possess irregularly shaped pores of broad size distribution. OMS pores, in contrast, are more uniform in shape with very narrow size distribution. The capability to "tailor" pore diameter during manufacture thereby allows close control of drug release rate. Decreased pore size leads to decreased drug release rate and vice versa [5, 9, 24]. Figure 10.1 exemplifies this effect. Depending on the pore size, drug release can occur in minutes, hours, or days.

Particle size/morphology may also significantly affect drug release rate [25], and several protocols have been published on how to modify OMS particle size/morphology [26, 27]. However, capability to control pore diameter may offer a better way of "designing" release rate. Drugs contained in OMS pores, being non-crystalline, can exhibit dissolution rates, and concentrations for absorption, not achievable by dissolution of low-energy crystalline forms. Stated otherwise, the release of poorly soluble drugs from OMS is associated with supersaturation. From a dosage form design perspective, this poses an interesting challenge based on the following possibilities:

- Increased concentrations at the site of absorption may – by virtue of Fick's First Law – enhance the flux of drug across the gastrointestinal epithelium or
- Supersaturation inherently poses the risk of the drug precipitating as an energetically more favorable but less soluble form, thereby reducing availability for absorption [28].

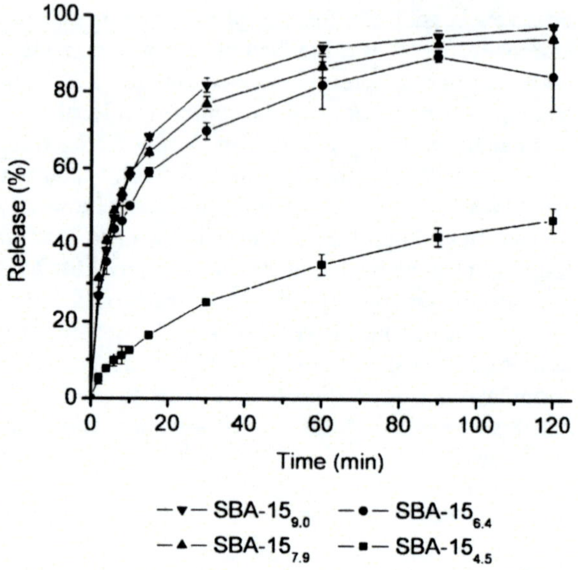

Fig. 10.1 Release profiles of itraconazole from SBA-15 in simulated gastric fluid supplemented with 1% of sodium lauryl sulfate. The release rate increases with increasing pore size. The *subscripts* denote the average pore diameter in nanometers

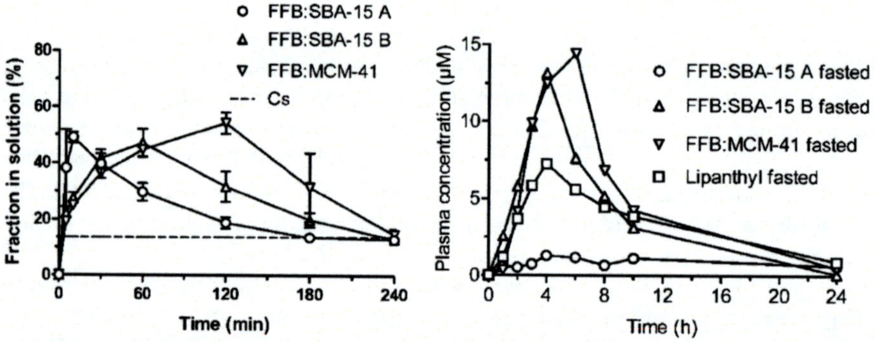

Fig. 10.2 In vitro release profiles of fenofibrate formulated with OMS materials with different pore diameter recorded under supersaturating conditions (*left panel*) and plasma concentration–time profiles of fenofibric acid (the active metabolite of fenofibrate) (*right panel*). The pore diameter of the materials amounts to: 7.3 nm for SBA-15 A, 4.4 nm for SBA-15 B, and 2.7 nm for MCM-41. The results are discussed in the text

OMS offers specific advantages over other supersaturating drug delivery systems in that the rate at which supersaturation occurs can be controlled. This is illustrated in a study on the effect of release rate from OMS on rate and extent of dissolution and absorption of the poorly soluble drug fenofibrate (Fig. 10.2) [23]. In vitro studies employing materials with the largest pore size showed an immediate, burst-like release, followed by rapid supersaturation of the dissolution medium. However, drug levels in solution then fell rapidly to equilibrium solubility levels. In contrast, materials with smaller pore diameter released fenofibrate more gradually and exhibited sustained supersaturation. This trend was reflected in vivo (rats, fasted state): the formulation with the smallest pore size (and thus, slowest release rate) exhibited the highest extent of absorption. Furthermore, the extent of absorption of the slow-releasing OMS-based formulations was higher than that of a commercially available product incorporating micronized drug (Lipanthyl®). OMS thus offers the potential to control the degree of supersaturation to maximize availability for absorption. For other supersaturating drug delivery systems, exerting such tight control over the drug release process may be much more complicated (solid dispersions, high-energy salt forms) or even impossible (lipid-based systems).

Capability to precisely design drug release rate has led many authors to conclude that OMS constitutes a valuable controlled release technology. At this time, there are no studies demonstrating that controlled release from OMS provides sustained blood levels in humans. Furthermore, various drug delivery systems exist that enable zero-order drug release. By contrast, drug release from OMS more closely resembles first-order kinetics. Although some authors have altered in vitro release kinetics by functionalizing the silica surface of an OMS material [29, 30], an in vivo proof of concept for such systems is still lacking.

However, zero-order release in vivo may not always be most appropriate. More rapid release after dosage (as could be obtained with a first order-type release profile) may be more suitable for drugs that are rapidly eliminated. It may be necessary to provide an effective therapeutic plasma concentration at the outset that is then maintained by the slower, later release (as would be provided by first-order kinetics). A modified release system has to take account of the pharmacokinetics of the drug and its therapeutic plasma window. What OMS-based systems may offer is very consistent release behavior that may translate to less intersubject variation, as would be important for medications with narrow therapeutic windows or rapid metabolism and clearance.

10.3 System Design

OMS evaluation and usage has to date been largely limited to laboratory studies. Commercial medications incorporating such technology have not yet been developed. Whenever in vivo (animal) studies were conducted, the drug-loaded silica powder was filled into hard gelatin capsules and administered as such, without further downstream processes. Thus, the design of OMS-based formulations is, as yet, very basic. However, a variety of OMS materials have been used for drug delivery applications and each of these materials exhibit specific features that can affect capability to formulate or pharmaceutical performance. This section gives a brief overview of how OMS materials are manufactured, and how differences in material characteristics might affect applicability.

OMS is manufactured by a sol-gel-derived process. Surfactants or polymers are included as a template for the polymerizing silica. After polymerization, the

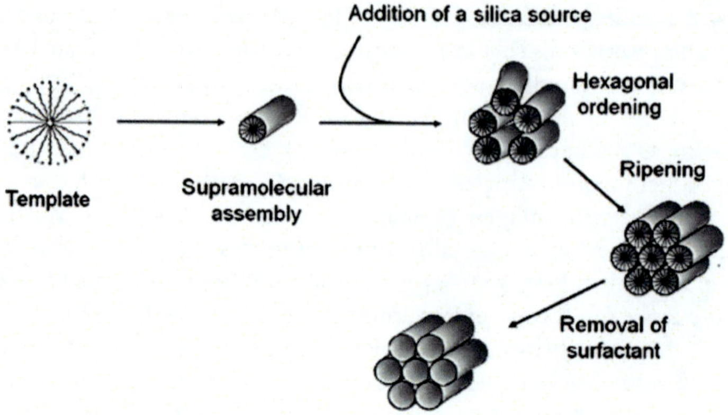

Fig. 10.3 Schematic for the synthesis of OMS

template is removed by chemical or thermal treatment, creating pores. The general pathway is depicted schematically in Fig. 10.3.

Template-assisted synthesis provides a material possessing an array of uniform pores that can be designed to diameters ranging from 2 nm to several tens of nanometers. The term "ordered" denotes that such materials exhibit long-range ordering (reflected in their X-ray diffraction patterns) [31, 32], although the pore walls comprise amorphous silica. Frequently used OMS materials for drug delivery are MCM-41 and SBA-15.

MCM-41 (after Mobil Composition of Matter) was first prepared in 1992 [33]. A quaternary ammonium surfactant is used as the structure-directing agent. MCM-41 possesses a regular array of uniform channels whose dimensions can be tailored (in the range 1.6–10 nm) through the choice of surfactant, auxiliary chemicals, and reaction conditions. The length of the alkyl chain of the surfactant determines pore diameter. Due to its relatively small pore diameter, MCM-41 has frequently been used in studies where slow drug release was envisaged.

SBA-15 (after Santa Barbara Amorphous) was first prepared in 1998. It employs amphiphilic poly(alkylene oxide) triblock copolymers as templates [34]. Appropriate polymer selection and processing conditions can provide pore diameters in the range of 5–30 nm. A transmission electron microscopy picture of an SBA-15 particle is depicted in Fig. 10.4. Its larger pore size makes SBA-15 the material of choice for fast drug release. In addition to uniform mesopores, SBA-15 possesses a complementary system of disordered micropores (diameter <2 nm) that protrude from the mesopore wall. Depending on the synthesis conditions, these micropores may or may not form interconnections between neighboring mesopores. The pore walls of SBA-15 are thicker (3–6 nm) than those of MCM-41 (1 nm), which is reflected in the higher hydrothermal stability of the former [34]. The relatively wide pore diameter of SBA-15, together with its high internal pore volume enable high drug loadings, reportedly up to 40% [35, 36].

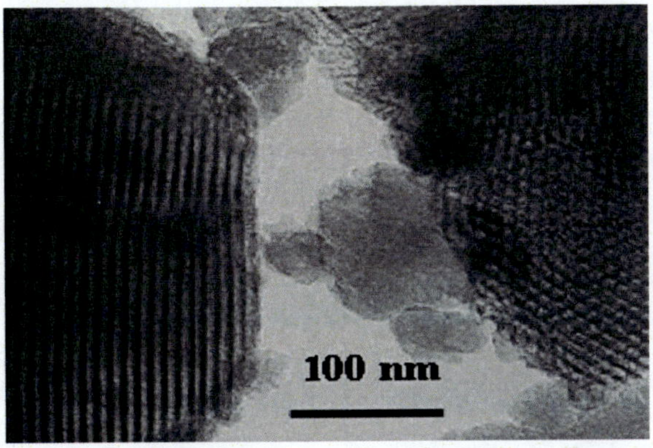

Fig. 10.4 Transmission electron microscopy picture of an SBA-15 particle, illustrating the array of uniform pores (*left hand side of the picture*) and the honeycomb type structure (*right hand side*)

Table 10.1 Mesoporous silica materials for drug delivery

	Pore diameter[a] (nm)	Surface area[a] (m^2/g)	Pore volume[a] (cm^3/g)	Structure	Literature example[b]
SBA-15	5–25	700–1,000	0.5–1.25	Hexagonally ordered, 1D	[9]
MCM-41	2–10	1,000–1,500	0.7–1.2	Hexagonally ordered, 1D	[5]
MCM-48	2–6	1,000–1,500	0.7–1.2	Cubic, 3D	[37]
TUD-1	2.5–25	500–1,000	0.6–1.7	Foam-like, 3D	[38]
FSM-16	1.5–3.2	700–1,000	0.3–0.8	Hexagonally ordered, 1D	[39]

[a] The range indicates the values that are typically encountered for drug delivery applications. Modification of template and synthesis parameters may result in more extreme values than those reported in this table

[b] Reference to an article where the material has been used in the context of drug delivery

Other mesoporous silica-based materials that have been used for drug delivery include:

- TUD-1 exhibiting a foam-like mesopore network.
- FSM-16 a highly ordered mesoporous silicate prepared via surfactant intercalation in a layered silica source.
- MCM-48 exhibiting a pore network comprising two independent and intricately interwoven systems of mesoporous channels.

The properties of these and other mesoporous materials are summarized in Table 10.1.

10.4 Release Mechanisms

Drug release from OMS-based formulations can be divided into four consecutive steps:

- Penetration of release medium into the pores.
- Drug dissolution/displacement from the silica surface.
- Diffusion of the drug through the solvent-filled pores.
- Diffusion of the released drug into the bulk release medium.

Several studies have indicated that drug release from OMS-based systems is via controlled diffusion. The most commonly used model to describe drug release kinetics is the Higuchi equation [40]. This assumes that drug release is diffusion-controlled; the fractional release (Q) of a drug from an insoluble carrier matrix is directly correlated to the square root of the immersion time (t)

$$Q = k_h \cdot t^{0.5},$$

where the release rate constant k_h comprises a variety of factors, including the fractional porosity (ratio pore volume/material volume), the matrix tortuosity, the solubility of the drug in the release medium, the initial drug load in the matrix, and the diffusion coefficient of the drug in the medium. The multitude of datasets that are well described by the classical Higuchi equation indicate that drug release from OMS is diffusion-controlled.

However, most studies with "poorly soluble" drugs have used ibuprofen as a model compound [5, 24, 36, 37]. Although ibuprofen is poorly soluble at low pH, its solubility increases sharply with increasing pH (ibuprofen is a weak acid with a pK_a of 3.9). Studies using ibuprofen were conducted in aqueous buffers of pH 5.5 or higher, i.e. under conditions where ibuprofen was virtually completely ionized, and thus exhibited high solubility. These studies invariably exhibited diffusion-controlled drug release. However, for truly poorly soluble drugs, the rate of water penetration may also be a limiting factor for drug release. The effect of water penetration on drug release is not clearly established, mainly due to experimental limitations. To maintain sink conditions, release media are usually supplemented with surfactants. In addition to creating a driving force for dissolution through micellar solubilization, the addition of surfactants also enhances wettability, facilitating water penetration. On the other hand, creating sink conditions for low solubility drugs without adding surfactants requires excessively high volumes of release medium (or alternatively: extremely low amounts of drug, which complicates analysis).

Several studies have focused on the functionalization of the silanol groups covering the silica surface in OMS materials, to promote interaction between drug and carrier and modify drug release. For such systems, drug release is not purely diffusion-controlled. The Korsmeyer–Peppas model may be used to describe release kinetics [41]. This is basically an extension of the Higuchi equation, incorporating a release exponent, *n*, which gives an indication of the drug release mechanism

$$Q = k \cdot t^n,$$

where *k* is the kinetic release constant incorporating structural and geometrical characteristics of the matrix, *t* the immersion time, and *n* the release exponent describing the drug release mechanism.

If $n = 0.5$, the Korsmeyer–Peppas equation is reduced to the Higuchi equation. If $0.5 > n > 1$, release rate is governed by anomalous diffusion. If $n = 1$, release is zero-order. The release exponent from OMS materials with a functionalized surface usually varies between 0.5 and 1, indicating that the release is governed by anomalous diffusion.

Electron microscopy mapping and optical single-molecule tracking experiments have been used to determine how a single luminescent dye molecule travels through linear or strongly curved sections of a mesoporous channel system [42]. This methodology provides detailed insights into the dynamics of molecules in OMS, which may help in optimizing the matrix parameters in order to obtain the desired release profile.

With the increasing importance of OMS in the context of drug delivery, efforts are also being made to model drug release from OMS. A recent study reported on a nonlinear diffusion model to predict drug release from two-dimensional ordered porous media. The model successfully reproduced release profiles obtained in an experimental study [43]. Such predictive models may significantly facilitate the development of OMS-based drug delivery systems.

10.5 Drug Candidates

Current knowledge suggests that adsorption on OMS can enhance and control dissolution rate of a wide variety of compounds. However, the number of in vivo studies using drugs incorporating OMS technology remains limited. Although release has been suggested to be pH-independent [44], the fate of released drug is not, and for most poorly soluble drugs, precipitation following release is likely. Thus, the in vivo performance of OMS may depend not only on the release rate, but also (and even more so) on the rate and extent of drug precipitation following release.

Mellaerts et al. compared the biopharmaceutical performance of OMS loaded with the poorly soluble weak base itraconazole [10], against that of the commercial solid dispersion product, Sporanox. In rabbits and dogs, there were no statistically significant differences in terms of rate and extent of absorption between the OMS formulation and Sporanox (Fig. 10.5).

In a subsequent study in rats, Van Speybroeck et al. demonstrated that the performance of the OMS-itraconazole formulation was enhanced by coadministering the polymeric precipitation inhibitor hydroxypropyl methylcellulose [35].

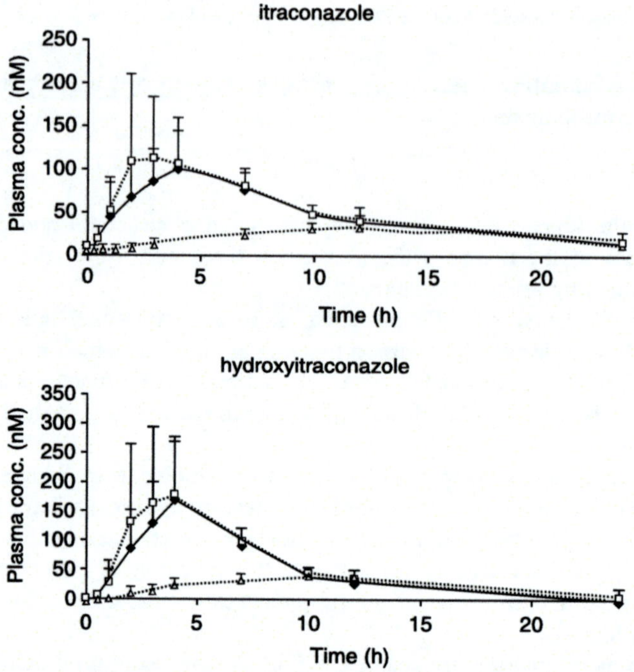

Fig. 10.5 Plasma concentration–time profiles of itraconazole and its major active metabolite hydroxyitraconazole in rabbits, after administration of crystalline drug (*open triangle*), an SBA-15-based formulation (*filled diamond*), and Sporanox (*open square*)

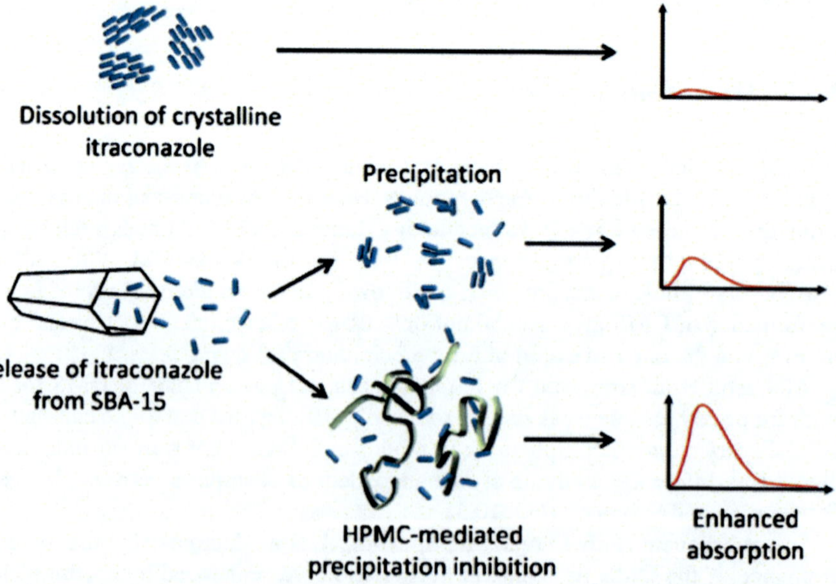

Fig. 10.6 Schematic illustrating the effect of HPMC addition on the biopharmaceutical performance of SBA-15

This polymeric component improved absorption by inhibiting itraconazole precipitation in the small intestine. Figure 10.6 illustrates this effect.

Fenofibrate, another poorly soluble compound has also been evaluated in vivo (rats) as described in Sect. 10.2.2, where enhanced absorption was ascribed to stabilization of released drug as supersaturated solution [23].

10.6 Formulation and Manufacturing

Several methods have been used to load drugs into OMS, but no studies have focused on the downstream processability of drug-loaded OMS material. The most frequently used approach for loading drugs is to soak the carrier in a concentrated solution. Typically, solvents such as methylene chloride or ethanol, which have a high solubilizing capacity for many poorly soluble drugs, are used for this purpose. Additionally and importantly, the vapor pressures of such solvents are high such that residual solvent can be readily removed. The time needed to reach maximal occupation of the silica surface can be determined from adsorption isotherms. Typically, the carrier is soaked in the drug-containing solution for several hours. The loaded carrier is then removed from the solvent and dried under vacuum at elevated temperature.

An alternative solvent-based technique involves impregnating the carrier with concentrated drug solution ("incipient wetness process"). Successive cycles of impregnation-drying can significantly enhance drug loading such that pore occupancy is virtually complete [8]. Such loading capacity is not feasible using the soaking procedure. Any excess drug on the external surface after impregnation may be removed by washing, provided that such treatment does not also extract drug from pores.

Melt-adsorption is also an effective loading technique. A physical mixture of drug and carrier is heated above the melting point of the drug, such that molten drug penetrates the mesopores [45, 46]. Solvent use is avoided. The technique is only applicable to thermally stable drugs viz. those not degraded on melting. Other techniques include adsorption using supercritical carbon dioxide [47], vapor phase adsorption [48], and comilling drug and carrier [49].

Little scale-up work has been reported for drug loading processes. However, operations, equipment and solvents hardly differ from those utilized during primary (drug substance) manufacture (mixing, liquid–solid separation, drying, etc.). Co-spray drying has also been used [50]. Co-spray drying SBA-15, dispersed in a concentrated solution of ibuprofen in ethanol, achieved drug loads as high as 50%. Nitrogen physisorption and transmission electron microscopy demonstrated that the drug was incorporated in the SBA-15 pores. However, there may be hazards associated with spray drying flammable organic solvents.

In general terms, there is no reason why some of the loading techniques cannot be used as part of the final stage of primary (active ingredient) manufacture, where equipment and operating systems for solvents processing and handling are more prevalent than for secondary (dosage form) manufacture.

Little information is available on downstream processing to dosage forms. Intuitively, it would seem unlikely that technical challenges are insurmountable or even complex. Most pharmaceutical secondary processes involve solids handling and OMS-based products should not be very different (except possibly their low density and (most probably) high inclusion levels). Pelletization of (unloaded) OMS and compression to compacts have been reported [51, 52]. These studies showed that the pressures typically applied in direct compression (i.e. around 100 MPa) may lead to a loss of surface area and pore volume due to pore collapse as well as pore blockage. Further work is required to explore the behavior of drug-loaded materials on compression, and how compression might affect properties such as drug release or disintegration. In particular granulation (melt or wet) of drug-loaded OMS may be critical, as molten binder or granulation solvents may extract drug from the mesopores.

10.7 Conclusion

Ordered mesoporous silicates are relatively newcomers to the field of enhanced and controlled oral absorption. They offer a number of advantages and possibilities for the plethora of poorly soluble drugs that now seem to emanate from Drug Discovery programs. These include:

- Capability to disperse the drug in the noncrystalline state over a wide surface area, thereby enhancing its propensity to pass from the solid to the solvated state.
- Retention (stabilization) in the noncrystalline form over time by virtue of geometric confinement of the adsorbed drug molecules.
- Capability to consistently define and control silicate pore size, possibly contributing to consistent drug release rate.

There are also promising indicators that, by appropriate use of adjuvants, released drug may remain in solution in the dissolution medium (supersaturation state) thereby enhancing and controlling absorption.

Many of these concepts and possibilities remain to be proven in human studies but data in animal studies are promising. Many facets of large-scale manufacture also remain to be explored. Nevertheless, findings to date suggest that ordered mesoporous silicates can become a valuable addition to the armamentarium of the scientist engaged in the design of systems to enhance and control oral drug delivery.

Acknowledgments The authors acknowledge the Flemish Government for financial support from the Industrial Research Fund (IOF). MVS acknowledges the Flemish Institute for the Promotion of Innovation through Science and Technology (IWT-Vlaanderen) for a PhD grant. RM acknowledges the Research Fund-Flanders (FWO- Vlaanderen) for a postdoctoral research fellowship. JAM acknowledges the Flemish Government for long-term structural funding (Methusalem).

References

1. Lipinski CA, Lombardo F, Dominy BW, Feeney PJ (2001) Experimental and computational approaches to estimate solubility and permeability in drug discovery and development settings. Adv Drug Deliv Rev 46:3–26
2. Monkhouse DC, Lach JL (1972) Use of adsorbents in enhancement of drug dissolution. I. J Pharm Sci 61:1430–1435
3. Monkhouse DC, Lach JL (1972) Use of adsorbents in enhancement of drug dissolution. II. J Pharm Sci 61:1435–1441
4. Vallet-Regi M, Ramilla A, del Real RP, Perez-Pariente J (2001) A new property of MCM-41: drug delivery system. Chem Mater 13:308–311
5. Horcajada P, Rámila A, Pérez-Pariente J, Vallet-Regi M (2004) Influence of pore size of MCM-41 matrices on drug delivery rate. Microporous Mesoporous Mater 68:105–109
6. Doadrio AL, Sousa EMB, Doadrio JC, Pérez Pariente J, Izquierdo-Barba I, Vallet-Regí M (2004) Mesoporous SBA-15 HPLC evaluation for controlled gentamicin drug delivery. J Control Release 97:125–132
7. Tozuka Y, Oguchi T, Yamamoto K (2003) Adsorption and entrapment of salicylamide molecules into the mesoporous structure of folded sheets mesoporous material (FSM-16). Pharm Res 20:926–930
8. Charnay C, Bégu S, Tourné-Péteilh C, Nicole L, Lerner DA, Devoisselle JM (2004) Inclusion of ibuprofen in mesoporous templated silica: drug loading and release property. Eur J Pharm Biopharm 57:533–540
9. Mellaerts R, Aerts CA, Van Humbeeck J, Augustijns P, Van den Mooter G, Martens JA (2007) Enhanced release of itraconazole from ordered mesoporous SBA-15 silica materials. Chem Commun 1375–1377
10. Mellaerts R, Mols R, Jammaer JAG, Aerts CA, Annaert P, Van Humbeeck J, Van den Mooter G, Augustijns P, Martens JA (2008) Increasing the oral bioavailability of the poorly water soluble drug itraconazole with ordered mesoporous silica. Eur J Pharm Biopharm 69:223–230
11. Leuner C, Dressman J (2000) Improving drug solubility for oral delivery using solid dispersions. Eur J Pharm Biopharm 50:47–60
12. Serajuddin AT (1999) Solid dispersion of poorly water-soluble drugs: early promises, subsequent problems, and recent breakthroughs. J Pharm Sci 88:1058–1066
13. Alcoutlabi M, McKenna GB (2005) Effects of confinement on material behaviour at the nanometer size scale. J Phys Condens Matter 17:461–524
14. Alba-Simionesco C, Coasne B, Dosseh G, Dudziak G, Gubbins K, Radhakrishnan R, Sliwinska-Bartkowiak M (2006) Effects of confinement on freezing and melting. J Phys Condens Matter 18:15–68
15. Kaneko K, Watanabe A, Iiyama T, Radhakrishan R, Gubbins KE (1999) A remarkable elevation of freezing temperature of CCl4 in graphitic micropores. J Phys Chem B 103:7061–7063
16. Sliwinska-Bartkowiak M, Dudziak G, Sikorski R, Gras R, Radhakrishnan R, Gubbins KE (2001) Melting/freezing behavior of a fluid confined in porous glasses and MCM-41: dielectric spectroscopy and molecular simulation. J Chem Phys 114:950–962
17. Sliwinska-Bartkowiak M, Dudziak G, Gras R, Sikorski R, Radhakrishnan R, Gubbins KE (2001) Freezing behavior in porous glasses and MCM-41. Colloids Surf A Physicochem 187–188:523–529
18. Azaïs T, Tourné-Péteilh C, Aussenac F, Baccile N, Coelho C, Devoiselle J, Babonneau F (2006) Solid-state NMR study of ibuprofen confined in MCM-41 material. Chem Mater 18:6382–6390
19. Mellaerts R, Houthoofd K, Elen K, Chen H, Van Speybroeck M, Van Humbeeck J, Augustijns P, Mullens J, Van den Mooter G, Martens JA (2010) Aging behavior of pharmaceutical formulations of itraconazole on SBA-15 ordered mesoporous silica carrier material. Microporous Mesoporous Mater 130:154–161

20. Van Speybroeck M, Barillaro V, Thi TD, Mellaerts R, Martens J, Van Humbeeck J, Vermant J, Annaert P, Van den Mooter G, Augustijns P (2009) Ordered mesoporous silica material SBA-15: a broad-spectrum formulation platform for poorly soluble drugs. J Pharm Sci 98:2648–2658
21. Janssens S, Van den Mooter G (2009) Review: physical chemistry of solid dispersions. J Pharm Pharmacol 61:1571–1586
22. Babonneau F, Yeung L, Steunou N, Gervais C, Ramila A, Vallet-Regi M (2004) Solid state NMR characterisation of encapsulated molecules in mesoporous silica. J Sol Gel Sci Technol 31:223
23. Van Speybroeck M, Mols R, Mellaerts R, Thi TD, Martens JA, Van Humbeeck J, Annaert P, Van den Mooter G, Augustijns P (2010) Enhanced absorption of the poorly soluble drug fenofibrate by tuning its release rate from ordered mesoporous silica. Eur J Pharm Sci 41(5):623–30
24. Andersson J, Rosenholm J, Areva S, Lindén M (2004) Influences of material characteristics on ibuprofen drug loading and release profiles from ordered micro- and mesoporous silica matrices. Chem Mater 16:4160–4167
25. Qu F, Zhu G, Huang S, Li S, Sun J, Zhang D, Qiu S (2006) Controlled release of Captopril by regulating the pore size and morphology of ordered mesoporous silica. Microporous Mesoporous Mater 92:1–9
26. Grün M, Unger KK, Matsumoto A, Tsutsumi K (1999) Novel pathways for the preparation of mesoporous MCM-41 materials: control of porosity and morphology. Microporous Mesoporous Mater 27:207–216
27. Kosuge K, Sato T, Kikukawa N, Takemori M (2004) Morphological control of rod- and fiber-like SBA-15 type mesoporous silica using water-soluble sodium silicate. Chem Mater 16:899–905
28. Brouwers J, Brewster ME, Augustijns P (2009) Supersaturating drug delivery systems: the answer to solubility-limited oral bioavailability? J Pharm Sci 98:2549–2572
29. Song S, Hidajat K, Kawi S (2005) Functionalized SBA-15 materials as carriers for controlled drug delivery: influence of surface properties on matrix-drug interactions. Langmuir 21:9568–9575
30. Rosenholm JM, Lindén M (2008) Towards establishing structure-activity relationships for mesoporous silica in drug delivery applications. J Control Release 128:157–164
31. Beck JS, Vartuli JC, Roth WJ, Leonowicz ME, Kresge CT, Schmitt KD, Chu CT, Olson DH, Sheppard EW, McCullen SB, Higgins JB, Schlenker JL (1992) A new family of mesoporous molecular sieves prepared with liquid crystal templates. J Am Chem Soc 114:10834–10843
32. Zhao D, Feng J, Huo Q, Melosh N, Fredrickson G, Chmelka B, Stucky G (1998) Triblock copolymer syntheses of mesoporous silica with periodic 50 to 300 angstrom pores. Science 279:548–552
33. Kresge CT, Leonowicz ME, Roth WJ, Vartuli JC, Beck JS (1992) Ordered mesoporous molecular sieves synthesized by a liquid-crystal template mechanism. Nature 359:710–712
34. Galarneau A, Nader M, Guenneau F, DiRenzo F, Gedeon A (2007) Understanding the stability in water of mesoporous SBA-15 and MCM-41. J Phys Chem 111:8268–8277
35. Van Speybroeck M, Mols R, Mellaerts R, Thi TD, Martens JA, Humbeeck JV, Annaert P, Van den Mooter G, Augustijns P (2010) Combined use of ordered mesoporous silica and precipitation inhibitors for improved oral absorption of the poorly soluble weak base itraconazole. Eur J Pharm Biopharm 75:354–365
36. Heikkilä T, Salonen J, Tuura J, Kumar N, Salmi T, Murzin DY, Hamdy MS, Mul G, Laitinen L, Kaukonen AM, Hirvonen J, Lehto V (2007) Evaluation of mesoporous TCPSi, MCM-41, SBA-15, and TUD-1 materials as API carriers for oral drug delivery. Drug Deliv 14:337–347
37. Izquierdo-Barba I, Martinez A, Doadrio AL, Pérez-Pariente J, Vallet-Regí M (2005) Release evaluation of drugs from ordered three-dimensional silica structures. Eur J Pharm Sci 26:365–373
38. Heikkilä T, Salonen J, Tuura J, Hamdy MS, Mul G, Kumar N, Salmi T, Murzin DY, Laitinen L, Kaukonen AM, Hirvonen J, Lehto V (2007) Mesoporous silica material TUD-1 as a drug delivery system. Int J Pharm 331:133–138

39. Tozuka Y, Wongmekiat A, Kimura K, Moribe K, Yamamura S, Yamamoto K (2005) Effect of pore size of FSM-16 on the entrapment of flurbiprofen in mesoporous structures. Chem Pharm Bull (Tokyo) 53:974–977
40. Higuchi T (1963) Mechanism of sustained-action medication. Theoretical analysis of rate of release of solid drugs dispersed in solid matrices. J Pharm Sci 52:1145–1149
41. Korsmeyer RW, Gurny R, Doelker E, Buri P, Peppas NA (1983) Mechanisms of solute release from porous hydrophilic polymers. Int J Pharm 15:25–35
42. Zurner A, Kirstein J, Doblinger M, Brauchle C, Bein T (2007) Visualizing single-molecule diffusion in mesoporous materials. Nature 450:705–708
43. Godec A, Gaberscek M, Jamnik J, Merzel F (2009) Nonlinear diffusion in two-dimensional ordered porous media based on a free volume theory. J Chem Phys 131:234106–234112
44. Mellaerts R, Mols R, Kayaert P, Annaert P, Van Humbeeck J, Van den Mooter G, Martens JA, Augustijns P (2008) Ordered mesoporous silica induces pH-independent supersaturation of the basic low solubility compound itraconazole resulting in enhanced transepithelial transport. Int J Pharm 357:169–179
45. Mellaerts R, Jammaer JAG, Van Speybroeck M, Chen H, Van Humbeeck J, Augustijns P, Van den Mooter G, Martens JA (2008) Physical state of poorly water soluble therapeutic molecules loaded into SBA-15 ordered mesoporous silica carriers: a case study with itraconazole and ibuprofen. Langmuir 24:8651–8659
46. Kinoshita M, Baba K, Nagayasu A, Yamabe K, Shimooka T, Takeichi Y, Azuma M, Houchi H, Minakuchi K (2002) Improvement of solubility and oral bioavailability of a poorly water-soluble drug, TAS-301, by its melt-adsorption on a porous calcium silicate. J Pharm Sci 91:362–370
47. Smirnova I, Mamic J, Arlt W (2003) Adsorption of drugs on silica aerogels. Langmuir 19:8521–8525
48. Konno T, Kinuno K (1989) Physical and chemical changes of medicinals in mixtures with adsorbents in the solid state. II. Application of reduced pressure treatment for the improvement of dissolution of flufenamic acid. Chem Pharm Bull (Tokyo) 37:2481–2484
49. Bahl D, Hudak J, Bogner RH (2008) Comparison of the ability of various pharmaceutical silicates to amorphize and enhance dissolution of indomethacin upon co-grinding. Pharm Dev Technol 13:255–269
50. Shen S, Ng WK, Chia L, Dong Y, Tan RBH (2010) Stabilized amorphous state of ibuprofen by co-spray drying with mesoporous SBA-15 to enhance dissolution properties. J Pharm Sci 99:1997–2007
51. Cassiers K, Linssen T, Mathieu M, Benjelloun M, Schrijnemakers K, Van Der Voort P, Cool P, Vansant EF (2002) A detailed study of thermal, hydrothermal, and mechanical stabilities of a wide range of surfactant assembled mesoporous silicas. Chem Mater 14:2317–2324
52. Hartmann M, Vinu A (2002) Mechanical stability and porosity analysis of large-pore SBA-15 mesoporous molecular sieves by mercury porosimetry and organics adsorption. Langmuir 18:8010–8016

Chapter 11
Geometric Release Systems: Principles, Release Mechanisms, Kinetics, Polymer Science, and Release-Modifying Material

Paolo Colombo, Gaia Colombo, and Christine Cahyadi

Abstract Geometrical features such as shape and surface area of a drug-releasing matrix affect drug release kinetics by changing diffusion rates across the matrix, lengthening the diffusion pathway through drug–matrix composition or simply presenting a different surface area to the dissolution medium. Such variables can change during the dissolution process, due to drug depletion or erosion or dissolution of release-modifying components.

Such phenomena can help the development of novel dosage units since they present opportunities to capitalize on shape and surface effects designing a matrix that optimally delivers drug at the required rate. The historical development and state-of-the-art of geometrically designed dosage forms are presented in this chapter.

11.1 Introduction

Hydrophilic matrix tablets, osmotic systems, and coated multiparticulates comprise the most prevalent modified release oral dosage forms. Materials, technologies, and release mechanisms for controlled drug delivery are selected depending on the site and desired rate or mode of drug release. Swellable matrices have many advantages, including simplicity of fabrication and drug delivery rate definition, development, and optimization. Furthermore, the availability of different grades of swellable excipients can help the design of appropriate release kinetics by capitalizing on non-Fickian diffusion behavior. These considerations can make swellable matrices attractive for controlling drug release.

P. Colombo (✉)
Dipartimento Farmaceutico, Via G.P.Usberti 27/A, 43100, Parma, Italy
e-mail: farmac2@unipr.it

C.G. Wilson and P.J. Crowley (eds.), *Controlled Release in Oral Drug Delivery*, Advances in Delivery Science and Technology, DOI 10.1007/978-1-4614-1004-1_11,

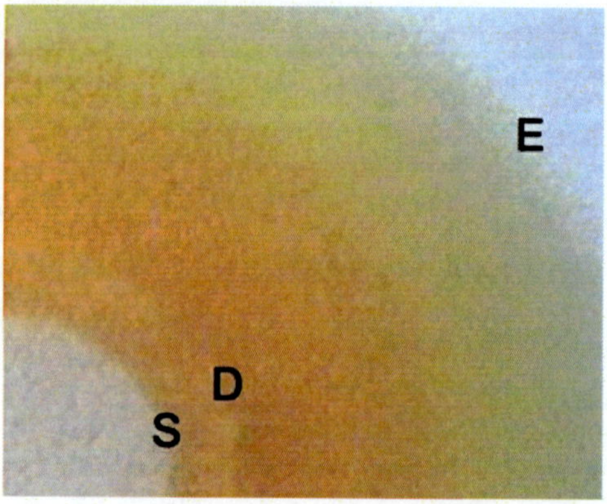

Fig. 11.1 Sector of a disk matrix made of HPMC and the orange colored drug Buflomedil pyridoxal phosphate, in which water penetrated from the lateral side only. *E* erosion front, *D* diffusion front, *S* swelling front. *Bottom left corner*: matrix glassy core; *Top right corner*: solvent

Swellable matrix compacts are moving boundary systems in which control of drug release is dictated by rate of hydration, swelling, and dissolution of the polymer/drug mixture. This sequence leads to boundary formation, i.e., fronts separating different physical regions inside the matrix during swelling (Fig. 11.1). These can determine the drug release kinetics. The swelling front separates glassy from rubbery polymer. In this location, physical stresses are created by polymer chain relaxation that determine matrix volume increase. Hence, swelling controls drug release.

Two other fronts determine release kinetics in a swellable matrix:

- The diffusion front, i.e., the boundary separating solid/undissolved drug from drug in solution; this can coincide with the swelling front.
- The matrix or erosion front, separating the swollen matrix from the solvent.

The distance between the swelling and erosion fronts defines the thickness of the gel layer. This in turn is a characteristic parameter of the diffusion layer for drug transport. Within the gel layer, concentration gradients (profiles) of polymer, water, and drug are established. Such profiles are generally depicted as linear since the swelling develops slowly. Consequent to this slow process, the local concentration may relax the variations in front position. This explains why in certain cases when the fronts move in a synchronized manner and the release area is constant, zero-order release can be obtained from such moving boundary systems. Thus, in swellable matrices the release area, i.e., the matrix surface exposed to the solvent, defines the release rate and kinetics. In general, the exposed area expands until erosion/dissolution phenomena counteract such increase, due to progressive disentanglement of the swollen polymeric chains.

This chapter illustrates how the swelling/release areas of matrices (geometry) can be used to design an appropriate release profile aligned with drug plasma profile requirements. Such control of the releasing area during swelling and dissolution determines release rate and kinetics without modifying the composition of the swellable system.

The concept of geometric manipulation to affect release rate and kinetics was introduced by Langer. Langer used inert matrices where diffusion layer thickness increases during release along with progressive increase of the diffusion front area [1]. Three examples of such geometric manipulation are presented in this chapter, reflecting research activity and experience of the authors.

11.2 Geomatrix®: Partially Coated Swellable Matrices

There have been many attempts to control release kinetics in hydrophilic matrices by manipulating and balancing diffusion and relaxation mechanisms. Zero-order release from a matrix was obtained by designing an appropriate matrix shape [1] or nonuniform drug distribution [2], by using ionic-exchange resins, hydrophobic porous materials [3, 4], hydrophilic soluble polymers capable of modifying the effective diffusivity of drug [5], or by surface cross-linking of the matrix [6] and others. Geometric design of zero-order release in swellable matrices resides in maintaining a constant release area during matrix swelling and drug diffusion. This releasing area is the area of the swollen matrix surface in contact with the dissolution medium. A partially coated matrix, providing a constant releasing area takes the form of a "core-in-cup" system, i.e., a "disk" of drug and swellable/soluble polymer coated with an impermeable film on the lateral surface and on one base (Fig. 11.2).

The coating film is impermeable to water and drug diffusion. On contact with aqueous media, the uncoated base undergoes swelling and erosion. If erosion/dissolution is sufficiently fast, core thickness inside the impermeable polymer cup is reduced maintaining a constant releasing area. Zero-order release kinetics are obtained if fronts are synchronized. Varying the type or amount of swellable/soluble

Fig. 11.2 Picture of different sizes of core-in-cup matrices obtained by coating one base and the size surface of the disk

polymer in the core enables the rate of release to be modulated, as does the area for release. In vitro release rate and in vivo absorption were directly related to releasing area [7].

Mechanisms governing drug release in such a system were explored using swellable polymers (PVA, HPMC, and NaCMC) that interact differently with water (swelling and dissolution). Diclofenac sodium was used as the model drug [8]. The unidirectional swelling induced by the core coating enabled the monitoring of the movement of the erosion and swelling fronts over time. When a soluble polymer such as polyvinylalcohol (PVA) was used, the synchronization of the movement of swelling and erosion fronts provided linear release kinetics of drug from such swelling-activated delivery systems. Such findings indicated that polymer swelling and dissolution in the matrix core governed front movement. Front synchronization was not attained with hydroxypropyl methylcellulose (HPMC) polymer; with sodium carboxymethylcellulose (NaCMC) front synchronization took place, but later than with PVA. Moreover, using drugs with different aqueous solubilities (diclofenac sodium, dyprofilline, and cimetidine) and PVA as polymer, the thickness of gel layer at synchronization increased with drug solubility [8]. Nevertheless, at front synchronization the release rates were the same since the concentration gradients of the differently soluble drugs in the gel were the same.

Hydrogel matrices may not always encounter environments that readily attain synchronization of the fronts, particularly when less soluble polymers are used. In this situation, during drug release, matrix swelling predominates over erosion/dissolution [9].

Geometric control of release rates during swelling can be obtained by applying impermeable coatings to different surfaces/areas of a disk matrix. This is illustrated in the sketch of Fig. 11.3. Case 0 matrices were not coated. Other cases are as follows [10]:

- Case 1: one base was coated
- Case 2: two bases were coated
- Case 3: lateral surface was coated
- Case 4: lateral surface and one base were coated

Applications of impermeable coatings on different surfaces of a matrix containing a water-soluble drug and HPMC as polymer do not alter the basic diffusion characteristics of drug within the matrix. This enables the design of dimensionality driven release systems in which the preferred dimension for release can be changed.

Hence, the matrix composition can remain unchanged, but release rate is altered by partial coating, thereby changing the dimensionality of swelling of the matrix. Swollen matrices present different shapes as a function of the location of the impermeable coating, as illustrated in Fig. 1 of Ref. [11]. This leads to the following observations:

- The uncoated cylindrical matrix (Case 0) exhibited isometric swelling with a propensity for thickness increase on hydration. This is typical of compressed swellable disk matrices.

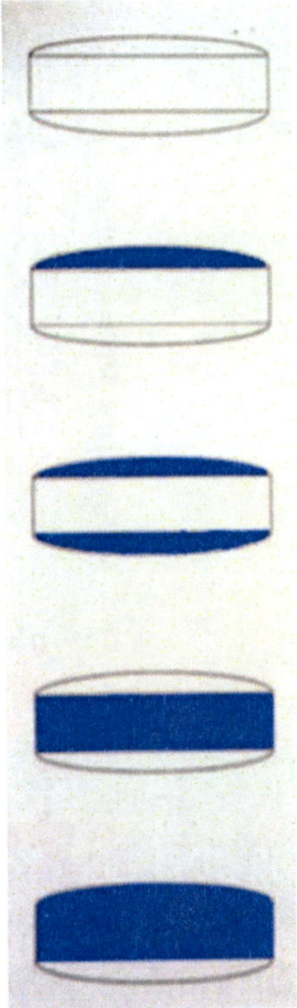

Fig. 11.3 Schematic representation of partially coated matrices. *From the top*: case 0; case 1; case 2; case 3; case 4

- The matrix with one face coated (Case 1) showed a less intense radial swelling with respect to the coated face compared to the uncoated face.
- The matrix with both faces coated (Case 2) exhibited the lowest axial increase in thickness and the largest diameter increase, indicating that swelling was mainly radially orientated.
- Increase in thickness was greatest in the matrix where the cylindrical sides of the compact were coated (Case 3), reflecting axial swelling.
- The matrix with the side and one face coated (Case 4) exhibited a one-direction axial swelling.

The findings indicated that coating applied to the disk faces changed matrix relaxation in axial or radial directions. The association between matrix swelling behavior and drug release was studied by changing the release areas for each of the partially coated matrices as shown in Fig. 11.3. The coating extension decreased in the order Case 4 > Case 2 > Case 1 > Case 3 > Case 0. Release rate decreased as the coat coverage increased. Plots of amount of drug released versus releasing area were linear in all cases (Fig. 11.4).

Surprisingly it was found that, as coating coverage was increased, a greater amount of drug was released per unit releasing area of swollen matrix. This indicated that swelling enhanced drug release by increasing the contribution of the relaxation mechanism to drug transport.

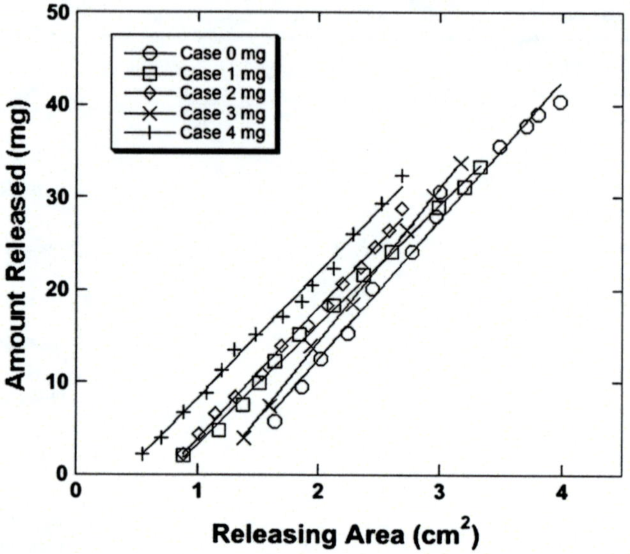

Fig. 11.4 Releasing area plotted versus the amount of drug released for the uncoated and partially coated matrices

Release rate profiles from the five systems were typical of swellable matrices. Release rate profiles over time were consistent for each system, being greatest with compacts where no coat was applied and least where all but one base surface were coated (Fig. 11.5).

The linear relationship between the swollen releasing area and the amount of drug released suggested that matrix swelling rates dictated release kinetics. Normalizing instantaneous release rates with the time-corresponding releasing area values revealed that swelling kinetics determined release kinetics. In fact, fluxes (amounts of drug released per unit area and time) for the five systems were practically the same (Fig. 11.6), despite the differing release rates from the complete units shown in Fig. 11.5. In conclusion, changes in releasing area due to matrix swelling determine rates and kinetics of delivery from swellable matrices.

Dimensionless numbers are frequent in drug transport analysis. A commonly used number is the Swelling Interface Number (Sw), defining anomalous release behaviors of swellable systems. The Sw, in analogy with the Peclet number, compares a pseudoconvective process with a diffusion-based process. The Sw value relates the contributions of penetrant transport (water) with the solute transport (drug) according to the expression:

$$\mathrm{Sw} = v\frac{\partial}{D},$$

where v is the penetrant front velocity, δ the swollen layer thickness, and D the drug diffusivity.

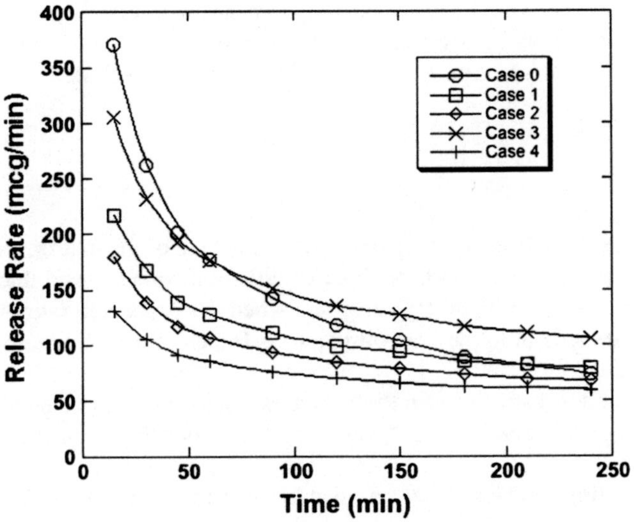

Fig. 11.5 Variation of release rate versus time for the uncoated and partially coated matrices

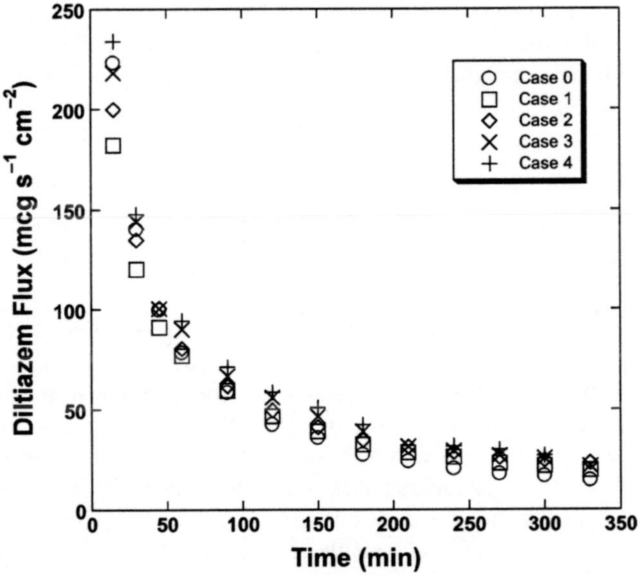

Fig. 11.6 Flux of the uncoated and partially coated matrices plotted versus time

A similar dimensionless number can be defined, based on the increase of matrix release area consequent to three-dimensional expansion due to swelling. This is relevant to release from matrices because the Sw value is inherently related to one-dimensional transport (as in thin disks or films). The new dimensionless number, in contrast, describes the behavior of three-dimensional systems such as matrices. The dimensional analogy between the releasing area rate and diffusion coefficient allows an improved Swelling Area Number, Sa, to be defined viz.:

$$\mathrm{Sa} = \frac{\mathrm{d}A}{\mathrm{d}t} \times \frac{1}{D},$$

where $\mathrm{d}A/\mathrm{d}t$ is the rate of increase of the releasing area of the matrix.

Sa values greater than 1 indicate Fickian diffusion; values lower than 1 indicate a relaxation-controlled (Case II) transport. When the expansion rate of the matrix drops to values similar to the drug diffusion coefficient, the release rate approaches a constant value.

The Sa numbers calculated for the five cases discussed here ranged from 30 to 10 as the coating extension increased, since the rate of development of releasing area decreased.

Matrix coating can also utilize permeable and semipermeable films, particularly for systems coated on the side and on one base, i.e., a core-in-cup system [13]. Permeable coating films can increase the delivery from partially coated matrices by adding other contributions to their swelling-dependent delivery mechanism. Semipermeable and permeable cups increase release rate, compared to an impermeable cup. However, the systems described so far require the application by casting of an impermeable film on a portion of the matrix. As an alternative, the application by compression of a polymeric barrier layer to both bases of the core disk was developed [12, 14].

11.3 Assembly of Release Modules: The "Click Technology" for Oral Drug Delivery Systems

Pharmacotherapy requires new platforms for controlling drug release, in particular systems that can deliver different drugs from the same dosage form at patient-specific or other desirable release rates. Such drivers stimulated the development of "release modules assembly" technology [15]. These systems are constructed by assembling two or more modules as a single dose unit. Components (modules) programmed to have the requisite release characteristics are first prepared. Two or more such modules, containing different drugs and/or differing release rates of the same drug, are then assembled as a single dosage form. Modules are manufactured as tablets, suitably shaped to facilitate interlocking or other such association with complementary-shaped units. Modules can be assembled in this way to delay, prolong, or otherwise control drug release over time. Assembly of combinations of modules provides one-piece delivery systems with appropriate delivery attributes for the drug payload and patient. The device may be a matrix or a conventional

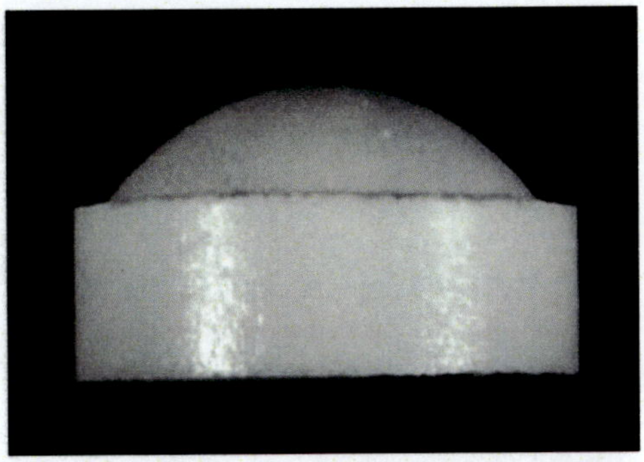

Fig. 11.7 Dome Matrix® module

disintegrating tablet shaped as a disk with convex and concave bases. The axial section of the module appears as a cupola, thus the name "Dome Matrix®" (Fig. 11.7).

This shape enables a number of modules to be assembled as one-unit systems by interlocking the convex base of one module with the concave base of another. Thus, modules containing different doses, units of the same dose or having different release behaviors or even different drugs can be assembled, providing the patient with "all in one" customized therapy.

Delivery kinetics from such systems reflects the differently shaped release surfaces of the individual modules and the surface/volume ratios of the assembled units (as well as reflecting matrix composition). It has also been shown that delivery from assembled units can be influenced by the way modules are oriented in single-unit format. Two basic configurations are:

- Inserting the convex base of one module into the concave base of another provides a system in "piled" or "stacked" configuration. Release rate and kinetics depend on the number of modules in the assembled unit.
- Inserting the concave base of one module with a concave base of a second module can provide a void within the assembled unit. This can cause the system to float when immersed in aqueous media. Two modules in such void configuration exhibited gastro-retentive behavior in humans (prolonged residence in the stomach) due to the unit buoyancy [16]. Average gastric retention time was 214.5 min (range 145–275 min), contrasting with an average time of 96.7 min (range 45–120 min) for a nonfloating system. Buoyancy force reflects the volume of inner empty space and the material density.

The effect of surface shape on swelling and release properties was studied, utilizing compacts with planar (flat), convex and concave bases, compacts being identical in composition. Figure 11.8 presents images of the swollen matrices, showing shape-related effects.

Measurements of the areas of gel in contact with the dissolution medium during swelling, i.e., the "releasing area," were concordant with the visual observations.

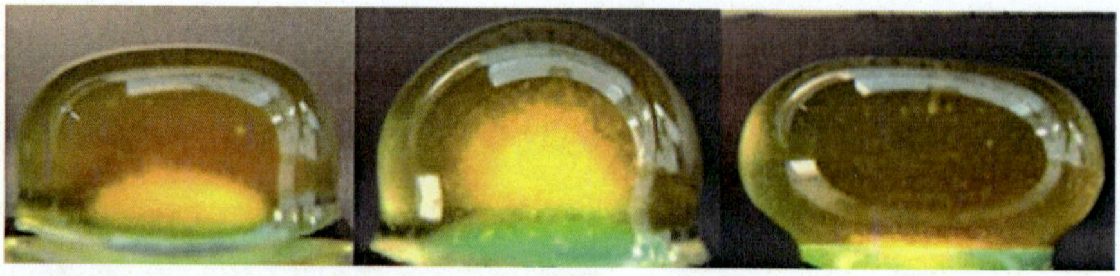

Fig. 11.8 Swollen matrices base at 300 min. *From left*: flat, convex, concave base

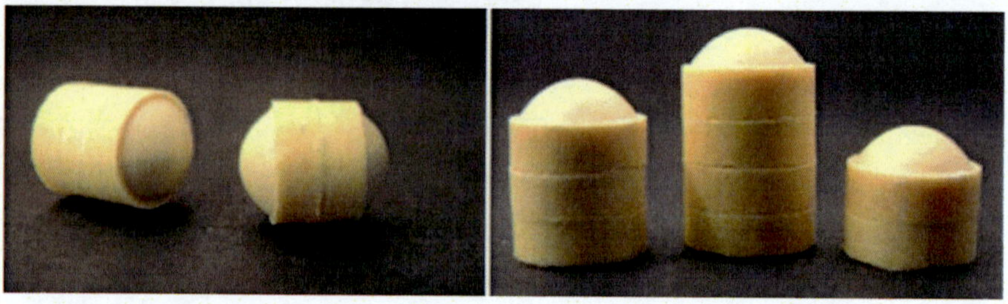

Fig. 11.9 Piled (*right*) and void configuration (*left*) of assembled modules

The convex base exhibited the highest increase in matrix surface area when exposed to dissolution medium (at the erosion front), the concave surface being lowest; the flat base area was intermediate. Observations of the shape of the swollen parts of the matrices suggested that the polymeric chains in the gel entangled differently in the extreme cases:

- The chains expanded along the outline of the curved surface of the convex base.
- For the concave base, the polymer chains were constrained to entanglement in the cavity.

Consequently, different release behaviors were evident from compacts with the three bases [15]. Although the two curved bases had the same initial area (in the dry state), the convex base released the water-soluble model drug buflomedil pyridoxal phosphate much faster than the concave surface. The release profile of the concave base was also lower than that of the flat base, which had the lowest initial area (Figure 6 in [15]). When drug release was expressed in flux units (amount released per unit surface area per unit time), flux values were practically identical in all cases. Such behavior again indicated that release rates were dictated by the swelling rates of the matrix bases.

Assembly of two swellable matrix modules can give delivery systems containing double amounts of drug but having a lower surface area/volume ratio relative to the two individual (nonassembled) modules. Type of assembly also plays a role. In void configuration, release rate was significantly slower compared to the nonassembled modules but was slightly higher than from modules assembled in "piled configuration" (Fig. 11.9) due to the different volume/area ratio.

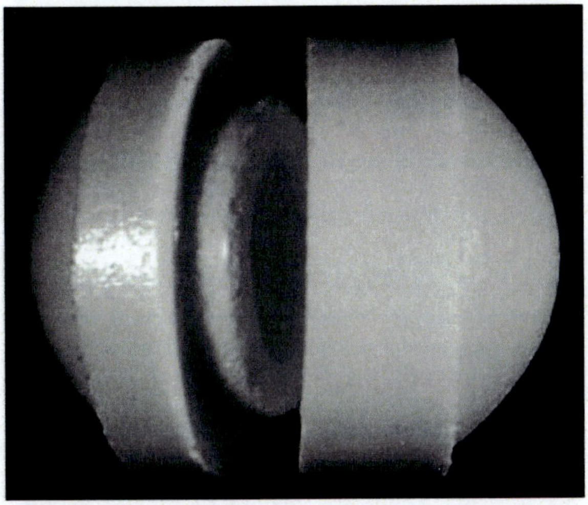

Fig. 11.10 Void assemblage of male and female modules

Figure 11.10 shows two uniquely shaped Dome Matrix® "male" and "female" modules designed to facilitate assembly in void, stacked, or mixed configurations by friction interlocking.

The concave female module can also fit the convex bases of both female and male modules. Clicking the concave bases of both male and female modules provides a unit with an empty inner space.

Dome Matrix® technology was successfully used to accommodate the antimalarials artesunate and clindamycin in one unit. Male and female modules containing clindamycin and formulated as swellable matrices for prolonged release were studied with particular attention to shape effects on drug release. Release profiles of clindamycin from male and female modules were typical for swellable matrices. The male module released drug more slowly than the female module, reflecting differing surface areas exposed to the dissolution medium. Rate profiles showed a rapid rate decrease of the initial delivery (first 100 min), followed by more steady release rate. Such behavior can be attributed to "burst" release, typical of hydrophilic matrices occurring before the formation of the rate controlling gel layer. The female module released the clindamycin faster than the male counterpart over the first 90 min. Release from the female module then decreased unexpectedly, falling lower than for the male module. Thereafter, the male module maintained a quasi-constant release rate, delivering more than 80% of drug load in 300 min [17].

Swelling/erosion behaviors were considered to explain such release rates. Photographic area measurements of swollen modules were compared with clindamycin release. Although the releasing areas between the two modules differed, fluxes were practically super-imposable. This suggests that modules could be designed with specific geometries to control release profiles from swellable matrices.

The delivery system for malaria treatment was completed by incorporation of two additional female modules, viz. one providing immediate (rapid) release of a fraction

Fig. 11.11 Clindamycin/artesunate combination delivery system obtained by assemblage of four modules

of the clindamycin dose and a module for immediate release of artesunate. This four-module assembly (Fig. 11.11) allows the patient to receive the artesunate and one-third of the clindamycin dose, formulated for immediate release, followed by the remainder of the clindamycin, formulated for prolonged retention/release under gastro-retentive conditions (80% in 8 h in vitro). Either assembly configuration (void or stacked) allowed site-specific delivery and alignment with kinetics and dynamics of both drugs in one unit. The void configuration, to enable floating and gastric retention, was obtained by joining the two prolonged release modules, having "male and female" shapes. The complete unit included the two additional immediate release (IR) modules stacked on either side of the sustained release modules assembled in void configuration. The four-module system was nonbuoyant (sank in aqueous systems), but rapid disintegration of the two fast releasing modules led to the assembled prolonged release modules attaining the appropriate density and floatation.

A bioavailability study in dogs (without the artesunate module) showed that the system sustained significant clindamycin plasma levels for up to 8 h, increased absorption (bioavailability), and suggested that dose frequency could be reduced [17]. The system is being proposed for delivery in one dosage unit the daily therapeutic dose suggested in WHO Guidelines. Other modules could be fitted to the system for a second "burst" of artesunate, for its sustained release, or for other drugs.

11.4 Further Geometric Considerations: The Butterfly Effect in Swellable Matrices

The "butterfly effect" was first observed when hypromellose-based matrix compacts partially separated as two "wings" during dissolution studies. Such splitting was pH-independent and the "halves" so generated remained attached to each other. Figure 11.12 illustrates such changes. Such a phenomenon, if reproducible, has potential applications for modifying drug release due to the changed release area consequent to splitting.

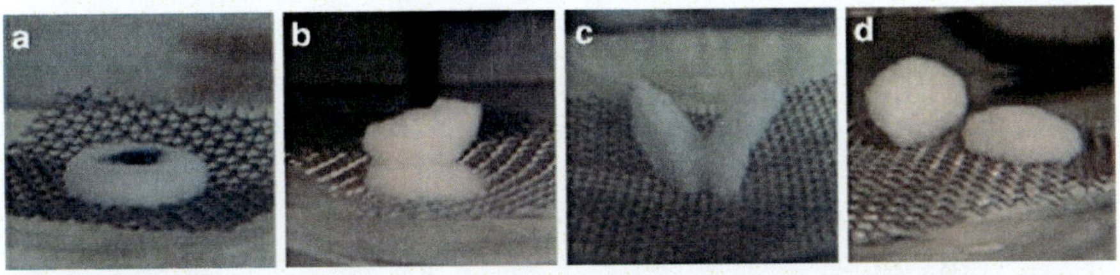

Fig. 11.12 Progression of events during disk matrix swelling leading to the manifestation of butterfly effect. (**a**) Hydration and swelling of the tablet disk matrix upon contact with dissolution medium. (**b**) Splitting of tablet at the radial side. The tablet edges began to curl outwards. (**c**) Formation of the butterfly shaped hydrated matrix. The two halves of the tablet remained attached at one end. (**d**) The two halves of the butterfly shaped tablet detached into two individual swollen matrices

The phenomenon is ascribable to the following behaviors:

- The dynamics of moving solvent and swelling fronts.
- The anisotropic expansion of materials in solution.

Hydration and swelling begins as the dissolution medium penetrates the compact. A solvent/swelling front is created as polymer swelling progresses. This moving boundary separates the glassy and rubbery states of polymer creating regions of high stress [18]. Stress relaxation occurs when the center of the matrix is plasticized by the penetrating dissolution medium, eliminating the glassy state [19]. In a disk-shaped matrix tablet, the swelling fronts move simultaneously from the axial and radial surfaces toward the center of the tablet. The glassy core tends to restrict the rubbery phase to one-dimensional swelling. Its elimination removes the swelling constraint [20]. This, coupled with the large axial swelling pressure causes sudden relaxation or volume increase of the swellable disk, splitting the tablet into a "butterfly"-shaped matrix.

The axial directional nature of compaction during tablet manufacture generates inherent mechanical anisotropy, resulting in axial and radial stresses and strains [21, 22]. Consequently, penetration of dissolution medium causes nonuniform swelling [23]. It is reported that hypromellose compacts swell predominantly in the axial rather than the radial direction on exposure to aqueous fluids [10]. Such more rapid release of stress in the axial direction would cause the tablet to split on the radial side. Dissolution medium penetrating the core through the fissure could cause greater swelling of the contact area. The outer surfaces, being hydrated much earlier, would have formed a viscous, more flexible layer which would "curl" outwards due to swelling pressure. This could also lead to the "butterfly" configuration.

Tablets exhibiting the "butterfly effect" were relatively thin, characterized by aspect ratios (2×diameter/thickness) ranging from 12.7 to 14.5. Thicker compacts did not display the effect. Such thicker units may be more likely to accommodate the swelling stresses generated as swelling fronts meet. Disappearance of the glassy core is slower, giving more time for the polymer chains to accommodate the swelling stresses. Moreover, the additional hypromellose possibly allows the matrix to be more strongly hydrated, thus maintaining its structural integrity.

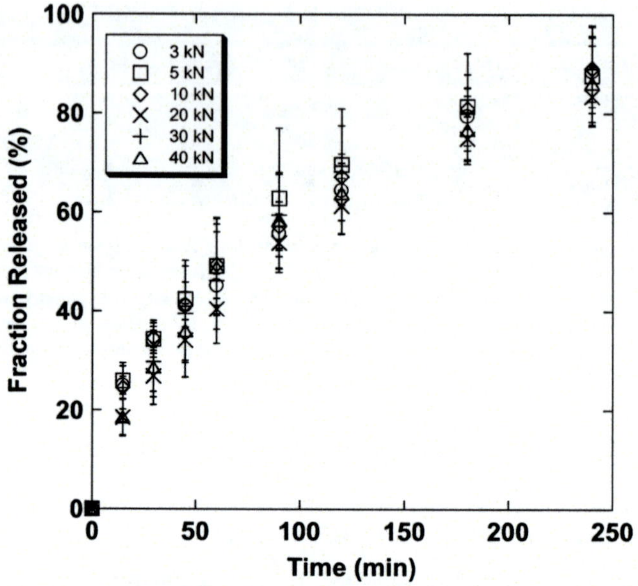

Fig. 11.13 Drug release profiles of matrix formulations containing unsieved hypromellose (Formula C) prepared at different compaction forces

The "butterfly" geometry causes the surface area for drug release to increase. The effect of shape change on release mechanism and kinetics was accordingly evaluated. Hypromellose particle size, applied compaction pressure and proportions of insoluble and soluble excipients were investigated. Findings were as follows:

- Compacts containing sieved (fine or coarse particles) hypromellose fractions did not manifest the "butterfly effect" during dissolution.
- Unsieved hypromellose (almost 70% of material less than 63 μm) contained a mix of coarse and fine particles which was most effective in manifesting the "butterfly effect" in compacts.
- The "butterfly effect" was only manifested when tablets were compacted at a suitable compaction pressure.

The proportions of soluble and insoluble excipients in the formulation affected "butterfly" shape formation and drug release.

Particle size distribution of hypromellose was key to "butterfly" shape formation.

Compacts incorporating predominantly fine hypromellose possess larger surface areas, facilitating greater polymer–water interactions. This would help preserve matrix structure by rapidly forming a gel barrier. Coarse hypromellose could have the opposite effect, with compacts disintegrating too rapidly.

Figure 11.13 shows the release profiles of compacts containing unsieved hypromellose prepared at different compaction pressures. There were no significant differences ($p > 0.05$) between dissolution profiles. However, the "butterfly

Table 11.1 Composition of tablets studied

		Formula A		Formula B		Formula C	
	Materials	mg	%	mg	%	mg	%
Granules	Flurbiprofen	37.50	33.13	37.50	33.13	37.50	33.13
	Mannitol	24.01		24.01		24.01	
	PVP K30	3.37		3.37		3.37	
Extragranular excipients	Mannitol	90.82	46.37	67.02	34.22	43.24	22.07
	MCC PH102	–	–	23.80	12.15	47.59	24.30
	Hypromellose K4M	39.17	20.00	39.17	20.00	39.17	20.00
	Magnesium stearate	0.98	0.50	0.98	0.50	0.98	0.50
	Total	195.85	100	195.85	100	195.86	100

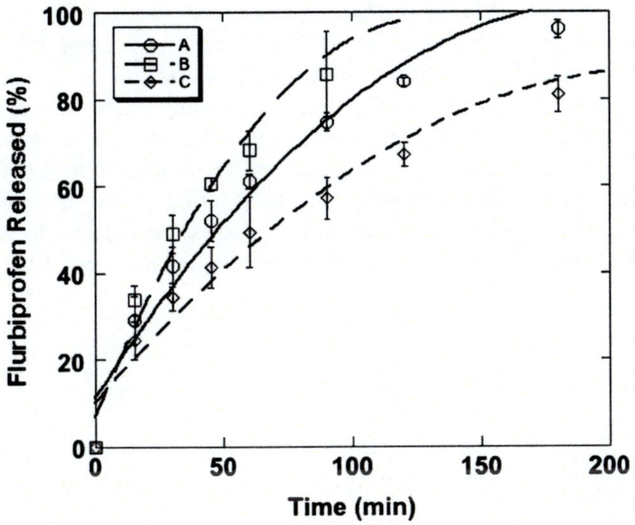

Fig. 11.14 Drug release profiles of matrices (Formula A, B, and C) prepared at compaction force of 10 kN

effect" was only manifested when compaction pressure exceeded 3 kN. At lower compaction pressures, greater compact porosity promoted disintegration due to enhanced penetration of dissolution medium and poorer establishment of swelling fronts. High compaction pressures might have been expected to reduce release rates. However, the "butterfly effect" counteracted this by increasing the surface area for release [11].

Three formulations (Table 11.1) prepared using unsieved hypromellose with different proportions of soluble (mannitol) and insoluble [microcrystalline cellulose (MCC)] excipients were studied.

Figure 11.14 shows release profiles of tablets produced using the different formulations prepared at a compaction pressure of 10 kN. Soluble excipients dissolve and diffuse out of the matrix, increasing porosity, facilitating drug release, and reduce strength of the matrix. The tablet containing the most soluble component

(Formula A) should theoretically display the fastest drug release. However, this was not the case. Release rates were in the following order:

Formula B (12.15% w/w MCC)>Formula A (0% w/w MCC)>Formula C (24.3% w/w MCC).

The "butterfly effect" was manifested in all three formulations. However, it was more extensive with Formula B and provided the fastest drug release. Even in the absence of MCC, Formula A tablets also manifested the "butterfly effect" but shape was quickly lost when the large amount of mannitol dissolved, thereby weakening the matrix. Collectively, these observations suggest that the shape of the "butterfly" disk matrix was influenced mainly by the unsieved hypromellose. MCC, by acting as a physical barrier, helped to promote the "butterfly effect" through an additional disintegrant effect and obstruction to polymer interaction and gelation. It was also noteworthy that Formula B showed the most extensive "butterfly" shape formation, not Formula C, which contained the most MCC. This further emphasizes the importance of formulation for generating the "butterfly effect."

In conclusion, the "butterfly effect" is a promising phenomenon observed in hypromellose matrices. Preliminary studies afforded new mechanistic insights on factors that promote or inhibit the phenomenon. The "butterfly effect" can increase drug release rates by increasing surface area, at a time in the dissolution process when rate may be slowing. However, further studies are required to better understand the phenomenon and utilize this effect in drug release control.

References

1. Hsieh DS, Rhine WD, Langer R (1983) Zero-order controlled-release polymer matrices for micro- and macromolecules. J Pharm Sci 72(1):17–22
2. Lee PI (1984) Effect of non-uniform initial drug concentration distribution on the kinetics of drug release from glassy hydrogel matrices. Polymer 25(7):973–978
3. Feely LC, Davis SS (1988) The influence of polymeric excipients on drug release from hydroxypropyl methylcellulose matrices. Int J Pharm 44(1–3):131–139
4. Swan EA, Peppas NA (1981) Drug release kinetics from hydrophobic porous monolithic devices. Proc Symp Control Release Bioact Mater 8:18–23
5. Korsmeyer RW, Gurny R, Doelker R, Buri P, Peppas NA (1983) Mechanisms of solute release from porous hydrophilic polymers. Int J Pharm 15(1):25–35
6. Colombo P, Conte U, Caramella C, Gazzaniga A, La Manna A (1985) Compressed polymeric mini-matrices for drug release control. J Control Release 1(4):283–289
7. Colombo P, Gazzaniga A, Caramella C, Conte U, La Manna A (1987) In vitro programmable zero-order release drug delivery system. Acta Pharm Technol 33(1):15–20
8. Conte U, Colombo P, Gazzaniga A, Sangalli ME, La Manna A (1988) Swelling-activated drug delivery systems. Biomaterials 9(6):489–93
9. Colombo P (1993) Swelling-controlled release in hydrogel matrices for oral route. Adv Drug Deliv Rev 11(1–2):37–57
10. Colombo P, Conte U, Gazzaniga A, Maggi L, Sangalli ME, Peppas NA, La Manna A (1990) Drug release modulation by physical restrictions of matrix swelling. Int J Pharm 63(1):43–48
11. Colombo P, Catellani PL, Peppas NA, Maggi L, Conte U (1992) Swelling characteristics of hydrophilic matrices for controlled release. New dimensionless number to describe the swelling and release behavior. Int J Pharm 88(1–3):99–109

12. Conte U, Maggi L, Colombo P, La Manna A (1993) Multi-layered hydrophilic matrices as constant release devices (Geomatrix™ Systems). J Control Release 26(1):39–47
13. Catellani PL, Colombo P, Peppas NA, Santi P, Bettini R (1998) Partial permselective coating adds an osmotic contribution to drug release from swellable matrixes. J Pharm Sci 87(6):726–731
14. Conte U, Maggi L (1996) Modulation of the dissolution profiles from Geomatrix multi-layer matrix tablets containing drugs of different solubility. Biomaterials 17(9):889–96
15. Losi E, Bettini R, Santi P, Sonvico F, Colombo G, Lofthus K, Colombo P, Peppas NA (2006) Assemblage of novel release modules for the development of adaptable drug delivery systems. J Control Release 111(1–2):212–8
16. Strusi OL, Sonvico F, Bettini R, Santi P, Colombo G, Barata P, Oliveira A, Santos D, Colombo P (2008) Module assemblage technology for floating systems: in vitro flotation and in vivo gastro-retention. J Control Release 129(2):88–92
17. Strusi OL, Barata P, Traini D, Young PM, Mercuri S, Colombo G, Sonvico F, Bettini R, Colombo P (2010) Artesunate-clindamycin multi-kinetics and site-specific oral delivery system for antimalaric combination products. J Control Release 146(1):54–60
18. Korsmeyer RW, Peppas NA (1984) Solute and penetrant diffusion in swellable polymers. 3. Drug release from glassy poly(HEMA-co-NVP) copolymers. J Control Release 1(2):89–98
19. Korsmeyer RW, Von Meerwall E, Peppas NA (1986) Solute and penetrant diffusion in swellable polymers. II. Verification of theoretical models. J Polym Sci Polym Phys 24(2):409–434
20. Lustig SR, Peppas NA (1987) Solute and penetrant diffusion in swellable polymers. VII. A free volume-based model with mechanical relaxation. J Appl Polym Sci 33(2):533–549
21. Malamataris S, Hatjichristos T, Rees JE (1996) Apparent compressive elastic modulus and strength isotropy of compacts formed from binary powder mixes. Int J Pharm 141(1–2): 101–108
22. Mullarney MP, Hancock BC (2006) Mechanical property anisotropy of pharmaceutical excipient compacts. Int J Pharm 314(1):9–14
23. Moe DV, Rippie EG (1997) Nondestructive viscoelastic analysis of anisotropy in compressed tablets. J Pharm Sci 86(1):26–32

Chapter 12
Extrudable Technologies for Controlling Drug Release and Absorption

Daniel Bar-Shalom, Matthew Roberts, and James L. Ford

Abstract Embedding a drug in a polymeric matrix in the molecular or dispersed state can enhance or control its rate of dissolution/release from the dosage form to modify its bioavailability or pharmacokinetics. The embedded drug may be dissolved in the polymer, be in a highly dispersed form, or in a stable high-energy state such as its amorphous form. Other materials added to the drug/polymer system may help provide a release and subsequent plasma profile or time-course that optimizes the drug's action.

The thermoplasticity of the polymer is a prerequisite for utility in such matrices. Functionality depends on its nature and level and on the presence and properties of other release-affecting additives. Thermal stability of the drug and compatibility with the other agents is vital.

Pharmaceutical textbooks have not traditionally provided detailed accounts of the equipment and technologies used for preparing thermoplastic matrices at industrial scale. Hence, a fulsome account is provided at the start of this chapter. The suitability and properties of the polymeric excipients are also summarized and discussed.

12.1 Introduction

Extrusion and injection moulding techniques can be used to produce controlled-release dosage forms, by homogeneously embedding drug in rate-controlling polymers. The technique can also be used to enhance the bioavailability of poorly

D. Bar-Shalom (✉)
Faculty of Pharmaceutical Sciences, Department of Pharmaceutics and Analytical Chemistry, Universitetsparken 2, DK-2980, Copenhagen Ø, Denmark
e-mail: dbs@farma.ku.dk

C.G. Wilson and P.J. Crowley (eds.), *Controlled Release in Oral Drug Delivery*, Advances in Delivery Science and Technology, DOI 10.1007/978-1-4614-1004-1_12,

soluble drugs. Products formulated in this way are not simple physical mixtures of drug and excipients. The drug/polymer association may be more complex and can be influenced by many factors. These are presented and explored in this chapter.

The provenance of equipment and techniques for the manufacture of controlled-release dosage forms lies largely in the industrial plastics industry. Pharmaceutics textbooks are largely bereft of accounts of equipment design and mode of operation. To rectify this deficiency, the chapter opens with an account of the technology, equipment, and component parts used to prepare extrudates.

Note: The terms Molding (US) and Moulding (UK) are purposefully used in this chapter to highlight their interchangeableness which reflects the country of origin of publication of the studies.

12.2 Historical Background

The concept of producing dosage units by molding is not new. Molded tablets or tablet triturates were introduced by Fuller in 1878 as unit oral dosage forms; a moistened powder comprising drug and excipients was pressed into a cavity, extruded, and dried (Fig. 12.1). Suppositories are also manufactured by casting a molten mass in a mold and allowing it to solidify by cooling (Fig. 12.2). Simple tablet and suppository molding techniques and equipment were prevalent in the nineteenth century. Earlier manufacture involved mixing with a spatula and "hand-rolling" on a glass plate [1]. Such "drug delivery" approaches did not encompass concepts of modifying release or otherwise enhancing drug performance. They were simply a means of providing an accurate dose of medication in as "user-friendly" a form as was possible at the time.

Manual and automatic machinery was developed for more efficient manufacture of molded tablets and suppositories, but as better unit dose presentations became

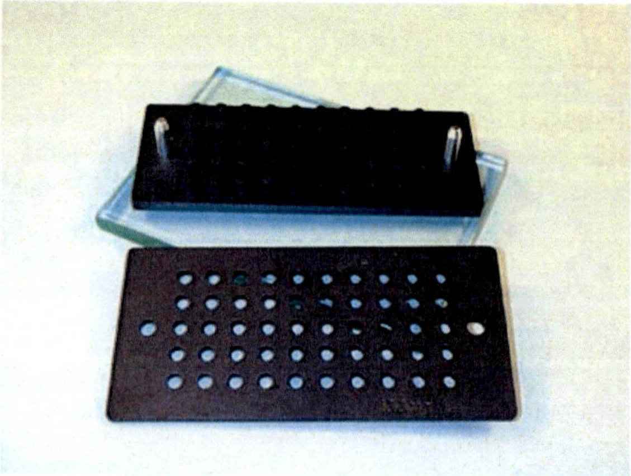

Fig. 12.1 Molded tablet or tablet triturate mold. The *cavities plate* is placed on the *glass plate*, the moistened mass is forced into the cavities and then the *pegs plate* is used to extrude the tablets

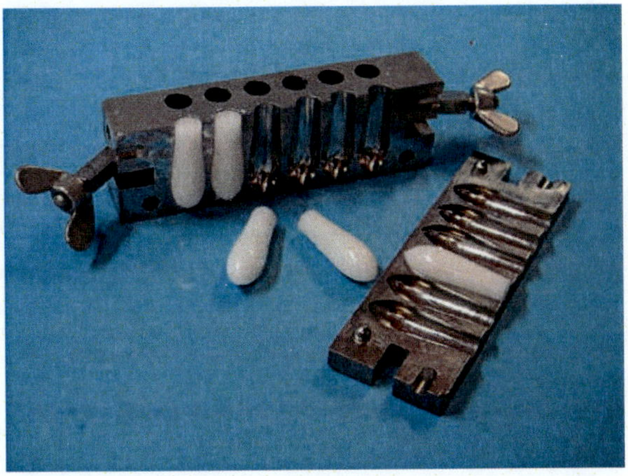

Fig. 12.2 Suppository form or mold. The molten mass is cast into the mold (held closed with the screws), the mass is allowed to solidify, the mold is opened by loosening the screws and the suppositories are ready for insertion

available (e.g. tablets and capsules) there was no significant evolution of technologies for pharmaceutical manufacture of "molded" dosage forms. However, as techniques for synthesizing different "plastic" materials were invented there were concomitant developments of technologies to process these to product forms. In 1855, Alexander Parkes developed a synthetic replacement for ivory made from cellulose treated with nitric acid and a solvent. Cellulose nitrate could be dissolved in alcohol and hardened into a transparent elastic material that could be molded to a desired shape or form when heated. Many other materials were subsequently invented. In 1868, John Wesley Hyatt developed a plastic, which he named "celluloid" that could be readily processed into product forms. He patented the first injection molding machine in 1872 [2], which was based on a plunger forcing the plastic through a heated cylinder into a mold.

The first plastic based on a synthetic polymer was made from phenol and formaldehyde (Bakelite) in 1909 by Leo Hendrik Baekeland [3]. In 1946, James Watson Hendry built the first screw injection molding machine (the screw is now termed the *extruder*), which afforded precise control over speed of injection and quality of product. It also allowed materials to be mixed before injection, for instance, colored or recycled plastic could be added to virgin material and mixed thoroughly before being injected. Today, screw injection technology is incorporated in the majority of machines. In the 1970s, Hendry developed the first gas-assisted injection molding process, permitting the manufacture of complex, hollow articles that cooled quickly. This greatly improved design possibilities as well as product strength and finish, while reducing production time, cost, weight, and waste [4].

For the production of objects made of plastic, thermoforming (when the plastic can be repeatedly melted and solidified) or thermosetting (when the plastic irreversibly cures) technologies were developed. Other technologies included calendaring, film casting, film blowing, extrusion, and injection. Many of these are used in the pharmaceutical industry, but for packaging or making patches and plasters and not

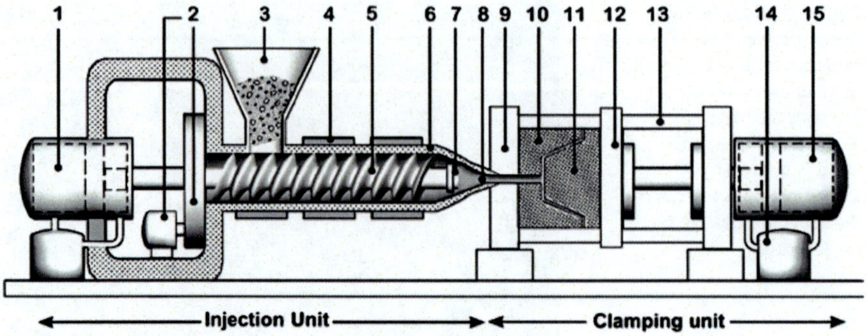

Fig. 12.3 Schematic section of an injection molding machine. The *numbers* in the picture correspond to the *numbers* in parenthesis in the text [#]

usually in dosage form design. However, injection molding has been used to manufacture implants [5], while screw extrusion and calendaring techniques are gaining momentum in melt-granulation. The next section describes equipment and mode of *extrudate* formation, as it pertains to dosage forms as well as other forms of product manufacture.

12.2.1 Mechanics of Injection Molding

This section gives an account of the injection molding, process, particularly those aspects that are important in a drug delivery/dosage form design context. The reader is encouraged to consult engineering textbooks on injection molding for more comprehensive accounts.

The following sequence of operations comprises the manufacturing cycle (Fig. 12.3):

- Materials are fed from a hopper *(funnel)* into a heated *barrel* with a reciprocating screw *(extruder)*.
- Materials are melted, mixed, and advanced into an *injection chamber* at the end of the barrel.
- The *screw-ram* pushes the *screw*; the *nonreturn valve* seals the *chamber*.
- The molten mix in the chamber is pushed through a nozzle into a mold where it solidifies under *hold pressure* to ensure complete filling and compensate for shrinkage.
- The mold opens and the product is ejected. The time span between mold openings is the *Cycle Time*, an important processing variable (the shorter the time, the more units are produced).

12.2.1.1 Feed Step

As represented in Fig. 12.3, material(s) in the *hopper* (*feeder*) (**3**) can be fed to the *barrel* (**6**) *passively* (gravimetrically) or by dosing screws, vibrators, grinders, regrinders, etc. A single *hopper* or an array of *hoppers* to feed more than one material concurrently may be employed. Ingredients may be also granulated in a screw-extruder and fed as pellets. Hoppers can be placed at different positions along the *barrel*, so that (for example) a heat-sensitive material can be added closer to the nozzle, minimizing the risk of thermal degradation. Accuracy of dosing is high, machine manufacturers claiming accuracies in the 0.1–1% range.

12.2.1.2 Melt Formation

The *barrel* (Fig. 12.3) can be heated by enveloping mantles (*heaters*) (**4**) with temperature usually rising toward the nozzle end (**8**). A motor and gears (**2**) rotate the reciprocating screw, advancing material toward the nozzle; a nonreturn valve (**7**) is fitted at the end of the screw. Material, possibly at its Glass Transition Temperature, is trapped in the space between the valve and the nozzle. A cylinder or motor (**1**) pushes the screw forward forcing the material into the cavity (or cavities). Material is cooled and solidifies while the cylinder maintains pressure (*retention*). This ensures complete filling of the cavity as many plastics shrink while cooling.

The path from the nozzle to the cavity (*runner*) results in formation of an extra, residual, part, the *cold runner*, in each cycle. This *cold runner* can be recycled but in applications where this could pose a problem (such as a risk associated with reprocessing a drug), a *hot-runner* might be used. As the name implies, it is heated so that the material remains fluid next to the cavity. Thus, when the mold opens, the plastic breaks leaving a smaller "scar" than if using cold runners. The *hot-runner* may also be gated so the break is cleaner and neater. The gating might be thermal (Fig. 12.4) (the heat shears the material after cooling but prior to the opening of the mold) or mechanical (Fig. 12.5) (a pin seals the opening). The benefits of hot-runners are many and can influence mold design, cycle time, product finish/elegance, etc., and usually outweigh the added cost.

12.2.1.3 Fusion of Melt

The product is formed when the molten material filling the cavity is allowed to solidify by cooling. Cavity shape defines the shape of the fused melt. There can be multiple cavities in the mold space, producing identical or different items. Molds must be *balanced* to ensure a good *fill pattern*. In the case of identical cavities (producing many identical items in each cycle), the number is a multiple of 2 to make the mold symmetric (Fig. 12.6) [6].

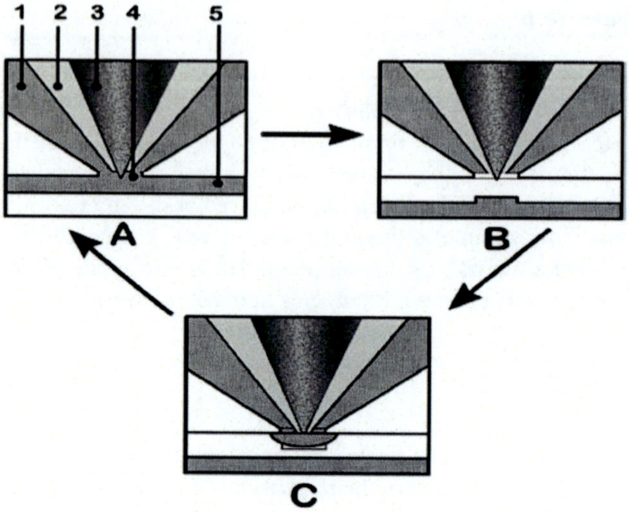

Fig. 12.4 Thermal Gating. The TIP is heated

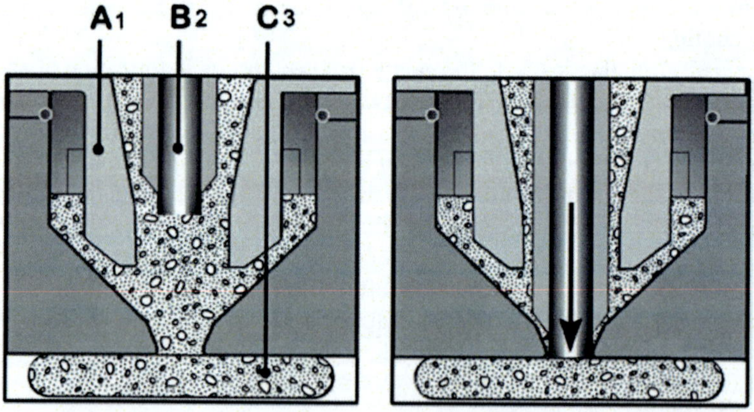

Fig. 12.5 Mechanical gating. The nozzle is sealed when the mold opens

12.2.1.4 Mold Design and Mode of Operation (Material Flow)

The simplest mold (Fig. 12.3) consists of two parts: The *Injection Mold* (**10**), which is attached to the *stationary platen* (**9**), and the *Ejector Mold* (**11**), which is attached to the *Movable Platen* or *Rear Platen* (**12**). The platens are connected by 4 (in some cases just 2) *Tie Bars* or *Tie Rods* (**13**). The *Movable Platen* is able to slide back and

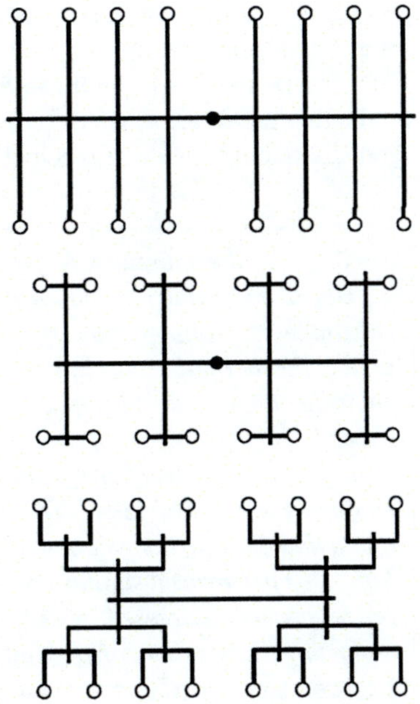

Fig. 12.6 Examples of fill patterns, note the symmetry

forth, driven by the *Clamping Cylinder* (**14**), and guided by the *Tie Bars*. The maximum mold size (high/wide) is limited by the distance between the *Tie Bars*. This distance and the *Clamping Force* (defined as the injection pressure multiplied by the total cavity projected area) are the parameters used for comparison of machines.

Molten material enters the mold through a *Sprue* where the nozzle docks. Channels transverse the mold for circulation of coolants (usually cooled or tempered water), thereby enabling appropriate solidification and optimum *cycle time*. When the content has solidified, the mold is opened (by moving the *Ejector* away from the *Injector*). Product remains on the *Ejector*, and the *Ejector Pins* push it (and *eventual cold runners*) out of the mold. As the molten materials are pushed into the cavity, the air present there either compresses and/or prevents complete filling, or, if the mold is well designed, escapes through air vents. Figure 12.6 provides a schematic of various "feed and fill" configurations.

Mold complexity reflects product complexity. Design modifications include lateral movement capability, transverse *Ejector Pins*, multilayer molds, *Slides*, etc. Cavities are usually engraved in changeable *Inserts*, which are mounted on the mold.

Some molds are designed to produce products made of two (or more) distinct materials, e.g. differing in color, chemical composition, or function. The Egalet® technology is an example, where a "Matrix-in-a-tube" is produced in one machine in one cycle. Such a machine has two (or more) injection units, and the molding process is repeated, often sequentially. In other cases, distinct injection processes are executed simultaneously.

Prototype molds can be made from mild steel, aluminum or nickel, and epoxy. Production molds are made of tool steel, hardened steel or beryllium–copper alloys. Production molds made of medical grade steel are also available.

Note: As the introductory paragraphs indicate, injection molding is a well-developed and mature technology in industrial operations other than pharmaceutical products. Consequently, specific terminologies have evolved and are widely used in the thermoplastics lexicon, some of which have equivalents in pharmaceutical technology. To avoid confusion, such terms are italicized throughout the text for clarity.

A good example of the problems related to terminology is the potentially confusing term "Extruder". In the minds of many pharmaceutical scientists it is a machine that forces wet material through a perforated screen ("wet granulation"), or performs a similar operation. In this chapter, only THERMOPLASTIC processes are addressed. An extruder as used in the plastics industry for thermoforming is termed a Hot-Melt Extruder in pharmaceutical operations, or a Compounder, if different materials are fed to form the extrudate (see: Pharmaceutical Extrusion Technology, Drugs and the Pharmaceutical Sciences, Vol. 133, 2003 I. Ghebre-Selassie and C. Martin Editors. Marcel Dekker).

12.3 Prototyping and Scale-Up

In recent years "tabletop" machines have been introduced where prototypes can be produced one at a time using minimal amounts of material (Fig. 12.7).

In the plastics industry, scale-up is often achieved by simply using additional machines rather than by using larger models. Hence, equipment used in development is often identical to industrial scale.

Most process parameters (temperatures, pressures, cooling temperature and rate of cooling, retention time, residence time, etc.) can be controlled accurately and precisely. Molds can sometimes be exchanged between machines of different models, or even makes.

12.4 Advantages and Disadvantages

From an engineering and operational perspective, injection molding technology can be low cost, highly precise (ensuring accurate drug content), and provide a wide range of product geometries. Scale-up can be less of an issue because the "unit operation" remains unchanged. It can also facilitate a largely continuous rather than "batch" mode of manufacture. Disadvantages are high initial set-up costs, with

Fig. 12.7 MiniJet injection molder (*left*) and MiniLab-compounder. Courtesy Thermo Scientific HAAKE

application limited to drugs and excipients that are thermostable. Moreover, injection molding technology may be unsuitable for materials where interactions (e.g. drug-excipient) can occur at high temperatures, but not at conditions encountered with more conventional modes of unit dose manufacture. As with all technologies, suitability is likely to be drug and dosage form specific and there will undoubtedly be "*niche*" areas where the technology may help surmount some barrier to development. Cleaning and validation might be an issue as machines are not usually designed specifically for medicinal product manufacture. Hence, it is important to consider cleaning capability in machine design and acquisition.

12.5 Injection Molding in Oral Drug Delivery

The possibility of utilizing injection molding for oral dosage form design and manufacture is mentioned as early as 1969 [7] and the first scientific article appeared in 1971 [8]. The 1980s saw a flurry of activity in the area.

Materials used in extruded dosage forms have been used in other more traditional pharmaceutical systems, and there are myriad examples of extrusion processes being used to produce granules, tablets, minimatrices, and pellets. Extrudates can be cut to a desired shape or size of tablet or matrix, or ground or milled to produce granules, or spheronized to pellets. Units may not require a coat to control drug release (although coating can be applied if desired). Release rate from the extrudate can be adjusted by polymer type, presence of pore-forming additives or hydrophilic polymers and pellet size [9].

The thermal sensitivity of a drug demands that matrix systems prepared by extrusion and/or injection moulding utilize a polymer that can be processed (is thermoplastic) at relatively low temperatures [10]. Polymer molecular weight, glass transition temperature, and sensitivity of the matrix or drug to heat and shear force can define suitability for extrusion [11]. Drug(s) and excipients may be amorphous or crystalline, depending on drug solubility in the molten polymeric carrier and vice-versa. Forster et al. [12] demonstrated that calculation of drug and excipient miscibility, based on solubility parameter could predict formation of amorphous solid solutions during melt extrusion. Quinten et al. [13] have studied matrix tablets produced by injection molding and confirmed that the drug, metoprolol tartrate, was partially dissolved in the polymer, but that clusters of solid-state drug were also present. Injection-molded composites are expected to be denser than extruded forms due to the pressure applied when filling the cavity and maintained while cooling [5]. This may be important if unit size is a consideration (e.g. high dose drugs). A valuable tool in predicting the behavior/compatibility is the Hansen Solubility Parameter [14], which is based on the "like dissolves like" concept.

In the 1980s, Snipes et al. [15] and Bar-Shalom et al. [16] independently proposed dosage units prepared by injection molding and comprising matrices of polyethylene glycol (PEG) or polyethylene oxide (PEO), encased in a cylinder that is impermeable to water but open at both ends. Furthermore, Bar-Shalom et al. suggested that the matrix could be divided in layers to enable pulsed release [17].

Clarke et al. described a process in which a compressed tablet is positioned in a mold and a coat is then injected-molded, leaving orifices for drug release [18].

12.6 Excipients for Extrusion Processes and Products

Extruded dosage forms are complex mixtures of active ingredient(s) and functional excipients, typically matrix carriers, release modifying agents and plasticizers. Carriers must be stable at the requisite processing temperatures, must deform readily when extruded and then solidify on cooling. They normally comprise a polymer and a low melting point wax or lipid material, or mixtures of these.

- The carrier has a significant influence on drug release mechanisms. Water insoluble materials (e.g. ethylcellulose) can act as diffusion-controlled systems. Hydrophilic polymers, such as hydroxypropyl cellulose, PEO and polyvinyl pyrrolidone, can control drug release via diffusion and erosion.
- Release-modifying excipients with various physical and chemical properties are usually incorporated to alter the porosity or tortuosity of the matrix, thereby influencing the release profile. Enteric-coating polymers and pH-adjusting agents (PVA phthalate and HPMC-AS) have also been used to produce extruded capsules with site-specific GI tract release [19].
- Plasticizers are typically low molecular weight compounds capable of reducing the T_g and melt viscosity of polymers, making them flexible. They may also

Table 12.1 Excipients used in extrudable and injection moulded delivery systems

Matrix carriers	Release modifiers	Plasticizers
Ethyl cellulose	Hydroxy propyl cellulose (HPC)	Citrate esters (e.g. triethyl citrate, acetyl tributyl citrate)
High MW Polyethylene glycols (PEG)	Hydroxypropyl methyl cellulose (HPMC)	Sebacate esters (e.g. Dibutyl sebacate)
Polyethylene oxide (PEO)	Xanthan gum	Phthalate esters (e.g. Diethyl phthalate)
Polyvinylpyrrolidone (PVP)	Chitosan	Low MW PEGs
Polyvinyl acetate (PVA)		
Acrylic copolymers (Eudragit®)		
Polyoxylglycerides (Gelucire®)		
Glycerol esters		

lower the shear forces needed to extrude a polymer. Incorporation may also lower the processing temperature for hot-melt extrusion and possibly reduce or obviate drug and carrier degradation [20]. Choice of plasticizer depends on polymer compatibility, stability, and volatility.

Table 12.1 lists commonly used materials for the formulation of extrudable and injection-moulded dosage forms. Examples of their application are discussed in the paragraphs that follow.

12.6.1 Cellulosics

Cellulose-based materials have been used in the manufacture of both immediate and modified release systems for many decades: a number have been utilized in extrudable dosage forms. These include:

- Water-insoluble ethylcellulose (EC) possesses excellent thermoplastic properties above its glass transition temperature (133°C) and is frequently used as a rate-controlling polymer [21].
- Hydroxypropyl cellulose (HPC) is the only water-soluble cellulose derivative that is thermoplastic. Its softening temperature is in the range of 100–150°C, depending on molecular weight [22]. Low-substituted hydroxypropyl cellulose (L-HPC) has also been studied in combination with EC [21].
- HPMC (hydroxypropyl methylcellulose), a nonthermoplastic hydrophilic cellulose ether, has been used to adjust release from extruded dosage forms. Inclusion level and viscosity grade can influence performance.

EC-based minimatrices were used to control ibuprofen release, with HPMC included as a hydrophilic rate-modifier by De Brabander [23]. Increasing HPMC concentration increased drug release rate, as did increasing its viscosity grade, due to increased matrix swelling. Drug release from formulations comprising 60% ibuprofen and low HPMC levels was diffusion-controlled (Fickian), while release tended toward anomalous transport with increasing concentrations and viscosity of the HPMC component.

Quinten et al. [13, 21] explored the use of hot-melt extrusion in combination with injection moulding to produce sustained-release matrix tablets. EC was used as the carrier and matrix former, while HPMC and L-HPC were used as hydrophilic-release modifiers. Plasticization with 20% w/w dibutyl sebacate was necessary to lower the T_g and melt viscosity of EC to obtain matrix tablets at lower temperatures. Increased process temperatures caused a significant decrease in release rate of metoprolol tartrate for all EC-based matrices. This effect was ascribed to a reduction in the polymer-free volume and better coalescence of the polymer fronts in the mould, promoting curing of the matrix [13]. Incorporation of 25 and 35% HPMC yielded faster and constant drug release rates by promoting diffusion. Formulations containing 50% HPMC displayed first-order profiles as drug release was controlled by a combination of diffusion and erosion mechanisms. Faster release of metoprolol tartrate was seen with higher viscosity grades of EC and HPMC due to increased water uptake and swelling [13]. Increasing the amount of L-HPC in the formulation increased release rates due to its swelling properties, with the mechanism of release changing from diffusion controlled to anomalous transport. A burst release component was observed for all formulations, which was more pronounced for high viscosity grades of EC [21].

Mehuys and coworkers developed a "matrix-in-cylinder" system comprising a hot-melt-extruded nonerodible EC tube, plasticized with 20%w/w dibutyl sebacate, surrounding a drug-containing, HPMC-Gelucire® matrix core. In aqueous media, the core formed a gel plug, which released drug through the open ends of the EC tube by erosion. Thus, erosion of the gel core was the most critical parameter affecting release rate. Sustained zero-order erosion-controlled release was obtained, which was independent of drug solubility. Rate could be changed by altering the length of the cylinder [24]. Although drug release was erosion-controlled, in vitro drug release was only slightly affected by hydrodynamics, mechanical stress, and composition of the dissolution medium. Furthermore, administration of the matrix-in-cylinder dosage form increased the bioavailability of propranolol in dogs, demonstrating the promise of this delivery approach for sustained release in vivo [25].

EC-based extrudates, 3 mm in diameter were manually cut to lengths of 2 mm, and release of ibuprofen modified by inclusion of hydrophilic components. Release rate could be changed by altering xanthan gum concentration and drug particle size [26]. In a later study [27], the effect of hydrophilic additives on such minimatrices was evaluated further. Metoprolol tartrate release was influenced by the inclusion of various PEG and PEO molecular weight grades at different concentrations. Slope and shape of the drug release profiles could be adjusted, without affecting drug crystallinity or homogeneity.

12.6.2 *Polyethylene Glycol*

PEGs are prepared by polymerization of ethylene oxide and are available over a wide range of molecular weights. High molecular weight PEGs have frequently been used as low-melting hydrophilic matrix carriers in hot-melt extrusion. Lower molecular weight grades are commonly used as plasticizers.

Quintavalle et al. [28] developed a cylindrical multilayered coextrudate comprising two concentric extruded matrices: a hydrophilic inner matrix (PEG 6000) and a hydrophobic outer matrix (microcrystalline cellulose wax). Drug release was successfully modified by selection of cylinder dimensions (hollow diameter and length), relative proportions, and composition of the inner and outer parts.

Zhang and McGinity [10] showed that PEG 3350 significantly decreased the processing temperature and torque exerted on PEO/drug blend during extrusion.

12.6.3 Polyethylene Oxide

Polyethylene oxide (PEO) resins are nonionic, water-soluble, high molecular weight (100,000–7,000,000 Da) homopolymers manufactured by the heterogeneous catalytic polymerization of ethylene oxide [10]. PEO is a thermoplastic, rapidly hydrating, semicrystalline polymer with a melting range of 57–73°C [29]. PEO is commonly used in the fabrication of matrices for controlling drug release through the formation of a hydrogel and subsequent diffusion and erosion. Polymer molecular weight can be used to achieve a desired erosion rate for sustained release. Low MW PEOs, which gel to a lesser extent may be advantageous with poorly soluble drugs due to reduced diffusion path length [29]. PEOs with molecular weights exceeding 300,000 Da are usually not suitable for hot-melt extrusion due to their high viscosity.

Release-retarding agents such as chitosan and xanthan gum have also been used to prepare PEO-based hot-melt-extruded tablets with pH and buffer species-independent sustained release [30].

12.6.4 Polyvinylpyrrolidone and Copolymers

Polyvinylpyrrolidone (PVP) is a water-soluble homopolymer made from the monomer *N*-vinyl pyrrolidone and is available in various molecular weight grades. It has been used in hot-melt extrusion processes to enhance solubility and dissolution rate of poorly soluble drugs. Hulsmann et al. showed that solid dispersions comprising 50% PVP K30 significantly enhanced drug release [31]. Similarly, melt-extrusion formulations with PVP K30 as the matrix carrier improved oral bioavailability of a poorly water-soluble drug [32].

Copolymers of vinylpyrrolidone and vinylacetate monomers have also been used for hot-melt extrusion applications.

12.6.5 Polyvinyl Acetate and Copolymers

Polyvinyl acetate (PVA) is a homopolymer synthesized from vinyl acetate monomer via free-radical polymerization. It is amorphous due to the presence of an acetate ester side chain and has a relatively low T_g due to the flexible backbone structure. Although PVA is water insoluble, it is slightly hydrophilic and can absorb water to some extent. Zhang and McGinity [33] demonstrated the excellent properties of PVA as a carrier for hot-melt extrusion systems, with processing conducted at 50–70°C. Suitable release profiles of theophylline were achieved by including the hydrophilic additives, PEO and HPMC in the extrudate.

Kollidon® SR, a controlled-release excipient based on PVA and PVP (4:1), was used to develop mini-matrices via hot-melt extrusion [34]. The low-melting point drug, ibuprofen, remained undissolved in the extrudates and its release rate was increased at higher processing temperatures, indicating a plasticizing effect of the drug. The higher-melting theophylline was dispersed in the matrix and release rate was inversely proportional to the processing temperature. Increased plasticizer (triethyl citrate) concentration formed a denser matrix and retarded theophylline release rate, whilst addition of a hydrophilic additive (HPC) as a pore former increased the release rates of both drugs.

12.6.6 Acrylic Copolymers (Eudragit®)

Poly (methyl) acrylates, commonly known under the trade name Eudragit®, are copolymers derived from esters of acrylic and methacrylic acid, whose physicochemical properties are determined by their functional groups. These versatile polymers have been used for various solid dosage form applications, including hot-melt extrusion.

- Eudragit S 100 and L 100 are anionic copolymers based on methacrylic acid and methyl methacrylate with 1:2 and 1:1 ratios of carboxyl groups to ester units, respectively.
- Eudragit® RL PO and RS PO are copolymers synthesized from acrylic and methacrylic acid esters, with the former having a greater molar ratio of ionizable quaternary ammonium groups than the latter, making it more permeable to aqueous media.
- Eudragit® RD 100 is comprised of 91% Eudragit® RL PO and 9% carboxymethyl cellulose, the inclusion of the latter increasing its permeability.

Zhu et al. [35] prepared tablets by hot-melt extrusion to control the release of the poorly soluble drug, indomethacin using the acrylic polymers Eudragit® RL PO, RD 100, L 100 or S 100, plasticized with triethyl citrate. The inclusion of Eudragit® L 100 or S 100 or a nonionic surfactant (Pluronic®) prior to processing increased release rate of indomethacin, as did an increase in media pH and decreased granule size.

Bruce et al. [36] prepared hot-melt extrusion tablets using Eudragit® S 100 as the polymeric carrier for delivery of 5-aminosalicylic acid to the colon. Drug release was shown to fit diffusion and erosion models at pH 7.4 and was influenced by the inclusion of lubricant and plasticizer. Citric acid functioned as a solid-state plasticizer at 10%w/w and delayed release of 5-ASA in phosphate buffer due to a lowering of micro environmental pH, which in turn suppressed ionization of

Eudragit® S 100. Heat-induced amide bond formation between citric acid and 5-ASA occurred during hot-melt processing.

Young et al. [37] developed acrylic-based controlled-release spherical matrix pellets using hot-melt extrusion and subsequent spheronization processes, demonstrating that release of theophylline was diffusion-controlled and influenced by the pH of the dissolution medium.

Schilling et al. [38] successfully prepared enteric matrix pellets with diameters <1 mm, containing up to 40% theophylline, by hot-melt extrusion, employing Eudragit® S 100, plasticized with either triethyl citrate or methylparaben as the matrix composition. The plasticizing efficiency and aqueous solubility of the plasticizer influenced drug release rate.

12.6.7 Lipids

Lipid-based systems for oral drug delivery offer an alternative approach to polymeric carriers due to their versatile physicochemical characteristics. They are also biodegradable and physiologically nontoxic. Solid lipid extrusion, where powdered glycerides are mixed with active pharmaceutical ingredient (API) and extruded below the melting point of the lipid, can avoid the polymorphic changes often associated with processing and storage of such materials [39]. The process has been described as "thermo-mechanical treatment by moderate temperature and pressure exposure, resulting in plastic mouldability of the lipid mass" [40].

Nonporous, sustained release matrices from formulations of glyceryl palmitostearate and glyceryl trimyristate, extruded at low temperatures (<50°C) provided sustained-release profiles [40]. Differences between the surface structures and fractured areas of the extrudates, which influence drug release properties, were attributed to friction-induced temperature increase during extrusion in the die plate. The initial burst release was minimized by reducing drug particle size and using intact extrudates [41].

Release profiles were controlled from extruded monoacid triglyceride (tripalmitin) matrices by incorporating increasing concentrations of the hydrophilic polymer, PEG 10000, which formed an interconnected pore network through dissolution, thus enhancing drug release [39].

Other applications of lipids for controlling drug release are presented in a separate chapter.

12.7 Active Pharmaceutical Ingredient Considerations

Although extrusion may avoid many of the issues associated with more traditional methods (e.g. hydrolytic degradation and the need for compressibility), the thermal, chemical and physical properties of the API can determine the feasibility of dosage form design via this process. The physicochemical properties of the API and other materials used in the formulation may determine whether the drug exists as undissolved particles, a solid solution or a combination of the two in the final dosage form. This can significantly impact processability and product stability [20].

The properties of the drug may also aid the functionality of other excipients through plasticization of the carrier. Examples include:

- Lidocaine lowers the T_g of Eudragit® E/HDPE films [42].
- Chlorpheniramine maleate lowers the T_g of HPC films [43].
- Indomethacin decreased the T_g of Eudragit® RL PO during extrusion, demonstrating that the drug exhibited a solid-state plasticizing effect on the polymer and was miscible with the polymer in the molten state [35]. These authors also showed that indomethacin was converted from the crystalline to a more soluble amorphous form in Eudragit® RD 100 following extrusion.

Conversely, a drug can be detrimental to the hot-melt extrusion process, as evidenced by the melting of oxprenolol hydrochloride, thereby decreasing extrudate viscosity [45] and fenoprofen calcium inhibiting the hardening of PEG–MCC matrixes [44].

Increased temperatures and pressures during hot-melt extrusion may increase drug solubility in the carrier; some drugs may melt or become solubilized in the polymer matrix. Recrystallization and nucleation of drug is retarded during cooling of the extrudate due to reduced solute migration in the highly viscous polymer medium. Miscibility of the API with the excipients is a key factor for successful hot-melt extrusion, with miscible components resulting in solid solution formation and immiscible components, possibly leading to amorphous drug being dispersed in crystalline excipients [12]. Solid solutions comprising drug and amorphous polymer are generally regarded as interstitial solutions with drug molecules occupying the interstitial space between the polymer chains. A solid dispersion of the drug may be susceptible to recrystallization during storage as the molecular dispersion is a metastable form.

The thermal stability of drugs influences the mechanical properties of extrudates. The instability of hydrocortisone at high temperatures, coupled with its hydrophobic nature, caused brittleness of HPC films [22]. A novel method of converting API to an amorphous form by solvent evaporation meant that subsequent melt-extrusion with PVP could proceed at a temperature below the melting point of the drug, and since the T_g of the drug was lower than that of the polymer the active substance served as plasticizer. Amorphous melt-extrusion formulations had higher bioavailability than formulations containing crystalline API [32].

References

1. University of the Sciences in Philadelphia (2005) Remington: the science and practice of pharmacy, Chap. 45, 21 edn. Lippincott Williams & Wilkins, Philadelphia
2. Painter PC, Coleman MM (2008) Essentials of polymer science and engineering. DEStech, Lancaster

3. U.S. patent 133229, 19 Nov 1872
4. Bryce DM (2001) Plastic injection molding: manufacturing process fundamentals. Society of Manufacturing Engineers, Dearborn
5. Rothen-Weinhold A, Besseghir K, Vuaridel E, Sublet E, Oudry N, Kubel F, Gurny R (1999) Injection-molding versus extrusion as a manufacturing technique for the preparation of biodegradable implants. Eur J Pharm Biopharm 48(2):113–121
6. Rosato DV, Rosato DV, Rosato MG (2000) In: Rosato DV (ed) Injection molding handbook. Springer, Heidelberg
7. Speiser P (1969) Injection molded oral medicament in solid form. US Patent 343 2592
8. Adel M et al (1971) Hot melt-extruded dosage forms. Pharm Acta Helv 46:31–52
9. Follonier N, Doelker E, Cole ET (1995) Various ways of modulating the release of diltiazem hydrochloride from hot-melt extruded sustained release pellets prepared using polymeric materials. J Control Release 36:243–250
10. Zhang F, McGinity JW (1999) Properties of sustained release tablets prepared by hot-melt extrusion. Pharm Dev Technol 4:241–250
11. Breitenbach J (2002) Melt extrusion: from process to drug delivery technology. Eur J Pharm Biopharm 54:107–117
12. Forster A, Hempenstall J, Tucker I, Rades T (2001) Selection of excipients for melt-extrusion with two poorly water-soluble drugs by solubility parameter calculation and thermal analysis. Int J Pharm 226:147–161
13. Quinten T, De Beer T, Vervaet C, Remon JP (2009) Evaluation of injection molding as a pharmaceutical technology to produce matrix tablets. Eur J Pharm Biopharm 71(1):145–154
14. Hansen C (2007) Hansen solubility parameters: a user's handbook, 2nd edn. CRC, Boca Raton, FL
15. Snipes WC, Wagner SJ (1986) Erodible matrix for sustained release bioactive composition. Appl No: 06/918,294 (14 Oct 1986)
16. Bar-Shalom D, Kindt-Larsen T (1995) Controlled release erodible composition. Appl No: 08/470,828 (6 June 1995)
17. Bar-Shalom D et al (2008) Development of the Egalet technology. In: Rathbone MJ et al (eds) Modified-release drug delivery technology drugs and the pharmaceutical sciences series, 2 edn. Marcel Dekker, New York
18. Clarke AJ, Glinike R, Raby R, Li C, Martini LG (2010) An injection molding process for the preparation of an oral delivery device comprising a core which contains a pharmaceutically active agent, having a coating with one or more openings leading to such a core. US Patent 2010/0221292
19. Mehuys E, Remon JP, Vervaet C (2005) Production of enteric capsules by means of hot-melt extrusion. Eur J Pharm Sci 24:207–212
20. Crowley MM et al (2007) Pharmaceutical applications of hot-melt extrusion: Part I. Drug Dev Ind Pharm 33:909–926
21. Quinten T et al (2009) Development of injection molded matrix tablets based on mixtures of ethylcellulose and low-substituted hydroxypropylcellulose. Eur J Pharm Sci 37:207–216
22. Repka MA, Gutta K, Prodduturi S, Munjal M, Stodghill SP (2005) Characterization of cellulosic hot-melt extruded films containing lidocaine. Eur J Pharm Biopharm 59:189–196
23. De Brabander C, Vervaet C, Remon JP (2003) Development and evaluation of sustained release mini-matrices prepared via hot-melt extrusion. J Control Release 89:235–247
24. Mehuys E, Vervaet C, Remon JP (2004) Hot-melt extruded ethylcellulose cylinders containing a HPMC-Gelucire® core for sustained drug delivery. J Control Release 94:273–280
25. Mehuys E, Vervaet C, Gielen I, Van Bree H, Remon JP (2004) In vitro and in vivo evaluation of a matrix-in-cylinder system for sustained drug delivery. J Control Release 96:261–271
26. Verhoeven E, Vervaet C, Remon JP (2006) Xanthan gum to tailor drug release of sustained-release ethylcellulose mini-matrices prepared via hot-melt extrusion: in vitro and in vivo evaluation. Eur J Pharm Biopharm 63:320–330
27. Verhoeven E, De Beer TRM, Schacht E, Van den Mooter G, Remon JP, Vervaet C (2009) Influence of polyethylene glycol/polyethylene oxide on the release characteristics of sustained release ethylcellulsoe mini matrices produced by hot-melt extrusion: in vitro and in vivo evaluations. Eur J Pharm Biopharm 72:463–470

28. Quintavalle U, Voinovich D, Perissutti B, Serdoz F, Grassi G, Dal Col A, Grass M (2008) Preparation of sustained release co-extrudates by hot-melt extrusion and mathematical modeling of in vitro/in vivo drug release profiles. Eur J Pharm Sci 33(3):282–293
29. Prodduturi S, Manek RV, Kolling WM, Stodghill SP, Repka MA (2005) Solid-state stability and characterization of hot-melt extruded poly(ethylene oxide) films. J Pharm Sci 94:2232–2245
30. Fukuda M, Peppas NA, McGinity JW (2006) Properties of sustained release hot-melt extruded tablets containing chitosan and xanthan gum. Int J Pharm 310:90–100
31. Hulsmann S, Backensfield T, Keitel E, Bodmeier R (2000) Melt extrusion: an alternative method for enhancing the dissolution rate of 17 β-estradiol hemihydrates. Eur J Pharm Biopharm 49:237–242
32. Lakshman JP, Cao Y, Kowalski J, Serajuddin ATM (2008) Application of melt extrusion in the development of a physically and chemically stable high-energy amorphous dispersion of a poorly water soluble drug. Mol Pharm 5:994–1002
33. Zhang F, McGinity JW (2000) Properties of hot-melt-extruded theophylline tablets containing poly (vinyl acetate). Drug Dev Ind Pharm 26:931–942
34. Ozguney I, Shuwisitkul D, Bodmeier R (2009) Development and characterization of extended release Kollidon® SR mini-matrices prepared by hot-melt extrusion. Eur J Pharm Biopharm 73:140–145
35. Zhu Y, Shah NH, Malick AW, Infeld MH, McGinity JW (2006) Controlled release of a poorly water-soluble drug from hot-melt extrudates containing acrylic polymers. Drug Dev Ind Pharm 32:569–583
36. Bruce LD, Shah NH, Malick AW, Infeld MH, McGinity JW (2005) Properties of hot-melt extruded tablet formulations for the colonic delivery of 5-aminosalicylic acid. Eur J Pharm Biopharm 59:85–97
37. Young CR, Koleng JJ, McGinity JW (2002) Production of spherical pellets by a hot-melt extrusion and spheronization process. Int J Pharm 242:87–92
38. Schilling SU, Shah NH, Malick AW, McGinity JW (2010) Properties of melt-extruded enteric matrix pellets. Eur J Pharm Biopharm 74:352–361
39. Windbergs M, Strachan CJ, Kleinebudde P (2009) Tailor-made dissolution profiles by extruded matrices based on lipid polyethylene glycol mixtures. J Control Release 137:211–216
40. Reitz C, Kleinebudde P (2007) Solid lipid extrusion of sustained release dosage forms. Eur J Pharm Biopharm 67:440–448
41. Reitz C, Strachan C, Kleinebudde P (2008) Solid lipid extrudates as sustained-release matrices: the effect of surface structure on drug release properties. Eur J Pharm Sci 35:335–343
42. Aitken-Nichol C, Zhang F, McGinity JW (1996) Hot melt extrusion of acrylic films. Pharm Res 13:804–808
43. Repka MA, Gerding TG, Repka SL, McGinity JW (1999) Influence of plasticizers and drugs on the physical-mechanical properties of hydroxypropyl cellulose films prepared by hot melt extrusion. Drug Dev Ind Pharm 25:625–633
44. Cuff G, Raouf F (1998) A preliminary evaluation of injection moulding as a technology to produce tablets. Pharm Technol 22:96–106
45. Follonier N, Doelker E, Cole ET (1994) Evaluation of hot-melt extrusion as a new technique for the production of polymer-based pellets for sustained release capsules containing high loadings of freely soluble drugs. Drug Development and Industrial Pharmacy 20:1323–1339

Chapter 13
Coated Multiparticulates for Controlling Drug Release

Brigitte Skalsky and Sven Stegemann

Abstract Coated particles offer a reliable way to delay, prolong, or pulse drug release. They are generally considered to perform more consistently than larger units with respect to GI tract transit. There are also potential safety advantages with potent drugs, if the dose is dispersed in many units. Individual unit failure does not pose an overdosing issue. This chapter reviews the various release-modifying materials used to coat such particles, and how they can be formulated to provide the requisite release characteristics. The technologies used for coat application are also discussed.

13.1 Introduction and Historical Background

Coating small particles is a well-established way to control drug dissolution and absorption to enhance therapeutic effect. One of the earliest of such multiparticulate systems was *Spansules*™ developed by SmithKline and French in the 1950s. Spherical particles were coated with wax-based materials to provide an array of release rates. More consistent and better performing materials have progressively replaced these earlier coatings such that a variety of coats with different performance attributes can now be used to control release. Drugs formulated as small particles that are coated with such materials may offer unique advantages over other modified release systems.

Dosage forms containing multiparticulates frequently comprise but are not constrained to loose-filled hard gelatin capsules. Multiparticulates can sometimes be compressed to tablets. Possibilities, coating materials, and associated technologies are discussed in this chapter.

B. Skalsky (✉)
Evonik Roehm GmbH, Darmstadt, Germany
e-mail: brigitte.skalsky@evonik.com

C.G. Wilson and P.J. Crowley (eds.), *Controlled Release in Oral Drug Delivery*, Advances in Delivery Science and Technology, DOI 10.1007/978-1-4614-1004-1_13,

13.2 Potential Benefits

Controlled release dosage forms, formulated as microparticles may offer the following advantages, relative to monolithic units:

- *Convenience*: Dysphagia and swallowing difficulties can complicate the use of large solid dosage forms especially in elderly patients. Multiparticulates, swallowed directly or dispersed in liquid are easier to swallow [1]. Such products can also be used to overcome potential swallowing issues in geriatric and pediatric patients by being "sprinkled" on food or beverages (http://www.dexilant.com). In severe cases multiparticulates can be dosed via intubation [2].
- *Reliability*: There is a large corpus of evidence that delivery/drug release in the GI tract is more consistent and reliable when multiparticulates are used.
- *Safety*: Controlled release dosage forms invariably contain higher levels of drug than conventional formulations. Failure of the release-controlling coat can accordingly lead to overdosage. Such "catastrophic" failure can be obviated with multiparticulates as coat failure is extremely unlikely in the totality of particles that constitute the dose.

Potential disadvantages that may be associated with microparticles concern dose of drug. If dose is high, a compressed single monolith occupies a smaller volume, there being no void spaces, as in a loose-filled multiunit capsule. In such cases it may be necessary to compress the particles or to fill them into sachets. There have also been allegations that capsule dosage forms (as used for most microparticulates) can lodge in the esophagus following dosage. These possibilities are addressed later.

13.3 Biopharmaceutical Considerations

13.3.1 Esophageal Transit

Capsules have been alleged to have a higher tendency to adhere to esophageal mucosa than do other solid dosage forms [3]. However, reports on esophageal injuries, ascribable to retention did not implicate a specific dosage form. Rather, the effect seemed to be compound specific [4, 5]. Hey et al. [6] used barium sulfate-containing tablets and capsules to monitor esophageal transit times of different solid dosage forms in healthy volunteers. Findings were that, with some dosage forms transit was mostly influenced by amount of coadministered liquid and position of the volunteer (upright or supine). The quantity of coadministered water in the supine position was most influential with tablets. Size or the amount of coadministered water did not influence capsule esophageal transit. The findings did not support the contention that capsules are more likely to lodge in the esophagus than small coated tablets. Other observations were:

- Longer transit times were evident with larger tablets in elderly subjects.
- Smaller coated tablets were easier to swallow than uncoated tablets.

Differences in esophageal transit between coated and uncoated tablets have been confirmed in other reports. Perkins et al. [7] compared transit times of small oval film coated with small round, flat-faced uncoated tablets. Mean esophageal transit times were 3.2±0.31 s for coated tablets and 65.2±32.8 s for uncoated tablets. Three out of the 31 subjects taking the uncoated tablet had transit times of more than 500 s and in some subjects the tablets persisted in the esophagus after 10 min.

Osmanoglou et al. [8] investigated esophageal transit of gelatin capsules using Magnetic Marker Imaging (MMI). Capsules were swallowed in supine and upright positions with 5 ml, 25 ml, and 50 ml of water. The study reaffirmed the importance of body position and amount of coadministered liquid that was noted with tablet dosage. Transit time was longest in the supine position with 5 ml of water (7.4±0.4 s) and shortest when administered in the upright position with 50 ml water (1.4±0.2 s). Assuming an in vivo capsule rupture time of 1.5–2.5 min [9] the risk of rupture during esophageal transit would seem to be low to nonexistent.

Gastrointestinal transit rates and GI tract location can be important determinants for dissolution, absorption, and consequent pharmacokinetic profile of controlled release dosage forms. Weitschies et al. studied an extended release, single-unit felodipine tablet to correlate gastrointestinal location, drug release, and fractional bioavailability in healthy volunteers under fasted and fed conditions [10]. Gastric residence times were highly variable, both fed and fasted; absorption was significantly influenced by GI tract release location. No absorption was evident when the tablet stayed in the proximal region of the stomach. Gastric residence time also influenced the pharmacokinetic profile, causing late and high plasma peaks. It was also shown that, in some subjects the tablet moved, not only from the proximal to the distal stomach but also in the reverse direction (distal to proximal).

Gastrointestinal transit time of solids is also dependent on nutritional status and unit size. Three distinct transit regions, viz., esophagus, stomach, and intestine need to be considered.

13.3.2 Gastric Emptying

Many controlled release units evince a sustained effect by releasing drug from the dosage unit while transiting the GI tract. Variable transit rates may influence onset, rate, and degree of absorption, with possible consequences for efficacy and safety. Consequently, the factors affecting gastrointestinal transit of solid oral dosage forms as single monoliths or multiple small units have been extensively investigated.

Gastric emptying times, in fasted conditions ranged from 6 to 60 min with an amoxicillin modified release tablet [11]. Other studies on tablet emptying reported average times of 42 min (range 12–126 min [12] and 25–155.5 min [13]). These findings illustrate the high intersubject variability for single unit dosage forms. Multiparticulates, in contrast empty more readily under both fasted and fed conditions. Digenis et al. found that erythromycin pellets were removed from the stomach gradually and within 90 min in fasted and fed conditions [14].

Gastric emptying time in fed conditions reflects caloric intake. For single monolithic units, a light meal (1,500 kJ) increased gastric emptying time from 0.3 (0.1–1.0) h (fasted) to 4.3 (1.7–5.0) h; a heavy meal (3,000 kJ) further increased transit time to 4.9 (1.9–18.0) h [12–14]. Borin et al. [15] reported that even a light meal (750 kJ) increased gastric emptying time in an unpredictable manner. Average emptying times in eight subjects taking a sustained release ibuprofen tablet was 4 h in three cases and 9 and 12 h in the others. Khosla et al. [16] investigated unit size effects on gastric retention following light (1,500 kJ) and heavy breakfast (3,500 kJ). The results did not confirm the usually accepted threshold requirement of 2 mm for particulates to pass through the pylorus in digestive mode. Lag (gastric retention) times were comparable to units of 5 mm in diameter after a light breakfast, being slightly longer following the heavier meal. A medium-calorie breakfast (2,300 kJ) gave results in between the light and heavy meal but some tablets of 5 and 7 mm remained in the stomach for more than 10 h in some subjects. Khosla and Davis [17] investigated the gastric emptying behavior of tablets with diameters of 7, 11, and 13 mm following a light breakfast (1,500 kJ). Findings indicated that tablets of up to 11 mm in diameter probably empty from the fed stomach during the phase 2 stage of the migrating myoelectric complex. Larger tablets would be expelled during phase 3 contractions. The authors concluded that gastric emptying from a fed stomach is not only controlled by pyloric sphincter diameter but also pressure gradient between antrum and duodenum. However, patient numbers were small and findings did not reach statistical relevance.

Gastric transit times of various preparations, dosed to volunteers in different nutritional states showed that solutions and small pellets were consistently emptied in fed conditions. This contrasted with variable transit times of larger single units, depending on caloric intake [18]. Similar findings were reported for gastric emptying of pellets containing tiaprofenic acid, following light (1,500 kJ) and heavy (3,600 kJ) breakfasts. Times for 50% emptying were consistent for each fed state, viz. (77.8 ± 8.2 min) after light breakfast and (170 ± 10.5 min) following the heavier meal [19]. Figure 13.1 compares transit rates of tablets and smaller pellets.

Overall, the corpus of evidence on small multiparticulate units is that gastric emptying is more consistent than for larger single unit dosage forms.

13.3.3 Intestinal Transit

Intestinal transit rates of solids are much more consistent than gastric transit. Passage is relatively independent of dosage form size or type (monolith or multiunit). Several studies have indicated that intestinal transit time is about 3 h [10, 12, 19, 21].

Multiparticulates have been found to have longer colonic transit time compared to single units [12]. Such extended residence could be advantageous for colonic drug delivery.

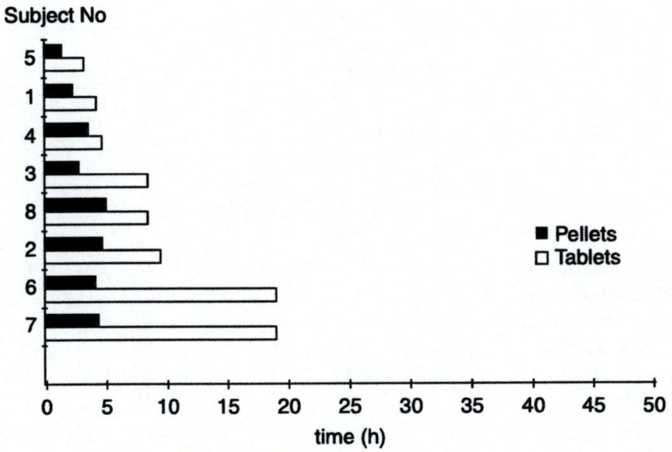

Fig. 13.1 Gastric transit time of pellets and single unit tablets after a breakfast of 2,800 kJ [20]

13.4 Functional Coats and Their Design

13.4.1 Fast Dissolving Coatings

Fast dissolving coatings may be used for taste masking, stabilization (e.g., if drug is photosensitive the coat can contain opacifying agents), differentiation during manufacture, to enhance physical stability (e.g., attrition during processing), or simply to improve appearance. Controlling drug release can mean protecting drug from an environment in which it is unstable. When the unit leaves the hostile environment, however, the coat needs to be removed or otherwise lose its protective functionality so that drug becomes available for absorption. Coats offering such "protection" must then dissolve or erode readily. Various low viscosity cellulose ethers, vinyl polymers, and basic methacrylic copolymers are used for such purposes (Table 13.1).

Several easy-to-use products containing these polymers are available in the market (Table 13.2). Commonly, they are powder mixtures that are dissolved or dispersed in aqueous or other vehicles before coating, without the need to add other excipients.

13.4.2 Enteric Coatings

Enteric coats are used to obviate gastric irritation, drug degradation at low pH, or to delay release until the unit reaches the small intestine or even the colon, possibly to evince a "local" effect or to ensure better absorption. They usually comprise acidic polymers as film formers. The pH at which the coat dissolves is determined by

Table 13.1 Fast dissolving coating polymers [22]

Polymer	Commercial products (selection)	Monographs	Manufacturer
HPMC (Hydroxypropyl methylcellulose, hypromellose)	Methocel® E3/E5/E6/E15	Ph.Eur, USP, JP	Dow Chemical
	Pharmacoat® 603/645/606/615		Shin-Etsu
	Walocel® HM 3 PA/HM 5 PA/HM 6 PA/HM 15 PA		Wolff Cellulosics
	Spectracel™		Sensient
Poly(butyl methacrylate-co-(2-dimethylaminoethyl) methacrylate-co-methyl methacrylate) 1:2:1	Eudragit® E PO/E 100	Ph.Eur, JP, DMF 1242 (USA)	Evonik
PVA–PEG Copolymer (Polyvinylalcohol–polyethylene glycol-copolymer)	Kollicoat® IR	Ph.Eur, USP, JP	BASF

Table 13.2 Commercial formulations for fast dissolution [22]

Commercial product (selection)	Polymer	Manufacturer
AquaPolish MS	HPMC/HPC	BioGrund
AquaPolish® TC	EUDRAGIT® E PO	
Kollicoat® IR/IR white	PVA–PEG copolymer	BASF
Kollicoat® Protect	Kollicoat® IR/PVA	
LustreClear™	Carrageenan/MCC	FMC
Opadry® II	PVA	Colorcon
Opadry® AMB	PVA	
Sepisperse™ Dry	HPMC/MCC	Seppic
Sepifilm™ LP/003/752		
Spectrablend™	HPMC/acrylate/shellac	Sensient
Spectrafilm™		
Spectracoat™		

PVA polyvinyl alcohol, *PEG* polyethylene glycol, *MCC* microcrystalline cellulose, *HPC* Hydroxypropylcellulose, *HPMC* Hydroxypropyl methylcellulose

polymer type or by using a mixture of polymers. Such pH-dependent solubility is related to polymer structure and content of acid groups. Enteric coating materials comprise anionic methacrylic copolymers or (less usually) cellulosic and vinyl monoesters of phthalic acid (Table 13.3).

Ready-made formulations of enteric coatings are commercially available, containing (besides the polymer) additives such as plasticizers, glidants, pigments, dispersing or wetting agents to facilitate suspension preparation and application. Table 13.4 lists some such commercial products.

It is also possible to coat different populations of multiparticulates with different polymers that dissolve at different intestinal pH values. Drug release may consequently be "pulsed" as the unit transits the stomach and small intestine. Such delivery can prolong an effect by sustaining plasma levels or can align release (and subsequent plasma levels) with "time of need" of the medication.

Table 13.3 Polymers for enteric coating [22]

Polymer	Commercial product (selection)	Monographs	Manufacturer
CAP (Cellulose acetate phthalate)	Aquacoat® CPD	EP, USP/NF, JPE	FMC
	Eastman C-A-P NF	EP, USP/NF	Eastman
CMC-Na (Carboxymethylcellulose-sodium)	Walocel® CRT A	EP, USP, JPE	Wolff Cellulosics
HPMCAS (Hydroxypropyl methyl-cellulose acetate succinate)	Aqoat®	JPE, NF	Shin-Etsu
HPMCP (Hydroxypropyl methylcellulose phthalate)	HP 50/HP 55	EP, USP, JPE	Shin-Etsu
Poly(methylacrylate-co-methyl methacrylate-co-methacrylic acid) 7:3:1	EUDRAGIT® FS 30 D	DMF 13941 (USA)	Evonik
Poly(methacrylic acid-co-ethyl acrylate) 1:1	EUDRAGIT® L 30 D-55/ L 100-55	EP, USP/NF, JPE	Evonik
	Kollicoat® MAE 30 DP/100 P		BASF
	Eastacryl 30 D NF		Eastman
Poly(methacrylic acid-co-methyl methacrylate) 1:1	EUDRAGIT® L 12,5/ EUDRAGIT® L 100	EP, USP/NF EP, USP/NF, JPE	Evonik
Poly(methacrylic acid-co-methyl methacrylate) 1:2	EUDRAGIT® S 12,5/ EUDRAGIT® S 100	EP, USP/NF EP, USP/NF, JPE	Evonik

Table 13.4 Commercial formulations for enteric coating [22]

Commercial products (selection)	Polymer	Manufacturer
Acryl-EZE®	EUDRAGIT® L 100-55	Colorcon
AquaPolish® E	Poly(meth)acrylate	BioGrund
Opadry® Enteric	Polyvinyl acetate phthalate	Colorcon
Sureteric®	Polyvinyl acetate phthalate	Colorcon

Polymers that dissolve at relatively high pH, e.g., 7.0 can be considered for drug delivery to the lower GI tract (colon) for local or even systemic effect. Polysaccharides degraded by colonic bacteria may also be appropriate coat components [23].

13.4.3 Time Controlled Release Coatings and Special Controlled Release Functionalities

Film formers for time controlled release are usually insoluble but swell under physiological conditions. Table 13.5 lists commonly used polymers. Drug release is controlled by diffusion through the swollen coat. Diffusion is influenced by polymer characteristics (hydrophilic and swelling properties), overall hydrophilicity of the applied coat, and coat thickness.

Table 13.5 Polymers for time controlled release [22]

Polymer	Commercial product (selection)	Monographs	Manufacturer
CA (Cellulose acetate)	Eastman CA	EP, USP	Eastman
CAB (Cellulose acetate butyrate)	Eastman CAB	EP, USP	Eastman
EC (Ethylcellulose)	Ethocel™	EP, USP	Dow Chemical
	Aquacoat® ECD		FMC
	Surelease® (ready-to-use)		Colorcon
Glyceride	GatteCoat™		Gattefosse
Poly(ethyl acrylate-co-methyl methacrylate) 2:1	EUDRAGIT® NE 30 D EUDRAGIT® NM 30 D	EP, JPE EP	Evonik
Poly(ethyl acrylate-co-methyl methacrylate-co-trimethyl ammonioethyl methacrylate chloride) 1:2:0.2	EUDRAGIT® RL 30 D EUDRAGIT® RL 100/ RL PO	USP/NF EP, USP/NF, JPE	Evonik
Poly(ethyl acrylate-co-methyl methacrylate-co-trimethyl ammonio ethyl methacrylate chloride) 1:2:0.1	EUDRAGIT® RS 30 D EUDRAGIT® RS 100/ RS PO	USP/NF EP, USP/NF, JPE	Evonik
PVAc (Polyvinyl acetate)	Kollicoat® SR 30 D		BASF
HPMC/CMC	Walocel® HM-PPA		Wolff Cellulosics

Strategies for time controlled release include:

- Combinations of soluble and insoluble film formers
- Use of pore formers
- Layering coats with different functionalities
- Specific film formulations which leverage the reactivity of polymeric functional groups as, e.g., salt formation of EUDRAGIT® RL/RS with anions that modify coat permeability [24]

Film coats that provide time-specific drug release need to confer drug release behavior that is aligned with the properties of the medication and the clinical condition. Hence most such time-controlled formulations are bespoke. An exception is Surelease® (Colorcon) which contains ethylcellulose as film former and is available in "ready-to-use" form.

13.4.4 Coat Composition

The film former, usually comprising a polymeric material or mixtures of polymers is the key component in a coat. It is also invariably present in greatest amount. However, other materials are also necessary to enhance performance, confer mechanical stability, or facilitate coat application. These include but are not limited to:

- A plasticizer to confer flexibility and ensure that coat cracking or other physical changes do not compromise drug releasing properties.

Table 13.6 Coating compositions [22]

Component	Poly(meth)acrylate film EUDRAGIT® NE 30 D 416.5	Ethylcellulose film Aquacoat® ECD 780.4
Film former (dry weight equivalent)	125.0	234.1
Plasticizer (dibutyl sebacate)	–	58.0
Antitacking agent (talc)	125.0	–
Vehicle (water)	458.5	161.6
Total weight	1,000.0	1,000.0

- Process aids such as antitacking agents to obviate particle aggregation during processing or storage.
- Dispersants, solvents, or wetting agents.
- Other materials to enhance appearance or provide differentiation (e.g., pigments, waxes). Some of these may be added as a "top coat."
- A pore former, where indicated to provide channels or ports for drug release.

Choice of formulation should not be based solely on functionality (release) requirements but should also consider incompatibilities between drug and coat ingredients and possibly impurities or residues in the coat components.

Coats can be applied from aqueous dispersions or are dissolved/dispersed in organic solvents. Constraints on organic solvent usage may well determine the type of vehicle employed. This in turn may influence coat component selection. Table 13.6 lists example compositions of aqueous dispersion formulations based on methacrylate or cellulose ether film formers.

13.5 Release Mechanisms

Multiparticulates can comprise pellets, granules, mini tablets, or mixtures thereof. They may be designed to release drug slowly, rapidly, or following a delay as discussed earlier. Mechanisms underpinning such release are illustrated in Figs. 13.2 and 13.3.

Figure 13.4 illustrates the potential flexibility of multiparticulates for delivering drug in a variety of modes from a single unit. In principle, different drugs with differing release requirements can also be incorporated in such units, allowing simultaneous, sequential, or other such delivery that can be aligned with mode of activity or clinical condition.

Matrix systems control drug release by diffusion- or erosion-mediated kinetics following matrix hydration. An outer coat can also be applied, e.g., to avoid release in the stomach. For example, a core with swelling properties can be coated with an insoluble but water permeable film that ruptures after a defined time for rapid release of drug, thereby providing time-controlled or pulsatile delivery [25].

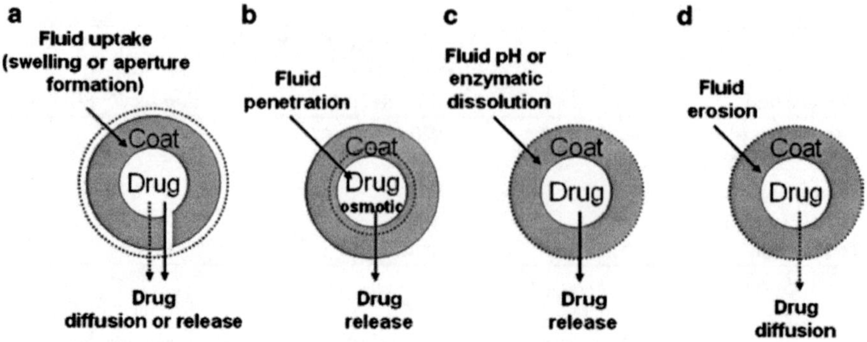

Fig. 13.2 Drug release from *coated particles*. (**a**) Release mediated by coat swelling or pore formation. (**b**) Osmotic *pressure*-driven release. (**c**) pH or enzymatic controlled coat removal. (**d**) Coat erosion-mediated release

Fig. 13.3 Drug release from *matrix pellets*. (**a**) Diffusion controlled, (**b**) erosion controlled

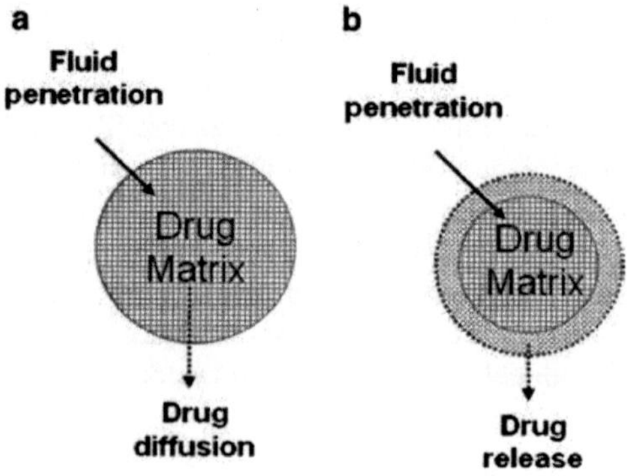

Fig. 13.4 Flexibility of multiparticulate technology

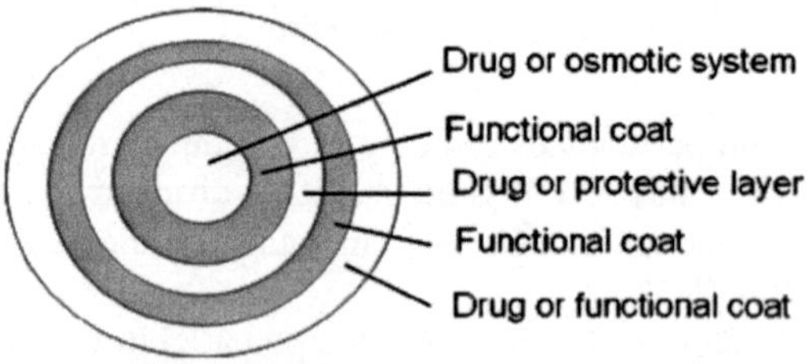

13.6 Factors Influencing Design and Development

13.6.1 API Characteristics

High dose drugs must of necessity comprise a major component in a microparticle. Such high inclusion level may require that a strong binder be used to ensure adequate particle robustness for processing. It may be possible, on occasion to apply the functional coat directly to the drug, no pregranulation being needed.

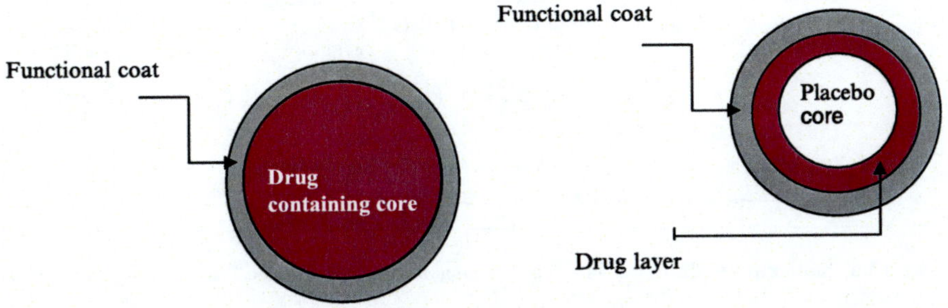

Fig. 13.5 Particle design depending on dose strength, *left*: high dose, *right*: low dose

Low dose drugs can be layered onto an inert core. The active can either be layered as solid, being fixed by a concentrated polymer binder solution (*powder layering*) or alternatively be applied as solution or suspension. Solution application can provide uniform dense API layers. It may also be possible to embed drug in the functional polymer.

Application from aqueous systems needs to take account of drug stability, viz., potential for hydrolytic degradation. However, propensity for hydrolysis may not mean that degradation is an issue, since contact time with water during coating is limited and only happens until the surface is sealed. Prudent process design and appropriate process controls may obviate hydrolysis enabling aqueous coating. Drug interactions with nonaqueous solvents or residues therein also need to be considered during process design.

13.6.2 Incompatibilities

All dosage form design programs must consider the potential and the consequences of any interactions between drug and materials with which it is partnered. Multiparticulates are no exception. Obviating such interactions requires knowledge of the drug, the excipients and, in the latter context the residues they may contain. Many release modifying materials, being polymeric are not "pure" in the strictest chemical sense. They comprise a mixture of polymeric chain lengths. Equally important other agents included in the coat or core can contain low levels of potentially "potent" residues such as aldehydes and peroxides. If such residues are "mobile" they may diffuse through the core or coat and interact with the active ingredient. Knowledge of residues or other impurities in excipients, and how they might vary over time or between batches can help judicious selection and evaluation of release modifiers and avoid stability issues that otherwise may only become apparent on long-term storage.

Ionic interactions between polymer and API can affect release characteristics [26]. High affinity between API and polymer may result in drug migration into the film, changing release characteristics. It may be possible to prevent such interaction by the application of a separate layer of neutral polymer like HPMC. Layer thickness

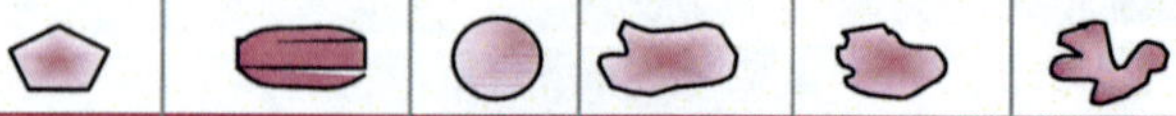

	Crystal	Mini tablet	Pellet	Dry granules	Wet granules	Fluid bed granules
Particle coating	-⇨+++	+++	+++	++	+	-

Fig. 13.6 Suitability of different shapes of particles for coating (*minus*: less suitable, *triple plus*: excellent)

of 10 μm is generally considered to be adequate for avoiding possible interactions. However, the diffusion propensity of a particular residue must be borne in mind and investigated.

13.6.3 Particle Core Design

Multiparticulate cores should preferably be dense, of uniform size and spheroidal shape, and have low porosity and sufficient hardness.

Pellets and mini- or microtablets intrinsically meet such requirements. Granules or milled products, in contrast, may require appropriate process controls such as screen classification. Needle-shaped particles are problematic as coating such shapes to consistent thickness is almost impossible. Furthermore, needles tend to break during coating exposing new surfaces and generating small particles. Both are risks for functionality. Pretreatment to confer more regular shape is warranted. Roller compaction, with subsequent size reduction and classification is one suitable technique (Fig. 13.6).

As release is invariably a function of the surface area interfacing with release medium it is important to ensure that surface area is consistent, between and within batches.

13.6.4 Coating Functionality

The polymer essentially determines the functionality of the coat. However, other ingredients can also impact release rate because of effects on properties such as hydrophilicity, porosity, and solubility. Such possibilities have to be considered during development and may be leveraged to achieve the desired release characteristics. Figure 13.7 illustrates how different antitack agents, present at functionality levels can influence release rates.

Reliable film functionality requires consistent coat thickness and integrity. Enteric coatings usually require thicknesses of 40–50 μm to confer gastroresistance.

Fig. 13.7 Influence of different antitacking agents on the release profile of phenylpropanolamine pellets coated with EUDRAGIT® RS 30 D

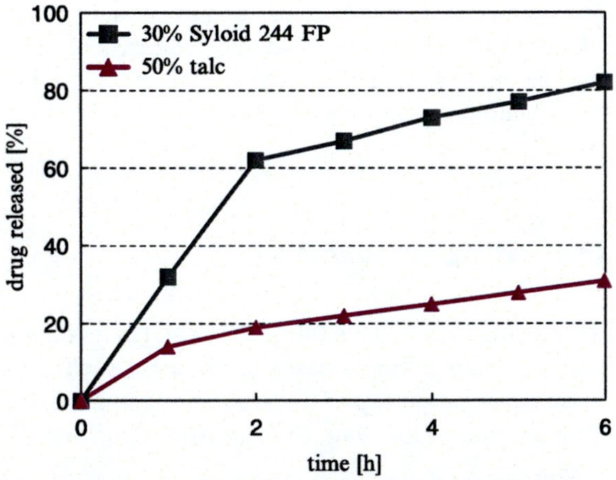

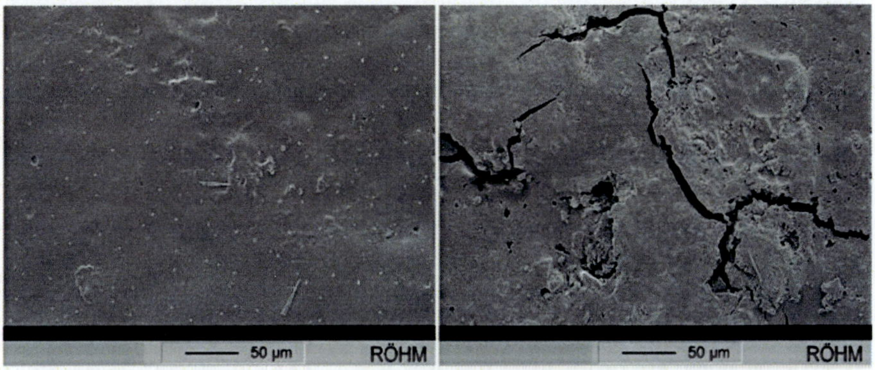

Fig. 13.8 SEM pictures: *left*: integer coating, *right*: cracks through insufficient plasticizer

For controlled release or protective applications, coat thicknesses must be designed to deliver the required release profile or protection. Film stability is assured by its flexibility but few polymers are intrinsically flexible. The majority require the presence of a plasticizer to avoid cracks developing during coating, subsequent processing, or storage. If coated particles are to be compressed to multiparticulate tablets, even higher film flexibility is required to withstand mechanical stress during tableting. Plasticizer needs to be matched with film forming polymer. Its impact can be checked by elongation-at-break measurements (ISO 527-3) or determination of Minimum Film forming Temperature (MFT) [27] (Fig. 13.8).

Pigments can affect coat porosity. Pigment binding capacity varies with polymer. Poly(meth)acrylates usually provide highest binding capacity, being capable of binding 300% of polymer weight. Cellulose-based polymers in contrast show only moderate binding capacities [27]. Loss or change of film functionality may be avoided by applying the pigments as a separate layer on top of the functional coat. Specific release profiles may require the sequential application of multiple coats.

An example is local treatment of inflammatory bowel diseases where a delayed release coat could ensure that drug release does not start before the formulation reaches the lower small intestine/colon. An inner insoluble coat can then prolong release during colonic transit [23].

13.7 Drug Candidates

Not all drugs are suitable for development in modified release form, nor is modification necessary in many cases. Drivers for such modification concern the various "biological" properties of the drug, the characteristics of the clinical condition, and the properties of the drug. Considerations may include:

- *Pharmacokinetic target*: The requisite plasma concentration, duration, and time of onset considerations (target plasma profile). This usually needs to be defined and validated by clinical experience. Lack of such information increases the risk of program failure.
- *Elimination (excretion) half life*: Drugs with half lives in the region 1–10 h are most suited to release modification, provided that plasma presence is directly related to therapeutic effect.
- *Therapeutic dose(s)*: If the aim of modifying release concerns prolonging duration of action this usually requires that a dose contain higher levels of drug. This can present unit size challenges where dose is high. It may be possible in such cases to utilize the active as the base particle for coating.
- *Dose response (therapeutic index)*: If dose response is steep, the target plasma concentration limits may be narrow. Performance standards are more stringent as a consequence. Multiparticulate formulations may be potentially advantageous in this context due to more consistent gastrointestinal transit and lower risk of "dose dumping" because of coat failure.
- *Location for drug absorption in GI tract*: The ideal drug for sustaining absorption is well and consistently absorbed throughout most of the GI tract, particularly the small intestine. In practice, pH–solubility effects, absorption windows, and susceptibility to metabolizing enzymes in the intestinal wall can affect absorption efficiency in the small intestine, regardless of technology used to modify release.
- *Physicochemical characteristics of the drug*: Absorption of drug and release from the dosage form requires that it be in solution. Solubility, however, can be affected by pH and possibly other environmental conditions in the GI tract. Multiparticulates may be advantageous with respect to release as their ready dispersion provides a larger surface area for dissolution than does a monolithic unit.

Multiparticulate systems offer an efficient platform for the development and manufacturing of products combining different drugs or one drug with different release profiles. For example, dual release can be achieved by the combination of a coated and uncoated fraction of the same multiparticulate composition. Dexlansoprazole, the R-enantiomer of the proton pump inhibitor lansoprazole, was developed using dual release to achieve a once-daily dosing regimen. The product combines enteric coated pellets releasing drug in the proximal duodenum and pellets that release drug in the distal small intestine [28]. Particles can also be "sprinkled" or otherwise dispersed in food or beverages to make for easier swallowing [29].

The same strategy has been adopted for treating Parkinson's disease patients. Dual release of a combination of two drugs (carbidopa and L-DOPA or benserazide-L-DOPA) was shown to prolong the therapeutic effect [30, 31].

Chronotherapy consists of aligning drug presence in the biosystem with circadian rhythm. Conditions that may be amenable to such therapy are hypertension, CNS disorders, cardiovascular disease, asthma, and cancer [32]. Multiparticulates have the potential, in certain cases to deliver drug(s) that provide such therapeutic advantage, being capable of providing appropriate release profiles [33]. Figure 13.9 shows a hypothetical design of a pulsatile releasing pellet.

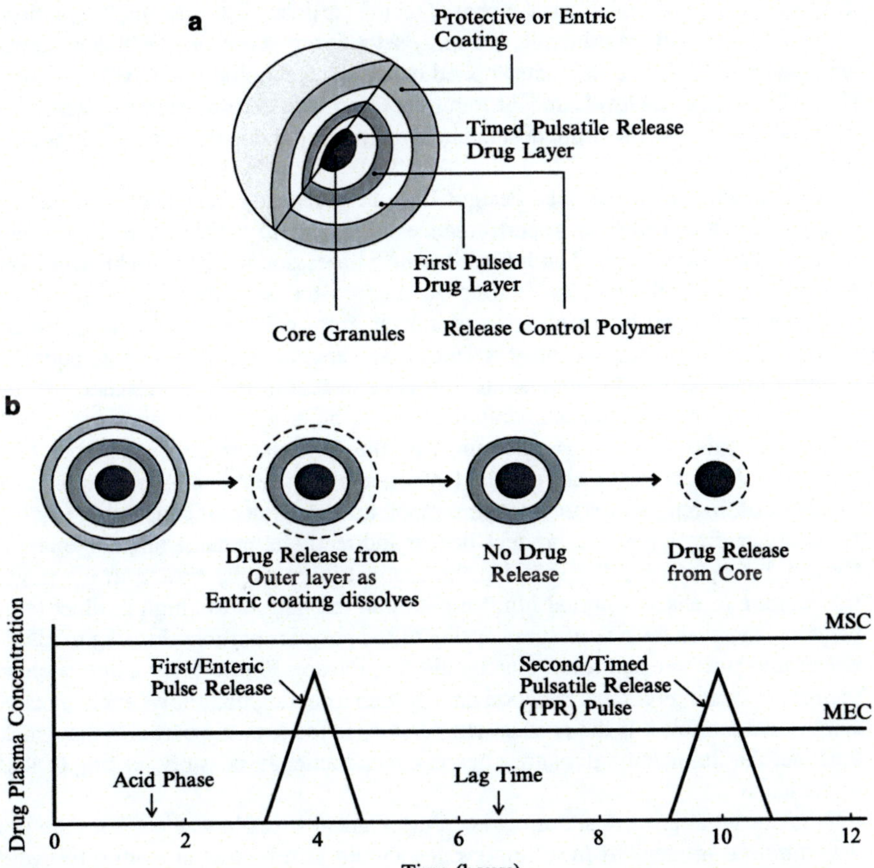

Fig. 13.9 Hypothetical design and plasma drug profile of a multiparticulate system. **(a)** Design of a pellet with multiple layers and **b**. **(b)** Subsequent bimodal plasma concentration profile [33]

Multiparticulate units can be advantageous for modified release formulations in early drug development programs where the therapeutic dose remains to be determined. Dose adjustment for dose–response studies can be readily made by filling different amounts of microparticles into capsules without any additional processing or formulation that could alter the release pattern.

13.8 Manufacture

13.8.1 Particle Coating

Particles of size 50 μm to 2 mm and mini tablets can readily be coated in conventional fluid bed equipment. The Wurster process is particularly suited to such layering and is frequently preferred setup due to its superior air streams providing longer residence of the coat in its "plastic phase" on the particles' surfaces and hence more uniform film quality. In principle, top spray setups with two-chamber filter cleaning are also feasible. However, coats applied using this technology may be less dense. Hence higher coating levels may be required. Coating in perforated pan coaters may be feasible with specific drum design but is not preferred due to suboptimal product movement patterns.

Various innovative fluid bed designs have been recently developed to optimize coating efficiency and process performance and to simplify scale up. Examples are the GEA Precision coater™ with FlexStream™ processor, the Oystar Huettlin Disk Jet® technology or the Innojet Ventilus®, and Aircoater® systems.

Films are formed by evaporation of solvent from polymer solutions or dispersions. Processing conditions must reflect the solvent and equipment. With aqueous polymer dispersions a homogeneous film is generated through coalescence of the polymer particles. Product temperature needs to be at least 10 K above the MFT (Minimum Film forming Temperature) of the formulation for acceptable film formation. Product temperature is the key process parameter, being a composite of contributions from spray rate, inlet air temperature, and inlet air capacity. It is measured either directly in the product bed or indirectly as exhaust air temperature. Usually before starting the spraying the product is gradually heated to the target temperature to ensure optimal film formation at the outset. Too high product temperatures increase the risk of spray drying of the polymer particles leading to insufficient film structure and level, thereby altering release. Figure 13.10 exhibits cross sections of coatings comparing good quality with a coating that shows enclosures of spray dried particles. If the product temperature is too low, the particles may stick together, the fluidization dynamics become dysfunctional, possibly leading to loss of the batch.

It is crucial at process startup to keep mechanical stress low as particles are not yet stabilized by applied film. Spray rates should also be low at startup to avoid penetration of water or solvent into the cores which may be subsequently difficult to remove due to the barrier in place, consequent to coat application.

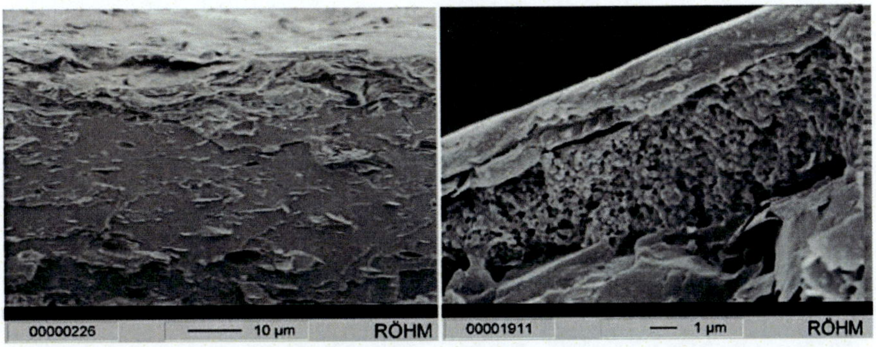

Fig. 13.10 Coating qualities: *left*: homogeneous and dense, *right*: enclosure of spray dried particles which may compromise film functionality

Some polymers applied as aqueous dispersions may require thermal treatment after spraying to enhance coalescence and film formation. This can be achieved by conventional tray drying or more efficiently by fluidization in the coating equipment. Polymer manufacturers usually provide polymer-specific recommendations for such processing.

Large-scale coating operations are usually more efficient than at laboratory scale. At the same time particles may encounter greater attrition at large scale due to higher material load and longer processing time.

In 2009, Vector corp. US launched a dry polymer coating fluid bed process [34] using the Granurex® conical rotor technology. Polymers that are available as micronized powders can be applied without organic solvents or other vehicle. An aqueous plasticizer solution is sprayed onto the pellets moving in a conical rotor. At the same time, the dry polymer powder is fed at a controlled rate from the powder feeder and dispersed onto the surface of the pellets via a powder spray gun. The dry polymer sticks to the surface of the pellet and the presence of the plasticizer in the spray causes the dry polymer to coalesce into a homogeneous film. Advantages for this process are much shorter process times, reduced or eliminated organic solvent usage, minimal exposure of the product to moisture, and decreased material preparation steps from the elimination of the need to prepare the polymer solution.

13.8.2 Encapsulation

The simplest way to provide a dose unit comprising microparticles is to fill into capsules. No special technology is required and coat rupture is not an issue as mechanical stress or high-pressure compaction is not employed. Neither are additional excipients necessary (except possibly a low level of lubricant). Volumetric or dosator filling devices are both suitable for encapsulation. A separate chapter in this book discusses capsules for use with controlled release preparations (Fig. 13.11).

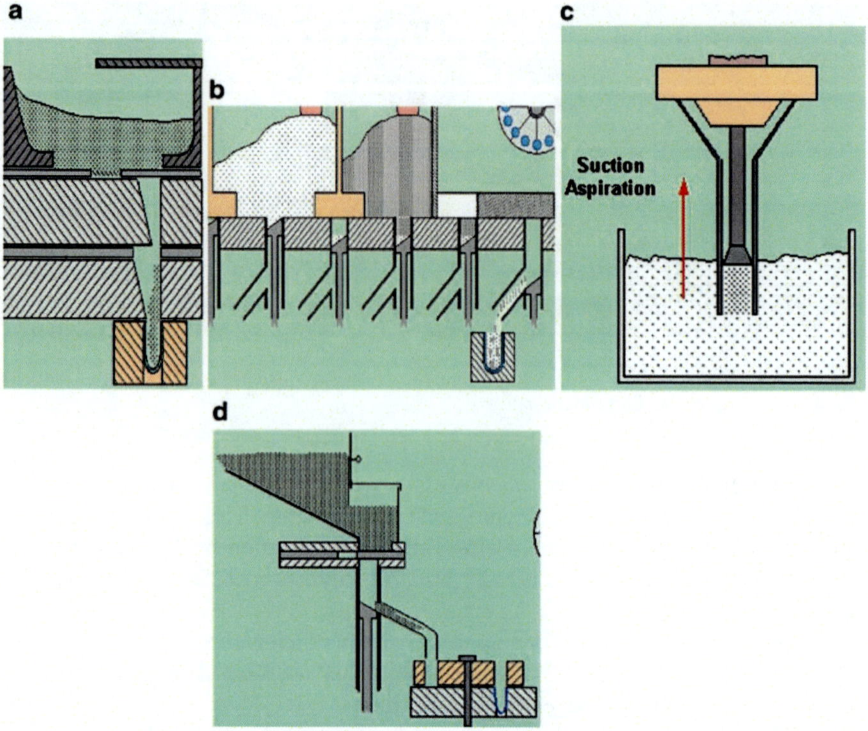

Fig. 13.11 Different capsule filling principles for multiparticulates. (**a**) Double slide filling, (**b**) dosage cylinder filling, (**c**) dosage tube filling, and (**d**) slide/piston dosing method

13.8.3 Tableting

Compressing coated particles requires that the coat withstand compaction stresses that might rupture coating and alter dissolution profile. Hence coatings need to be more flexible, possibly by adding higher levels of plasticizer. Concave, rather than flat-faced punches may be better aligned with particle shape, avoiding high local compression forces. Uniformity of mass and content may pose challenges as coated particles are usually larger than excipients, possibly posing segregation risks. Such potential hazards may explain why most dosage forms incorporating multiparticulates are capsules.

References

1. Gonzalez MA, Golub AL (1983) Theo-Dur® and Theo-Dur Sprinkle™ controlled-release delivery systems for theophylline. Drug Dev Ind Pharm 9(7):1379–1396
2. Zimmermann A, Walters JK, Katona B, Souney P (1997) Alternative methods of proton pump inhibitor administration. Consult Pharm 9:990–998

3. Bott S, Prakash C, Callum RW (1987) Medication induced esophageal injury: survey of the literature. Am J Gastroenterol 82(8):758–763
4. O'Neill JL, Remington TL (2003) Drug-induced esophageal injuries and dysphagia. Ann Pharmacother 37:1675–1684
5. Carlborg B, Densert O, Lindqvist C (1983) Tetracycline induced esophageal ulcers. A clinical and experimental study. Laryngoscope 93(2):184–187
6. Hey H, Jorgensen F, Sorensen K, Hasselbalch H (1982) Oesophageal transit of six commonly used tablets and capsules. Br Med J 285:1717–1719
7. Perkins AC, Wilson CG, Frier M, Blackshaw PE, Dansereau RJ, Vincent RM, Wenderoth D, Hathaway S, Li Z, Spiller RC (2001) The use of scintigraphy to demonstrate the rapid esophageal transit of the oval film-coated placebo risedronate tablet compared to a round uncoated placebo tablet when administered with minimal volumes of water. Int J Pharm 222:295–303
8. Osmanoglou E, Van der Voort IR, Fach K, Kosch O, Hartmann V, Stenzke A, Weitschies W, Wiedenmann B, Trahms L, Mönnikes H (2004) Oesophageal transport of solid dosage forms depends on body position, swallowing volume and pharyngeal propulsion velocity. Neurogastroenterol Motil 16:547–556
9. Weitschies W, Karaus M, Cordini D, Trahms L, Breitkreutz J, Semmler W (2001) Magnetic marker monitoring of disintegrating capsules. Eur J Pharm Sci 13:411–416
10. Weitschies W, Wedemeyer R-S, Kosch O, Fach K, Nagel S, Söderlind E, Trahms L, Abrahamsson B, Mönnikes H (2005) Impact of the intragastric location of extended release tablets on food interactions. J Control Release 108:375–385
11. Gottfries J, Svenheden A, Alpsten M, Bake B, Larsson A, Idström J-P (1996) Gastrointestinal transit of amoxicillin modified-release tablets and a placebo tablet including pharmacokinetic assessments of amoxicillin. Scand J Gastroenterol 31(1):49–53
12. Hardy JG, Healey JNC, Lee SW, Reynolds JR (1987) Gastrointestinal transit of an enteric-coated delayed-release 5-aminosalicylic acid tablet. Aliment Pharmacol Ther 1:209–216
13. Goodman K, Hodges LA, Band J, Stevens HNE, Weitschies W, Wilson CG (2010) Assessing gastrointestinal motility and disintegration profiles of magnetic tablets by a novel magnetic imaging device and gamma scintigraphy. Eur J Pharm Biopharm 74:84–92
14. Digenis GA, Sandefer EP, Parr AF, Beihn R, McClain C, Scheinthal BM, Ghebre-Sellassie I, Iyer U, Nesbitt RU, Randinitis E (1990) Gastrointestinal behavior of orally administered radiolabeled erythromycin pellets in man as determined by gamma scintigraphy. J Clin Pharmacol 30:621–631
15. Borin MT, Khare S, Beihn RM, Jay M (1990) The effect of food on gastrointestinal (GI) transit of sustained-release ibuprofen tablets as evaluated by gamma scintigraphy. Pharm Res 7(3):304–307
16. Khosla R, Feely LC, Davis SS (1989) Gastrointestinal transit of non-disintegrating tablets in fed subjects. Int J Pharm 53:107–117
17. Khosla R, Davis SS (1990) The effect of tablet size on the gastric emptying of non-disintegrating tablets. Int J Pharm 62:R9–R11
18. Davis SS, Hardy JG, Fara JW (1986) Transit of pharmaceutical dosage forms through the small intestine. Gut 27:886–892
19. Davis SS, Khosla R, Wilson CG, Washington N (1987) Gastrointestinal transit of a controlled-release pellet formulation of tiaprofenic acid and the effect of food. Int J Pharm 35:253–258
20. Abrahamsson B, Alpsten M, Jonsson UE, Lundberg PJ, Sandberg A, Sundgren M, Svenheden A, Tölli J (1996) Gastro-intestinal transit of a multiple-unit formulation (metoprolol CR/ZOK) and a non-disintegrating tablet with the emphasis on the colon. Int J Pharm 140:229–235
21. Davis SS, Hardy JG, Taylor MJ, Whalley DR, Wilson CG (1984) The effect of food on the gastrointestinal transit of pellets and an osmotic device (Osmet). Int J Pharm 21:331–340
22. Specht F (2006) APV basics, Praktikum Filmcoating, Kurs no 6022, Darmstadt 2006, by courtesy of F Specht and the APV
23. Bott C, Rudolph MW et al (2004) In vivo evaluation of a novel pH- and time-based multi-unit colonic drug delivery system. Aliment Pharmacol Ther 20:347–353

24. Wagner KG, McGinity JW (2002) Influence of chloride ion exchange on the permeability and drug release of EUDRAGIT® RS 30 D films. J Control Release 82(2–3):385–397
25. Arora S, Ali J, Ahuja A, Baboota S, Qureshi J (2006) Pulsatile drug delivery systems: an approach for controlled drug delivery. Indian J Pharm Sci 68:295–300
26. Ravishankar H, Patil P et al (2006) Modulated release metoprolol succinate formulation based on ionic interactions: in vivo proof of concept. J Control Release 111:65–72
27. Bauer KH, Lehmann K, Osterwald HP, Rothgang G (1998) Coated pharmaceutical dosage forms. Medpharm Scientific, Stuttgart, pp 82–98
28. Fock KM, Ang TL (2008) Dexlansoprazole, a modified release formulation of an enantiomer of lansoprazole, for the treatment of reflux esophagitis. Curr Opin Invest Drugs 9(10): 1108–1115
29. Metz DC, Vakily M, Dixit T, Mulford D (2009) Dual delayed release formulation of dexlansoprazole MR. A novel approach to overcome the limitations of conventional single release pump inhibitor therapy. Aliment Pharmacol Ther 29:928–937
30. Rubin AA (2000) Improvement in the treatment of Parkinson's disease and related disorders by novel formulations of the combination carbidopa-L-dopa. Patent WO2000015197
31. Gasser UE, Crevoisier C, Ouwerkerk M, Lankhaar G, Dingemanse J (1998) Comparative single- and multiple-dose pharmacokinetics of L-dopa and 3-O-methyldopa following a new dual release and a conventional slow release formulation of L-dopa and benserazide in healthy subjects. Eur J Pharm Biopharm 46:223–228
32. Khan Z, Pillay V, Choonara YE, du Toit LC (2009) Drug delivery technologies for chronotherapeutic applications. Pharm Dev Technol 14(6):602–612
33. Roy P, Shahiwala A (2009) Multiparticulate formulation approach to pulsatile drug delivery: current perspectives. J Control Release 134:74–80
34. Kucera S, Li J, Engels S, Jensen B (2009) Utilization of a novel rotor-granulator powder layering technique to taste mask acetaminophen pellets with EUDRAGIT® E PO. AAPS Annual Meeting, Los Angeles, 2009

Chapter 14
Capsules as a Delivery System for Modified-Release Products

Sven Stegemann

Abstract Capsules can offer particularly unique features for controlling drug delivery. They can accommodate multiparticulate compositions comprising pellets, beads, or even small tablets. Such units can be coated to control release, or drug may be embedded in a release-controlling matrix.

A capsule-based delivery system can have a number of advantages. The units need not be compressed, as is the case with tablets. Hence release-modifying coats are not in danger of being ruptured nor are release rates from matrices altered by compression forces. There is also a strong body of evidence that transit behaviors of small particles in the gastrointestinal tract are more consistent than for single larger units. Hence release, designed to be pH or time-dependent, and subsequent drug plasma profile is likely to be more consistent. Coat failure and consequent dumping of total dose is also far less likely with multiparticulates than with monolithic systems where the total dose is contained in a single unit. This is an important safety feature as controlled-release dosage forms usually contain a greater dose than units where release is not modified. Capsules can also readily accommodate units with different modes of release or different drugs.

Capsules have traditionally been formed from gelatin. Hydroxypropyl methylcellulose (HPMC) capsules have been made available in recent years as well. These HPMC capsules contain lower levels of residual moisture that may be beneficial with moisture-sensitive drugs or formulations. With these two types of capsules available, the possibilities in development and manufacturing of the various modified-release dosage forms has been broadened.

S. Stegemann (✉)
Capsugel, Rijksweg 11, B-2880 Bornem, Belgium
e-mail: sven.stegemann@capsugel.com

C.G. Wilson and P.J. Crowley (eds.), *Controlled Release in Oral Drug Delivery*, Advances in Delivery Science and Technology, DOI 10.1007/978-1-4614-1004-1_14,

14.1 Introduction

Gelatin capsules were one of the first drug delivery systems for pharmaceutical products. What started as a simple gelatin mold filled with the product and closed by a drop of gelatin has evolved into two different types of capsules:

- Soft gelatin capsule (SGC). This capsule is manufactured and filled in one step. It is usually used for immediate-release liquid products and for drugs requiring solubility-enhancing formulations such as nonaqueous solvents or complex emulsion systems.
- Hard capsules (HC) are premanufactured as two-piece units that are filled with the formulated drug and closed on separate high-speed filing machines. These are most commonly used for immediate-release and modified-release products.

Recent advancements in capsule-sealing technology have resulted in HC being employed to accommodate liquids, formulated to enhance drug solubility, dissolution and absorption or possibly a combination of a liquid and solid, the solid being capable of modifying the release of drug if required. Such dual release and combination products are being increasingly considered for designing individualized medicines.

This chapter focuses on the two-piece hard capsule as a drug delivery system and technology platform for modified-release products.

14.2 Capsules as a Technology Platform

14.2.1 Capsule Sizes, Materials, and Properties

Different design options exist for HC capsules, depending on the active ingredient (API) and fill composition. They may accommodate solids such as powder, granules, pellets, mini-tablets, or combinations thereof in standard two-piece capsules (e.g. ConiSnap® capsules: Fig. 14.1). Liquid, semisolid, or formulations comprising liquid and solid (e.g. pellets, capsules) can be contained using capsules designed for liquid fillings (e.g. Licaps capsule) and readily sealed by microspray technology (e.g. LEMS sealing) or band sealed with gelatin.

Units of appropriate volume are used to accommodate the required drug dose (Table 14.1). Dose can be adjusted by altering fill weight, as in the case of individualized medicines or where different doses of a novel entity are being evaluated in a blinded clinical trial.

Two-piece HC were traditionally manufactured from gelatin due to its unique gelling and film-forming properties within a very narrow temperature window.

More recently, capsules made from hypromellose (hydroxypropyl-methylcellulose, HPMC) became available. "First-generation" HPMC units also contained a gelling agent [gellan gum (e.g. Vcaps® capsules) or carrageenan (QualiV® capsules) and a

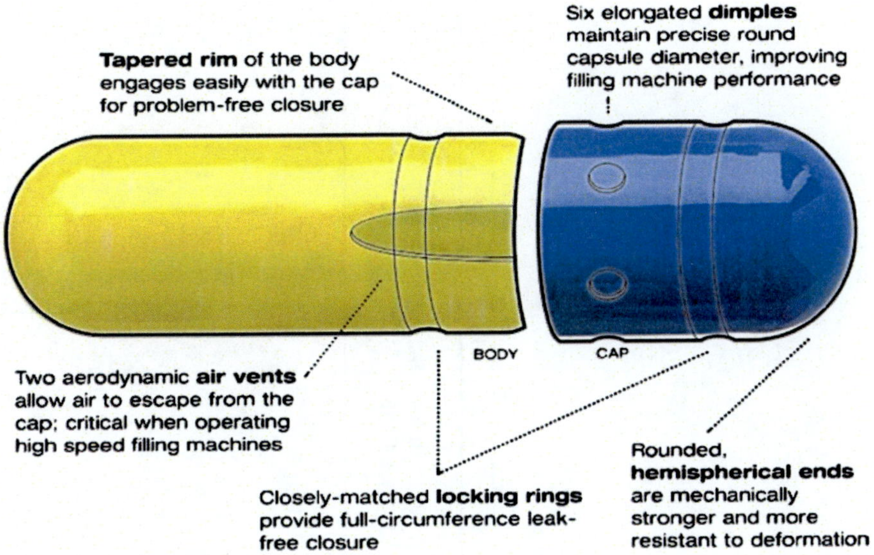

Fig. 14.1 ConiSnap® capsules

gel promoter (e.g. potassium chloride or potassium acetate)]. The gelling systems were found to affect drug delivery as capsule shell dissolution depended on pH and ionic strength of the dissolution medium.

A "second generation" of hypromellose capsules has also been developed. These comprise a "hypromellose-only" capsule (e.g. Vcaps Plus® capsules). Capsule shell dissolution is independent of pH and ionic strength providing the same predictability of release as gelatin capsules.

Table 14.2 summarizes the characteristics and properties of two-piece capsules.

Gelatin capsules have an equilibrium water content of 13–16% at 35–60% rh and 15–25°C. Capsules can be dried to below 13% water content if beneficial for moisture-sensitive drugs but decreasing water content can lead to brittle shells that may crack during product manufacture. Generally however, reduction in shell moisture to about 12% water does not cause excessive brittleness. Hypromellose capsules have an equilibrium moisture content of 4–9% under normal processing conditions. Such capsules can be dried to <1% water content if beneficial for moisture-sensitive drugs without losing their mechanical properties. Figure 14.2 compares equilibrium moisture contents for gelatin and hypromellose capsules. It is evident that both materials have comparable propensity at each specific relative humidity to donate moisture to materials with lower equilibrium relative humidity, less would be transferred in the case of HPMC because of its lower equilibrium moisture content.

Hypromellose, unlike gelatin does not interact with aldehyde residues that may be present in some drugs and excipients.

Table 14.1 Capsule sizes (in millimeter) and capsules volumes (in milliliter)

Size	000	00el	00	0el	0	1el	1	2el	2	3	4el	4	5
Volume	1.37	1.02	0.91	0.78	0.68	0.54	0.50	0.41	0.37	0.30	0.25	0.21	0.13
Closed length	1.029	0.995	0.917	0.953	0.854	0.804	0.765	0.760	0.709	0.626	0.621	0.563	0.437
External diameter	0.390	0.336	0.336	0.301	0.300	0.272	0.272	0.250	0.250	0.229	0.209	0.209	0.193

Table 14.2 Characteristics of two-piece capsules made of gelatin, hypromellose with gelling system (first generation) and pure hypromellose (second generation)

Polymer	Gelatin	HPMC First generation	HPMC Second generation
Material	Protein	Cellulose	Cellulose
Origin	Animal derived	Plant derived	Plant derived
Pharmacopoeial inclusion	EP, USP/NF, JP	EP, USP/NF, JP	EP, USP/NF, JP
Moisture content	13–16%	4–9%	4–9%
Oxygen permeability	Low	High	High
Cross-linking Propensity	Yes	No	No
pH-independent dissolution	Yes	Depends on the co-gelling agent	Yes

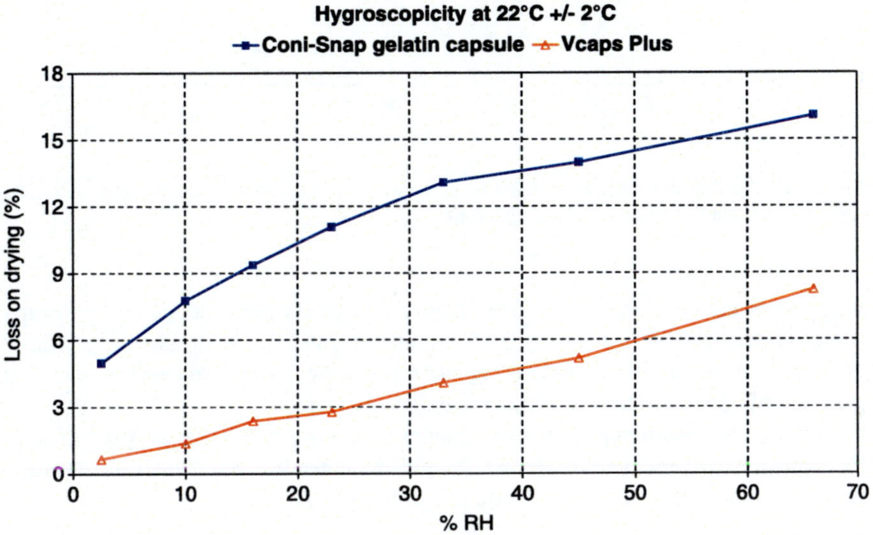

Fig. 14.2 Equilibrated moisture content of gelatin and hypromellose capsules after 1 week at different RH [1]

14.2.2 *In-Vitro Release from Capsules*

Gelatin capsules usually start to open at the cylindrical/hemispherical interface ("shoulder") within 2 min in aqueous media at 37°C. Liquid penetration causes wetting and disintegration/release of the contents while the capsule shells continue to dissolve [2]. Figure 14.3 shows the dissolution profile of a caffeine/lactose/croscarmellose formulation in USP apparatus 2 dissolution test in simulated gastric fluid pH 1.2 and simulated intestinal fluid pH 6.8.

As mentioned earlier, two generations of hypromellose capsules exist with different in vitro release profiles. First-generation hypromellose capsules have a pH and ionic strength-dependent dissolution profile that is related to the gelling system used. Carrageenan-based hypromellose capsules show a slow dissolution in media

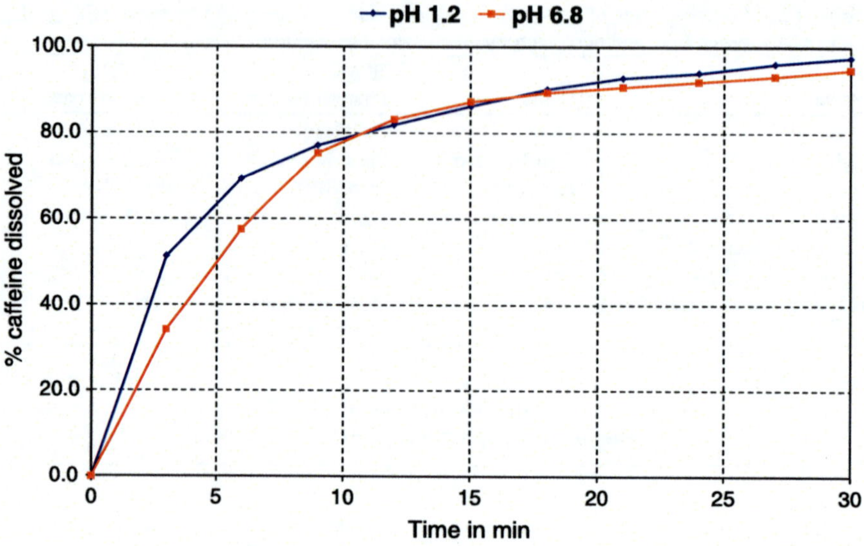

Fig. 14.3 Dissolution profile of caffeine from a hard gelatin capsule in simulated gastric fluid (pH 1.2) and simulated intestinal fluid (pH 6.8)

with pH 6.8 and high ionic strength [3]. In contrast, gellan-gum containing hypromellose capsules dissolve slowly at pH 1.2. Figure 14.4 shows the dissolution profile of an ibuprofen lactose blend in a carrageenan-based hypromellose capsule (viz. first generation) compared to release from a gelatin capsule.

Second-generation hypromellose capsules exhibit lag times that are pH and ionic strength-independent. Compared to the gelatin capsules, hypromellose capsules have a 3–4 min longer lag time as illustrated in Fig. 14.5.

14.2.3 Capsule Technology Platform

During the past decade, attitudes within regulatory agencies have changed from rigid procedure-based requirements and "mandatory guidelines" for dosage form design and manufacture to risk and science-based approaches incorporating an ethos of continuous improvement and quality by design (QbD). At the same time pharmaceutical manufacturers and R&D units seek to streamline activities, enhance efficiency, and reduce development timelines. The adoption of platform technologies is one component in such initiatives. Such technologies must suit a wide range of medicinal agents, whether dose or physicochemical properties if they are to be viable. Such requirements apply to modified-release dosage forms as well as more conventional units. Features of an ideal technology platform are summarized in Table 14.3.

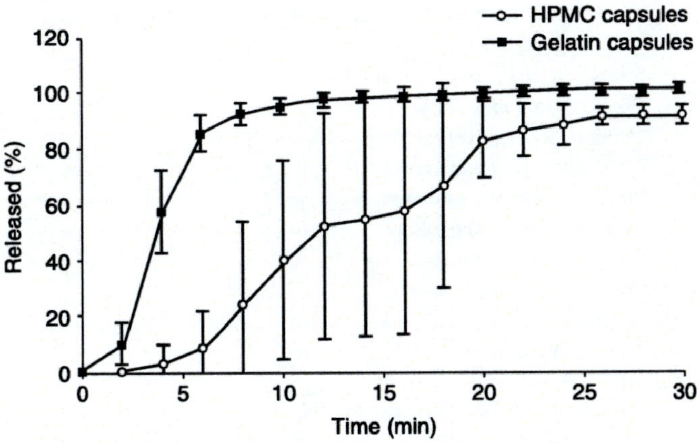

Fig. 14.4 Ibuprofen dissolution profile of a gelatin and a hypromellose capsule containing carrageenan as a gelling agent in potassium phosphate buffer pH 7.2 at 150 rpm [4]

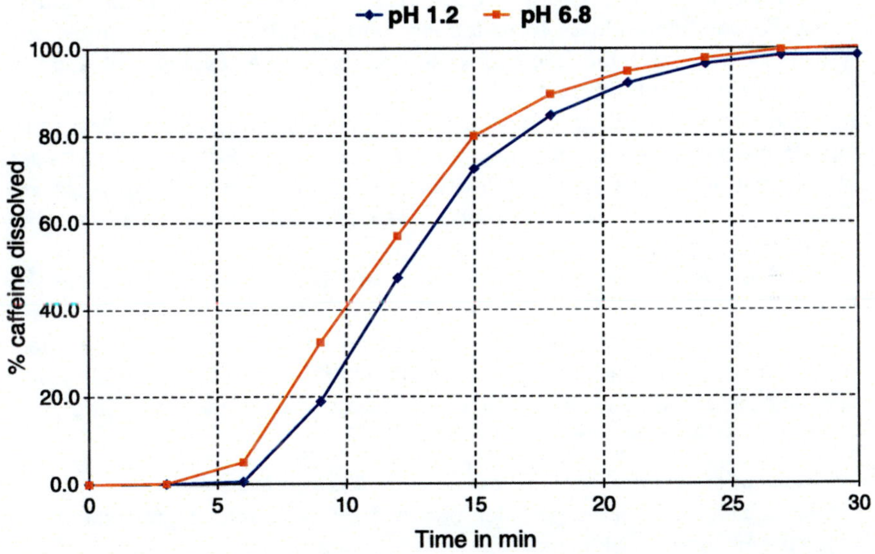

Fig. 14.5 Dissolution profile of caffeine in hypromellose gelling agent free capsules (Vcaps Plus® capsules) in simulated gastric fluid (pH 1.2) and simulated intestinal fluid (pH 6.8)

Capsules have become more attractive as a technology platform mainly in the light of their versatility and processing efficiency. Capsules (if suitably coated) can carry the drug or drug delivery system to the site of release in the gastrointestinal tract and then provide fast disintegration and dispersion of the API or formulation if that is what is desired. Further to this, HC are a flexible technology

Table 14.3 Features of an ideal technology platform for dosage form manufacture

Simplicity and ease of operation
Cost and time efficiency
Multiple drug delivery options
Flexibility in manufacturing
Product and process knowledge
Familiarity with technology
Robust and controllable
Cover small to large scale
Reusable
Quick change-over
Transferability
Suitable for worldwide manufacturing
Potential for innovation

platform as they can be used in a variety of different immediate-release and modified-release drug delivery formulations from development through to manufacturing. For modified-release formulations, hard capsules are a well-suited drug delivery system due to the ease of dose adjustment during development, and later the manufacturing of different dose strengths of commercial product from a single "platform intermediate." Several major products are modified-release hard capsule products, for example, omeprazole (Losec® and Prilosec®) and its second-generation esomaprazole (Nexium®), tamsulosin (Omnic®), venlafaxin (Effexor®) or the most recent introduction of a dual release form of dexlansoprazole (Dexilant®) [5].

In the field of controlled-release delivery, capsules can have certain advantages over single units where the formulation is compacted. Compression can rupture release-modifying membranes or coats and alter release rate. Capsules contents are not usually compacted to the same extent as tablets (if at all) and can readily accommodate a variety of multiparticulate systems without attendant compression or other release-altering stresses, e.g.:

- Coated mini- or microparticulates that control drug release.
- Matrix modified-release particles prepared by extrusion or granulation with release-modifying polymers.
- A number of "mini-tablets" providing different rates or modes of release.
- Mixtures of particles with differing release rates.
- Mixtures of "immediate" and controlled-release units.

Compression of such units to a tablet carries the risk of coat rupture or that compression force has variable effects on drug release. Filling into a capsule does not pose such a risk or other release-altering effects. Such possibilities can also be advantageous during new product development as it provides flexibility for evaluating different release modes in clinical studies (in "look-alike" units) without separate formulation programs for each and every dose variable, and for the rapid preparation of such units when trial findings warrant modification (Fig. 14.6).

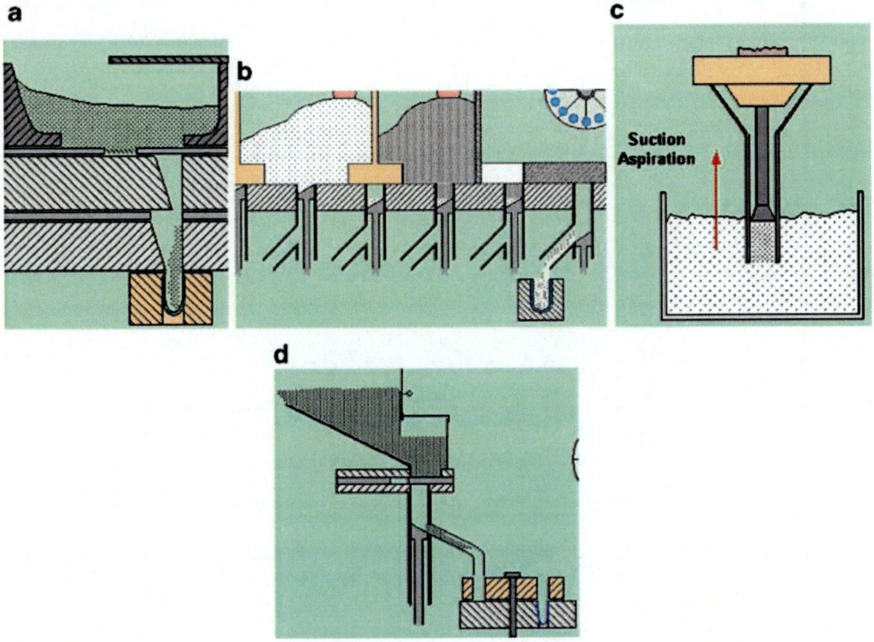

Fig. 14.6 Different capsule filling principles for multiparticulates. (**a**) Double slide filling, (**b**) Dosage cylinder filling, (**c**) Dosage tube filling and (**d**) Slide/piston dosing method

14.3 Capsules for Drug Delivery

14.3.1 In-Vivo Release from Capsules

In vivo disintegration and release of drug from capsules has been investigated using gamma scintigraphy and Magnetic Marker Imaging (MMI). Disintegration of a fast-release sumatriptan tablet that was over-encapsulated in a gelatin capsule shell was comparable to that observed with nonencapsulated tablets in fasted subjects, both in terms of shell dissolution and tablet disintegration. Individual values are shown in Fig. 14.7 [6].

Brown et al. [7] confirmed such fast in vivo disintegration in the fasted state, even when the gelatin was cross-linked and capsules failed to meet pharmacopoeial standards for disintegration time.

Digenis et al. [8] used gamma scintigraphy to study the effect of formaldehyde-induced cross-linking on disintegration time and shell rupture in fasted and fed volunteers. The bioavailability of the contained drug, amoxicillin, was also determined in a conventional bioavailability study. Findings are summarized in Table 14.4.

In vivo disintegration time (onset of capsule rupture) was prolonged in both fasted and fed subjects with cross-linked capsules. It was significantly longer with severely cross-linked capsules in fed subjects. Onset of absorption was also prolonged by

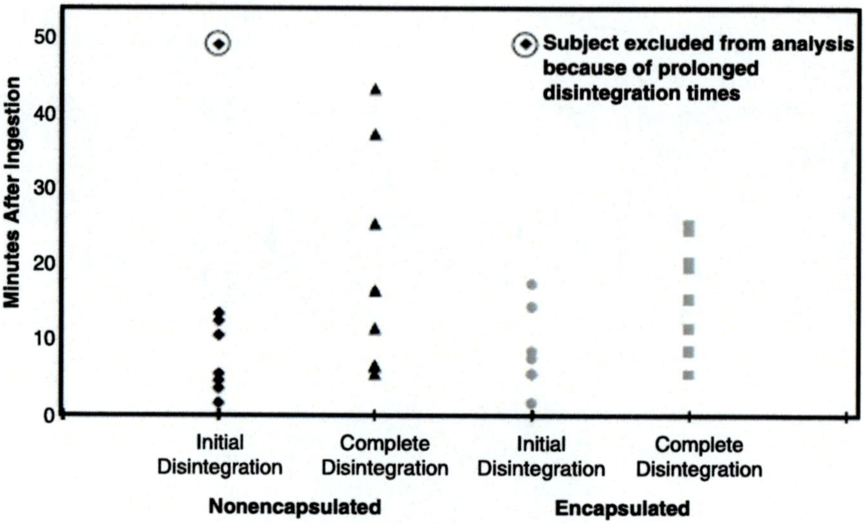

Fig. 14.7 Disintegration profiles of nonencapsulated and encapsulated sumatriptan 100 mg tablets [6]

Table 14.4 Effect of cross-linking on in vivo performance of gelatin capsules

Degree of cross-linking	In vivo disintegration (min)		Bioavailability*					
			Fasted			Fed		
	Fasted	Fed	AUC	C_{max}	T_{max}	AUC	C_{max}	T_{max}
None (Control)	7	11	18.03	7.78	1.17	18.86	7.02	1.5
Moderate	22	23	18.08	6.42	1.62	18.32	6.35	1.6
Severe	31	71	15.93	5.77	1.85	18.56	6.42	2.55

PK parameters are the usual ones viz. AUC (0-inf) = μg h ml^{-1}, C_{max} = μg ml^{-1}, T_{max} = hours

about 1 h in the same group. Numeric differences were also evident for the other PK parameters but were less striking and not statistically significant. Hence, for the usually more clinically relevant values of peak plasma concentration and area under the curve the impact of cross-linking, even with a compound-like amoxicillin, whose absorption window is in the upper small intestine, was hardly significant.

Cole et al. [9] also used scintigraphy and pharmacokinetic determinations (using ibuprofen) to characterize behavior of gelatin and 1st generation hypromellose capsules. Hypromellose capsule opening was hindered in acid conditions and in vivo disintegration (onset and complete) was significantly longer with hypromellose capsules. However, there were no numerical nor statistical differences in the pharmacokinetics of the ibuprofen, suggesting that the differences in disintegration were not biopharmaceutically relevant. Esophageal transit times were also comparable with both types of capsule.

MMI is also used to characterize dosage form behavior in vivo by means of magnetically marked dosage forms. The technique affords continuous measurement of location and trafficking of a dosage form in the GI tract. Hard gelatin capsules marked with iron oxide as a magnetic marker were evaluated in eight healthy volunteers for in vivo disintegration. In vitro disintegration corresponded well with in vivo performance, capsule rupture time being between 1.5 and 2.5 min [10].

These studies allay putative concerns suggested by in vitro studies that delayed shell disintegration might compromise in vivo performance.

14.3.2 Coating Gelatin Capsules

Capsules coated with materials preventing gelatin dissolution at low pH can be used to obviate gastric degradation of acid-unstable drugs. Such a dosage form may be particularly useful for acid-unstable drugs that are sensitive to compaction or need to be formulated as liquids or semisolids. Coating may also be utilized to delay delivery until a dosage form is further along the GI tract (e.g. colonic delivery) or to provide a mix of rapidly released (uncoated) and delayed release units to prolong plasma levels. pH or enzymatically sensitive coatings can be utilized to effect such delay in drug release and ensure local delivery [11]. Appropriately coated (pH or time-mediated release) units may also be useful during drug or new product development to establish whether drug is degraded or absorption is optimal in specific locations of the GI tract.

Enteric coating of hard gelatin capsules has been intensively reviewed by Thoma and Bechtold [12]. Commonly used polymers are:

- Polymethacrylates (copolymer of methacrylic acid and either methylmethacrylate or ethyl acrylate) (e.g. Eudragit®).
- Cellulose-based polymers such as cellulose acetate phthalate (e.g. Aquateric®).
- Polyvinyl derivatives such as polyvinyl acetate phthalate (e.g. Coateric®).

Plasticizers such as triacetin, triethyl citrate, diethyl phthalate, silicon oil, and acetyltriethyl citrate may also be present. Materials such as talc, magnesium stearate, and Aerosil® may be included as antiadhesion agents. Titanium dioxide or iron oxides can provide color.

The amount of film coat to be applied per capsule can be calculated based on capsule surface per capsule size according to the formula

$$A = \pi \times d \times h (\mathrm{mm}^2) \tag{14.1}$$

where by A is the area, d the diameter, and h the heights or length of the capsule.

A precoating may be applied to improve coat adhesion and reduce water uptake (softening) or water losses (brittleness) in gelatin capsules. Typically a 4 mg cm^{-2} precoat as an aqueous suspension of hydroxypropylcellulose (HPC) is used before applying the functional coat to gelatin capsules [13].

For colonic drug delivery, biodegradation actions by the colonic microflora are regarded as the most selective and hence most suitable to deliver drug to the colon [14]. The limitation with polysaccharide coatings is their hydrophilicity, solubility, and swelling properties in aqueous media [11]. To overcome these issues, attempts have been made to add a film-forming material (e.g. ethylcellulose) or applying a top coat with enteric properties [15]. Colonic delivery is the subject of a separate chapter in this book.

Pulsatile delivery systems offer the potential to deliver discrete sequential doses of drug that may be aligned with onset of action (delayed) or duration of effect by maintaining therapeutic plasma levels. Such delivery could be beneficial for drugs with short half-lives if associated with short duration of action and for chrono-pharmacotherapeutic drugs, where timed, pulsatile release could be aligned with time of greatest need or of optimal effect (e.g. Parkinson's disease, rheumatoid arthritis, asthma). Different populations of multiparticulate formulations, each population designed to provide the relevant release (time or place) contained in a capsule could be appropriate for such medications. Bussemer et al. investigated pulsatile drug delivery by a rupturable system comprising a coated capsule with a swellable polymer layer and a water-insoluble but water-permeable top coat. Croscarmellose sodium with PVP as a binder was incorporated in the swellable subcoat [16]. Ethylcellulose with hypromellose as pore former have been identified as the best coating combination to achieve predictable time-dependent rupture and pulsatile release [17, 18]. The same approach was successful with soft gelatin capsules [19]. The differences between soft and HC pulsatile delivery was ascribed to the shorter lag time for the soft gelatin capsules. Superior strength prevents a partial swelling discharge inwards as was observed with under-filled hard gelatin capsules.

14.3.3 Coating HPMC Capsules

The advent of hypromellose capsules has greatly simplified capsule coating using aqueous systems due to good polymer to polymer interaction and adhesion. Furthermore, hypromellose does not soften nor become brittle when aqueous film coating is applied. Cole et al. [20] reported the successful coating of hypromellose capsules with two types of pH-sensitive polymers viz. Eudragit® L 30 D-55 or Eudragit® FS 30D and confirmed the in vivo release using gamma scintigraphy. Optical microscopy showed a contiguous coat of uniform thickness around the capsule shell (Fig. 14.8).

Dissolution tests confirmed that the capsule coated with either 6 or 8 mg cm^2 Eudragit® L 30 D-55 did not release the drug acetaminophen over 2 h at pH 1.2, but released rapidly at pH 6.8. Capsules that were coated with either 6, 8 or 10 mg cm^2 Eudragit® FS 30 D did not release the drug at either pH 1.2 for 2 h nor pH 6.8 for an additional hour but released at pH 7.4. Gamma-scintigraphy studies in human volunteers confirmed that:

- Neither type of coated capsule disintegrated in the stomach.
- Eudragit® L 30 D-55-coated capsules disintegrated completely in the small intestine.
- Eudragit FS 30 D-coated capsules disintegrated between the mid distal small intestine and the proximal colon.

This study illustrated the potential for intestinal targeting through coated hypromellose capsules [20].

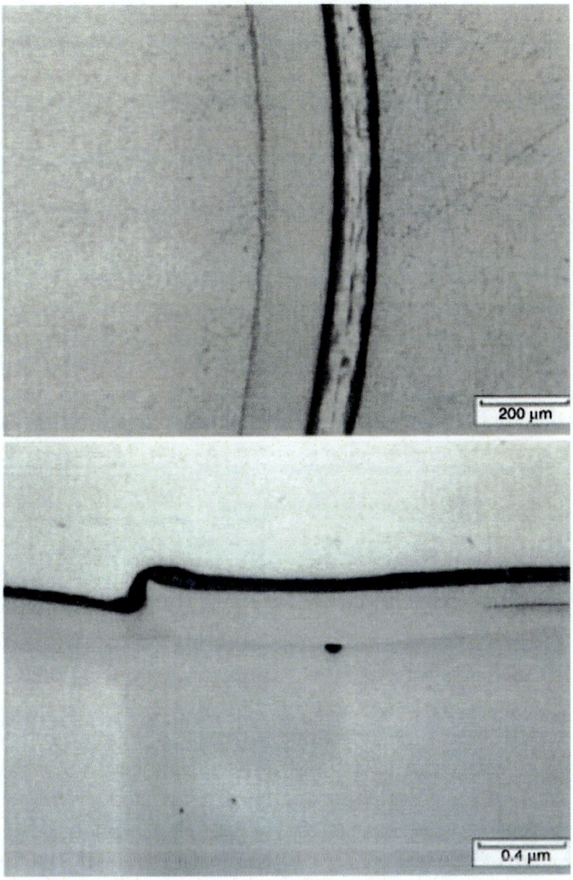

Fig. 14.8 Optical micrographs of HPMC capsules coated with Eudragit® L 30 D-55. (*Top*) Cross-section of domed end of capsule coated with 10 mg cm^2 Eudragit® L 30 D-55. (*Bottom*) Longitudinal cross-section through a capsule coated with 6 mg cm^2 Eudragit L 30 D-55 [20]

14.4 Capsules for Modified Release

14.4.1 Single Release

Multiparticulates in capsules received substantial attention in the mid-1980s as a way of controlling drug release, when the first proton pump inhibitor (PPI) omeprazole was developed. Omeprazole is extremely pH-sensitive, with a degradation half-life of less than 10 min below pH 4.0. Conventional immediate-release dosage forms were not appropriate so an enteric-coated multiparticulate pellet formulation was developed [21]. The pellet contained an alkaline-reacting component, a basic salt of omeprazole together with high pH-buffering materials in a subcoat and an enteric film top coat [US4786505, Nov 22, 1988 and US4853230, Aug 1, 1989]. Pellets were contained in two-piece hard gelatin capsules. The formulation protected the omeprazole from acid degradation in the stomach and released drug rapidly in the small intestine (Losec®). This technology has subsequently been applied to other acid labile compounds.

Roxithromycin is an oral macrolide antibiotic with half-life in simulated gastric fluid of about 14 min. Enteric-coated roxithromycin pellets were developed and compared in an in-vivo human study with a dispersible tablet formulation. Bioavailability was enhanced by 143% with the pellet formulation. T_{max} of the pellet formulation was delayed slightly consequent to the delayed release provided by the coat (2.83 h vs. 1.43 h) [22].

Multiparticulates, from which drug release is controlled have also been effective in treating chronic pain, with potent analgesic compounds having short biological half-lives and consequently limited duration of action. Tramadol, a drug for treating chronic pain has an elimination half-life of 5–6 h and requires 4–6 times a day dosing for pain control. A sustained-release pellet formulation was compared to the marketed SR tablet formulation in a bioequivalence study in 24 subjects. Findings were:

- Both formulations had similar AUC values.
- Pellets had lower C_{max} (148 vs. 183 μg l^{-1}).
- T_{max} was longer with pellets (5.9 vs. 4.9 h).
- Elimination half-life was significantly longer with pellets (13.4 vs. 10.4 h).

Moreover, the inter- and intrasubject variability in terms of rate and extent of absorption was much less for the pellet formulation. Hence, pellets provided a less variable and more prolonged plasma profile [23].

Morphine sulfate is also used for chronic pain, particularly in terminal illness. Elimination half-life is short (2.2 h) as is its duration of action (about 4 h). An extended-release capsule formulation comprising coated pellets provided plasma levels that sustained the analgesic effect for at least 12 h with dose proportionality over the dose range of 30–100 mg for C_{max} and AUC. Thus, the pellet formulation in capsules had equivalent performance to the twice-a-day SR tablet formulation [24].

Hydromorphone hydrochloride is an opioid analgesic with a short elimination half-life (2.3 h). Immediate-release formulations need to be dosed every 3–6 h to sustain analgesia. An ER pellet formulation was compared with an IR tablet for food effect, dose proportionality, and steady-state behavior. The pellet formulation demonstrated dose proportionality and provided more constant steady-state plasma levels with lower C_{max} and higher C_{min} plasma concentrations (Fig. 14.9). Adverse drug reactions were fewer in the pellet cohort, possibly ascribable to the smoother plasma profile [25].

Formulation of diclofenac as multiparticulates reduced in vivo disintegration lag time compared to coated monoliths. This lead to faster release, more rapid drug absorption, and increased bioavailability [26, 27].

Valproate is used to treat epilepsy. A relatively constant plasma concentration is necessary for optimal efficacy and safety, but its rapid absorption and fast elimination rate causes plasma concentrations to fluctuate. Wangemann et al. compared the pharmacokinetic profile of a multiparticulate capsule formulation with an enteric-coated tablet formulation, both containing 300 mg sodium valproate. Under a twice-daily

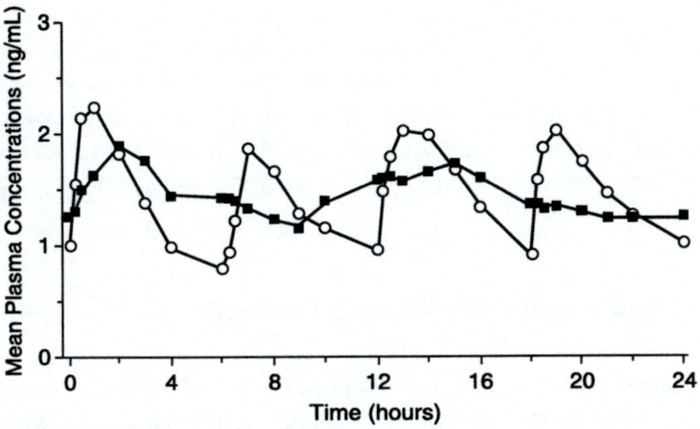

Fig. 14.9 Mean steady-state plasma concentration of hydromorphone hydrochloride following the administration of 12 mg ER pellet capsules once daily (*closed circle*) compared to every 6 h 3 mg IR tablets (*open circles*) [25]

regimen, both formulations showed comparable extent of absorption at steady state. However, the plasma fluctuations on dosage of the multiparticulate capsule formulation were only one-third of those seen on dosage of the enteric-coated tablet formulation [28].

Pulsatile delivery may enable the design of chronotherapeutic delivery systems [29]. Single-unit approaches using coated capsules have been described and discussed earlier [16–19]. Multiparticulate pulsatile drug delivery systems in capsules might reduce the unpredictability of gastric transit that is evident with monolithic units. They also offer the possibility of combining multiparticulate populations with distinct time-release profiles in a single capsule [30].

14.4.2 Dual Release

Dexlansoprazole is the R-enantiomer of Lansoprazole, sharing its proton pump acid inhibition properties and short elimination half-life (ca 1 h). A dual-release pellet formulation provides fast release of one population of pellets that are enteric-coated to release drug in the proximal duodenum. A second population is enteric-coated with polymer that dissolves at higher pH so that the remaining drug is released in the distal small intestine. Pharmacokinetic evaluation showed two plasma peaks after 1–2 h and 4–5 h, respectively, thereby providing sustained plasma levels of drug and potential for once-daily dosage [31]. A Phase III clinical study in erosive esophagitis and heartburn confirmed efficacy with once-daily dosing [32]. The dual-release, once-daily product is marketed under the trade name Dexilant® [33].

Tamsulosin is a selective α_{1a}-selective adrenoreceptor antagonist for treating benign prostate hyperplasia. A capsule product containing two populations of pellets coated with pH-sensitive polymers viz. either Eudragit® NE30D an insoluble low permeable film with time-controlled release or Eudragit® L30D-55 dissolving at pH >5.5 was developed to provide once-daily treatment and reduce cardiovascular adverse reactions associated with excessive α blockade [34]. Coreg CR® is another capsule-based product that contains three populations of pellets viz. an uncoated portion for fast release, a second population that releases drug at pH 5.5 and a third that releases at pH 6.4. Eudragit coats are utilized to provide the delayed-release profiles.

14.4.3 Combinations of Different Products

Chronic clinical conditions such as hypertension and diabetes usually require management with more than one medication (drug). Patient convenience can determine compliance which in turn can influence efficacy. Multiple unit and differently timed dosage are best avoided. At the same time better knowledge on the molecular biology of drug action (and clinical condition) and of temporal aspects of drug action teaches that dose frequency and time of dosage can be important. Additionally, genomic considerations and the increased interest in biomarkers points to personalized medicines and to products that combine different drugs [35].

It will be evident from earlier discussions that multiple units possibly with differing performance characteristics controlling drug delivery can be clinically beneficial. The same potential exists for dosage forms containing more than one drug viz. combination drug products. Different release profiles, or times, or location of release may be feasible to reflect the performance requirements of the individual drugs. Different doses and release (of the same drug) to reflect biological activity or requirements are also feasible.

Duodart® combines a small immediate-release dutasteride soft-gel capsule and sustained-release pellets of tamsulosin filled into one HC [GSK Media Press Release March 31, 2010].

14.5 Future Perspectives on Capsule Drug Delivery

14.5.1 Capsule Technology Platforms in Development and Manufacturing

Quality by Design (QbD) philosophies of generating a true product and process understanding encourage the concept of intrinsic reliability and reproducibility of manufacturing processes [36]. This is particularly relevant to modified-release preparations where too rapid or too slow release of drug can compromise safety or efficacy. It will be evident from previous comment in this chapter and information elsewhere in this book that multiparticulate-controlled delivery systems offer many possibilities for designing a variety of release modes. Multiparticulates can be converted to tablets by compression or can be filled directly into capsules. The latter operation carries little or no risk that mode or rate of drug release is altered. By contrast compaction stress carries the risk of coat rupture, or of altering the porosity of a matrix pellet. At the same time, compaction can reduce unit size due to the greater compact density. Processes or operating parameters and variabilities need to reflect such considerations.

Principles, processes, and associated equipment and monitoring systems for development and manufacture of coated pellets are well established [37–39]. Recent advances in pellet formation strategy as suggested by Roblegg using a single-component pellet system formed by calcium stearate and ibuprofen [40] and innovative dry powder coating technology [41] provide opportunities for additional process simplification and efficiency enhancements.

Tableting of pellets can be complex and multifaceted. Various factors have been shown to impact product quality and performance. Optimization may be difficult due to the complex nature of pellets, the need for excipients to aid tableting as well as the compression process per se. Bodmeier [42] reviewed the tableting of coated pellets and concluded that "the challenges of preparing tablets from coated pellets are evident." Later articles confirmed the challenge of compressing coated pellets to a tablet [43–46].

Abbaspour et al. investigated pellet load and compression force on friability, hardness, and disintegration of tableted pellets. Disintegration times of 7 min, being equivalent to usual capsule disintegration times, could only be achieved with a maximum pellet load of 60% and low compression force (5 kN). Hardness of resulting tablets was only just above the minimum requirement [47]. Friability can also be poor [48].

Encapsulation of multiparticulates into HC is a single-unit process allowing pellets to retain integrity during processing. A capsule technology platform allows filling of many kinds of modified-release drug delivery systems, like pellets, granules, mini-tablets and combinations thereof. In contrast to tableting, the encapsulation process is only dependent on a few parameters like pellet size, shape and the tendency to form aggregates. These do not normally compromise content uniformity and can be predetermined using computer simulation [49–51]. Fill-weight monitoring during encapsulation can accordingly be utilized as a Process Analytical Technology (PAT) tool for content uniformity assurance.

Many controlled-release products are marketed in several dose strengths. The above-mentioned morphine sulfate extended-release product (Kadian®) is available in eight different strengths from 10 to 200 mg in capsule presentations [24]. Due to the variety of different drug delivery technologies and combinations thereof, the number of available capsule sizes, their diversity in terms of color, color combinations or imprints which is critical for product safety, especially in polypharacy conditions [52] and the ease of dose adjustment on filling machines, a capsule technology platform provide flexibility and efficiency in development and manufacturing (Fig. 14.10).

A lean and simplified process is directly related to the number of unit operations and hence resources and input materials needed. Fewer unit operations and fewer excipients contribute to cost-efficient manufacturing [53].

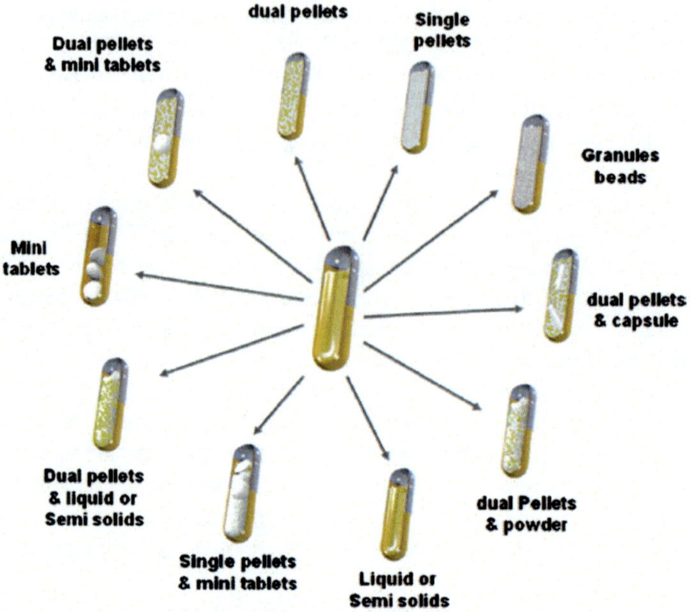

Fig. 14.10 Controlled drug delivery options provided by capsule technology

14.5.2 Individualized Medicines

Increasing understanding of disease mechanisms, the increasing prevalence of genotyping together with enhancements in diagnostic capabilities, including biomarkers have all contributed to the personalization of medicines (e.g. Herceptin®, Glivec®). Changing demographics and other societal changes worldwide are also likely to challenge traditional ways of product design, development, and manufacturing. There is likely to be demand for "individualized medicines" taking account of the disease and individual patient. Polypharmacy will enable the treatment of more than one condition (e.g. in elderly patients), but will pose challenges such as drug–drug-interactions. Dosing of different drugs must be tailored to the individual. This may require the availability or capability to dispense several strengths and combinations of medication products. To reduce pill burden and enhance compliance modified release and combination of different drugs in one product will be require a flexible dosage form and technology platform for its manufacturing [54].

Patient demographics are also a consideration, in particularly the increasing number of elderly patients. This patient group can suffer from dysphagia and have difficulty swallowing larger solid dosage forms. Capsules, suitably formulated may provide help overcome such difficulties. Multiparticulates can be sprinkled into drinks or on food on opening or even taken directly from the capsule, provided that accuracy of dose is not compromised (as could be the case with simple dry powder fills) [55, 56].

Multiparticulates have also been successfully administered via nasogastric tubing (if of appropriate dimensions). Esomeprazole enteric-coated pellets were removed from capsules and tested for their stability in different soft foods and beverages. With one exception (milk) pellets remained stable for up to 2 h in the food and beverages mixed with hydrochloric solution to mimic the gastric content [57]. Morphine sulfate pellets were investigated in a similar way and found to be stable in all tested foods and beverages. Pellet passage through two different tube systems, a gastrostomy tube with a 16 French size and a nasogastric tube with a 12 French size, was also assessed. Pellets passed through the 16 French tube, but were too large to pass the 12 French tube [58], emphasizing the need to evaluate each medication/tube system prior to adoption.

Looking ahead to the future of controlled drug delivery, it is apparent that encapsulated products will continue to play an important role, particularly with respect to manufacturing efficiency and the envisaged need for individualized medications.

14.6 Conclusion

Capsules can have a number of advantages for controlled drug-release delivery. These include:

- They can be readily formulated as single or multiple unit systems or contain more than one medication having different release modes.
- Capsule coating is a mature technology. Furthermore, the advent of hypromellose-based capsules offer additional or alternative functionality for controlled or otherwise targeted drug delivery because of ease of coating, in vitro–in vivo comparability of disintegration and freedom from residues that might interact with and destabilize some active ingredients.
- Coated capsules are far less likely to be damaged by the encapsulation process such that release is altered as a consequence.
- MMI and gamma scintigraphy have shown that capsules rapidly and consistently pass through the esophagus and disintegrate in the stomach within 5–10 min to release their contents. Furthermore, capsules do not have higher tendency to adhere to the esophagus than tablet dosage forms. They transit the esophagus in the same way as smaller oval coated tablets and possibly better than uncoated or large round tablets.
- In contrast to single-unit systems, multiple-unit systems have more predictable and less variable gastric retention times that can lead to more consistent plasma profiles.
- Capsule technology platform is readily amenable to Quality-by-Design (QbD) and Process Analytical Technology (PAT) control tools in accordance with the ICH Q 8, Q 9 and Q 10 guidelines (http://www.ich.org/cache/compo/276-254-1.html).

References

1. Ku MS, Li W, Dulin W, Donahue F, Cade D, Benameur H, Hutchison K (2010) Performance qualification of a new hypromellose capsule: Part 1. Comparative evaluation of physical, mechanical and processability quality attributes of Vcaps Plus®, Quali-V® and gelatin capsules. Int J Pharm 386(1–2):30–41
2. Ludwig A, Van Oetegham M, Delva A (1979) Disintegration of hard gelatin capsules. Part 1: Composition and structure of the capsule wall. Pharm Ind 41:796–798
3. Chiwele I, Jones BE, Podczeck F (2000) The shell dissolution of various hard capsules. Chem Pharm Bull 48(7):951–956
4. Honkanen O, Seppä H, Eerikäinen S, Tuominen R, Marvola M (2001) Bioavailability of ibuprofen from orally administered hydroxypropyl methyl cellulose capsules compared to corresponding gelatin capsules. S.T.P. Pharma Sci 11(2):181–185
5. Olsen KM, Hitzeman ML (2009) Dexlansoprazole MR in the management of gastroesophageal reflux disease. Clin Med Ther 1:1641–1652
6. Wilding IR, Clark D, Wray H, Alderman J, Muirhead N, Sikes CR (2005) In-vivo distegration profiles of encapsulated and non-encapsulated sumatriptan: gamma scintigraphy in healthy volunteers. J Clin Pharmacol 45:101–105
7. Brown J, Madit N, Cole ET, Wilding IR, Cade D (1998) The effect of cross-linking on the in-vivo disintegration of hard gelatin capsules. Pharm Res 15(7):1026–1030
8. Digenis GA, Sandefer EP, Page RC, Doll WJ, Gold TB, Darwazeh NB (2000) Bioequivalence study of stressed and nonstressed hard gelatin capsules using amoxicillin as a drug marker and gamma scintigraphy to confirm time and GI location of in-vivo capsule rupture. Pharm Res 17(5):572–582
9. Cole ET, Scott RA, Cade D, Connor AL, Wilding IR (2004) In-vitro and in-vivo pharmacoscintigraphic evaluation of ibuprofen hypromellose and gelatin capsules. Pharm Res 21(5):793–798
10. Weitschies W, Karaus M, Cordini D, Trahms L, Breitkreutz J, Semmler W (2001) Magnetic marker monitoring of disintegrating capsules. Eur J Pharm Sci 13:411–416
11. Chourasia MK, Jain SK (2003) Pharmaceutical approaches to colon targeted delivery systems. J Pharm Pharm Sci 6(1):33–66
12. Thoma K, Bechtold K (1992) Enteric coated hard gelatin capsules. Capsugel Library BAS 146 E
13. Degussa (2005) Enteric coating of hard gelatin capsules with Eudragit® L 30 D-55. Pharma Polymers 01/2005
14. Sinha VR, Kumria R (2001) Polysaccharides in colon-specific drug delivery. Int J Pharm 224:19–38
15. Singh BN (2007) Modified-release solid formulations for colonic delivery. Recent Pat Drug Deliv Formul 1:53–63
16. Bussemer T, Peppas NA, Bodmeier R (2003) Evaluation of the swelling, hydration and rupturing properties of the swelling layer of a ruturable pulsatile drug delivery system. Eur J Pharm Biopharm 56:261–270
17. Bussemer T, Peppas NA, Bodmeier R (2003) Time-dependent mechanical properties of polymeric coatings used in rupturable pulsatile release dosage forms. Drug Dev Ind Pharm 29(6):623–630
18. Bussemer T, Dashevsky A, Bodmeier R (2003) A pulsatile drug delivery system based on rupturable coated hard gelatin capsules. J Control Release 93:331–339
19. Bussemer T, Bodmeier R (2003) Formulation parameters affecting the performance of coated gelatin capsules with pulsatile release profiles. Int J Pharm 267:59–68
20. Cole ET, Scott RA, Connor AL, Wilding IR, Petereit H-U, Schminke C, Beckert T, Cade D (2002) Enteric coated HPMC capsules designed to achieve intestinal targeting. Int J Pharm 231:83–95
21. Pilbrant A, Cederberg C (1985) Development of an oral formulation of omeprazole. Scand J Gastroenterol 20(suppl 108):113–120

22. Sun J, Zhang T, Qui F, Liu Y, Tang J, Huang H, He Z (2005) Impact of a pharmaceutical dosage form on the pharmacokinetics of roxithromycin in healthy human volunteers. J Antimicrobiol Chemother 55:796–799
23. Cnota PJ, Nowak H, Tagarro I, Erb K, Schürer M, Schulz H-U, Maus J (2005) Tramadol SR formulations: pharmacokinetic comparison of a multiple-unit dose (capsule) versus a single-unit dose (tablet). Clin Drug Invest 25(7):435–443
24. Nicholson B (2008) Morphine sulfate extended-release capsule for treatment of chronic, moderate-to-severe pain. Expert Opin Pharmacother 9(9):1585–1594
25. Vashi V, Harris S, El-Tahtaway A, Wu D, Cipriano A (2005) Clinical pharmacology and pharmacokinetics of once-daily hydromorphone hydrochloride extended-release capsules. J Clin Pharmacol 45:547–554
26. Walter K, von Nieciecki A (2001) Relative bioavailability of diclophenac after a single administration of a new multiple-unit formulation with enteric-coated pellets. Drug Res 51(2):643–650
27. Hosny EA, El-Mahrouk GM, Gouda MW (1998) Formulation and in vitro and in vivo availability of diclofenac sodium enteric-coated beads. Drug Dev Ind Pharm 24(7):661–666
28. Wangemann M, Retzow A, Vens-Cappell B (1999) Pharmacokinetic characteristics of a new multiple unit sustained release formulation of sodium valproate. Int J Clin Pharmacol Ther 37(2):100–108
29. Shidhaye SS, Lotlikar VM, Ghule AM, Phutane PK, Kadam VJ (2010) Pulsatile delivery systems: an approach for chronotherapeutic deiseases. Sys Rev Pharm 1(1):55–61
30. Roy P, Shahiwala A (2009) Multiparticulate formulation approach to pulsatile drug delivery: Current perspectives. J Control Release 134:74–80
31. Metz DC, Vakily M, Dixit T, Mulford D (2009) Review article: dual delayed release formulation of dexlansoprazole MR, a novel approach to overcome the limitations of conventional single release proton pump inhibitor therapy. Aliment Pharmacol Ther 29:928–937
32. Metz DC, Howden CW, Perez MC, Larsen L, O'Neil J, Atkinson SN (2009) Clinical trial: dexlansoprazole MR, a proton pump inhibitor with dual delayed-release technology, effectively controls symptoms and prevents relaps in patients with healed erosive esophagitis. Aliment Pharmacol Ther 29:742–754
33. Abel C, Desilets AR, Willett K (2010) Dexlansoprazole in the treatment of esophagitis and gastroesophageal reflux disease. Ann Pharmacother 44:871–877
34. Zhang X, Tang X, Yang R (2009) Development of a tamsulosin hydrochloride controlled release capsule consisting of two different coated pellets. Drug Dev Ind Pharm 35:26–33
35. Rubenstein K (2008) Systems biology: a disruptive technology. Insight Pharma Reports. Insight Pharmaceuticals, Langhorne
36. Potter CJ, Berridge JC et al (2010) PQLI roadmap: product design, development, and realization, a science- and risk based approach to implementation – an overview of ISPE's first PQLI guide. Pharm Eng 30(2)
37. Tang ESK, Chan LW, Heng PWS (2005) Coating of multiparticulates for sustained release. Am J Drug Deliv 3(1):17–28
38. Rhodes CT, Porter SC (1998) Coatings for controlled-release drug delivery systems. Drug Dev Ind Pharm 24(12):1139–1154
39. Prabhakaran L, Prushothaman M, Sriganesan P (2009) Pharmaceutical micropellets: an overview. Pharm Rev 7(4)
40. Roblegg E, Ulbing S, Zeissmann S, Zimmer A (2010) Development of lipophilic calcium stearate pellets using ibuprofen as a model drug. Eur J Pharm Biopharm 75:56–62
41. Terebesi I, Bodmeier R (2010) Optimised process and formulation conditions for extended release dry polymer powder-coated pellets. Eur J Pharm Biopharm 75:63–70
42. Bodmeier R (1997) Tableting of coated pellets. Eur J Pham Biopharm 43:1–8
43. Türko lu M, Varol H, Celikok M (2004) Tableting and stability evaluation of enteric-coated omeprazole pellets. Eur J Pharm Biopharm 57:279–286
44. Ando M, Kojima S, Ozeki Y, Nakayama Y, Nabeshima T (2007) Development and evaluation of a novel dry-tablet technology for pellet as a substitute for conventional encapsulation technology. Int J Pharm 336:99–107

45. Abbaspour MR, Sadeghi F, Afrasiabi Garekani H (2008) Design and study of ibuprofen disintegrating sustained-release tablet comprising coated pellets. Eur J Pharm Biopharm 68:747–759
46. Dwibhashyam M, Ratna JV (2008) Key formulation variables in tableting of coated pellets. Indian J Pharm Sci 70:555–564
47. Rudnic E, Schwartz JB (1990) Oral solid dosage forms. In: Gennaro AR (ed) Remington's pharmaceutical sciences, 18th edn. Mack, Easton, PA, USA, pp 1633–1665
48. Gohel MC, Parikh RK, Brahmbhatt BK, Shah AR (2007) Improving the tablet characteristics and dissolution profile of ibuprofen by using a novel coprocessed superdisintegrant: a technical note. AAPS PharmSciTech 8(1):E94–E99
49. Rowe RC, York P, Colbourn EA, Roskilly SJ (2005) The influence of pellet shape, size and distribution on capsule filling – a preliminary evaluation of three-dimensional computer simulation using Monte-Carlo technique. Int J Pharm 300:32–37
50. Ali AM, de Matas M, York P, Rowe RC (2009) Influence of pellet aggregate populations on the variability of pellet filling into hard shell capsules: a comparison of experiments and computer simulation. Eur J Pharm Sci 38:197–2005
51. Ali AM, de Matas M, York P, Rowe RC (2010) Role of pellet size, shape, and filling method in achieving fill weight uniformity for encapsulated pelletized systems: a comparison of experiment and computer simulation. J Pharm Sci 99(3):1299–1308
52. Stegemann S (2005) Colored capsules – a contribution to drug safety. Pharm Ind 67(9): 1088–1095
53. Nichols J (2008) Reducing the cost of manufacturing through targeted processes, process intensification, and lean/continuous manufacturing. ISPE Knowledge Brief, http://www.ispe.org
54. Stegemann S, Ecker F, Maio M, Kraahs P, Wohlfart R, Breitkreuz J, Zimmer A, Bar-Shalom D, Hettrich P, Broegmann B (2010) Geriatric drug therapy: neglecting the inevitable majority. Aging Res Rev 9:384–398
55. Gonzalez MA, Golub AL (1983) Theo-Dur® and Theo-Dur Sprinkle™ controlled-release delivery systems for theophylline. Drug Dev Ind Pharm 9(7):1379–1396
56. Zimmermann A, Walters JK, Katona B, Souney P (1997) Alternative methods of proton pump inhibitor administration. Consult Pharm 9:990–998
57. Johnson DA, Roach AC, Carlsson AS, Karlsson AAS, Behr DE (2003) Stability of esomeprazole capsule content after in vitro suspension in common soft foods and beverages. Pharmacotherapy 23(6):731–734
58. Jones R, Hale E, Talomsin L, Philips R (1996) Kapanol™ capsules pellet formulation provides alternative methods of administration of sustained-release morphine sulfate. Clin Pharmacol 12(2):88–93

Chapter 15
Lipids in Oral Controlled Release Drug Delivery

Ben J. Boyd, Tri-Hung Nguyen, and Anette Müllertz

Abstract Lipids can perform many useful functions that modify or otherwise enhance drug absorption or disposition. They may influence release/absorption-controlling processes such as gastro retention and muco-adhesion. They can facilitate lymphatic transport of lipophilic drugs, and make them less susceptible to "first pass effects" by avoiding or reducing the hepatic portal route. Understanding lipid digestion and structure formation in vivo is key to exploiting possibilities for their use in controlling absorption or disposition. In particular, liquid crystalline structures are being increasingly investigated for their roles in the aforementioned processes. Such phenomena and possibilities are discussed in this chapter.

15.1 Introduction

15.1.1 Lipids and Lipid Self-Assembly in Oral Drug Delivery

There are a number of considerations associated with the use of lipids in dosage form design. In particular, the following properties and behaviors can play key roles, as well as forming the basis for lipid classifications:

- Lipids can act as solvents, leading to drug being present in the gastrointestinal tract (GIT) (at least initially) in solution thereby overcoming the drug dissolution step [1].
- Lipids may have amphiphilic structures that determine their capability to self-assemble in aqueous environments. Such behavior can have a critical effect on drug disposition kinetics in the GIT.

B.J. Boyd (✉)
Drug Delivery, Disposition and Dynamics, Monash Institute of Pharmaceutical Sciences, Monash University (Parkville Campus), Parkville, VIC, Australia
e-mail: Ben.Boyd@monash.edu

C.G. Wilson and P.J. Crowley (eds.), *Controlled Release in Oral Drug Delivery*, Advances in Delivery Science and Technology, DOI 10.1007/978-1-4614-1004-1_15,

- Lipids may or may not be digestible. Digestion of dietary and formulation lipids can lead to generation of colloidal structures in the GIT, providing transient solubilization of drug, and reducing the propensity for precipitation prior to absorption.

Understanding these issues can provide valuable insights into potential applications and choice of lipids for controlled drug delivery. These are discussed in the following sections.

15.1.1.1 Classification of Lipids

Structurally, lipids may be assigned to nine distinct groups viz. fatty acids, glycerolipids, glycerophospholipids, sphingolipids, sterolipids, prenol lipids, lycolipids, saccharolipids, and polyketides [2]. Fatty acids, glycerolipids, glycerophospholipids, and sterolipids are present in most human foods and are therefore most likely to be encountered in the GIT. Glycerolipids (acylglycerols) comprise mono-, di- or triesters with glycerol, with either short ($<C_8$), medium (C_8–C_{12}) and long ($>C_{12}$) fatty acid side chains. The majority of dietary and formulation lipids are glycerolipids of varying chain length and degree of branching.

Small et al. classified lipids based on their propensity for self-assembly with water [3]. Polar Class I lipids are usually present in animal tissue, are consumed in the diet, and are frequently components of lipid-based formulations. They include tri- and di-acylglycerides, protonated long-chain fatty acids and alcohols, waxes, cholesterol and fat soluble vitamins which neither swell nor dissolve in the bulk phase. However, the presence of hydrophilic groups allows these amphiphilic structures to spread at the air/water interface, differentiating them from nonpolar lipids. Class II lipids comprise monoacylglycerides and lecithins, which are also insoluble in water, but are capable of swelling, depending on the temperature, and aliphatic chain length, saturation and branching. To minimize the thermodynamically unfavorable interactions between the hydrophobic aliphatic chains and water Class II lipids can self-assemble, forming liquid crystal phases in bulk liquids and/or monolayers at air/water interfaces. Such assemblies are often formed during lipid digestion where digestion products such as free fatty acid salts and monoglycerides form liquid crystals prior to solubilization by endogenous amphiphilic Class III lipids.

Free fatty acids may be introduced to the GIT as formulation components, products of digestion, or impurities in formulation lipids. Their ionization influences lipid self-assembly, which in turn may influence in vivo release relative to in vitro performance. Protonation by gastric acid may cause formation of "inverse" structures by modifying the curvature of lipid assemblies, favoring smaller aqueous domains. Conversely, ionization in the small intestine favors micelle formation.

15.1.1.2 Lipid Dispersion and Digestion

Lipids encounter a number of biological and mechanical processes in the GIT. These include emulsification, lipolysis, and intestinal absorption. Preliminary emulsification

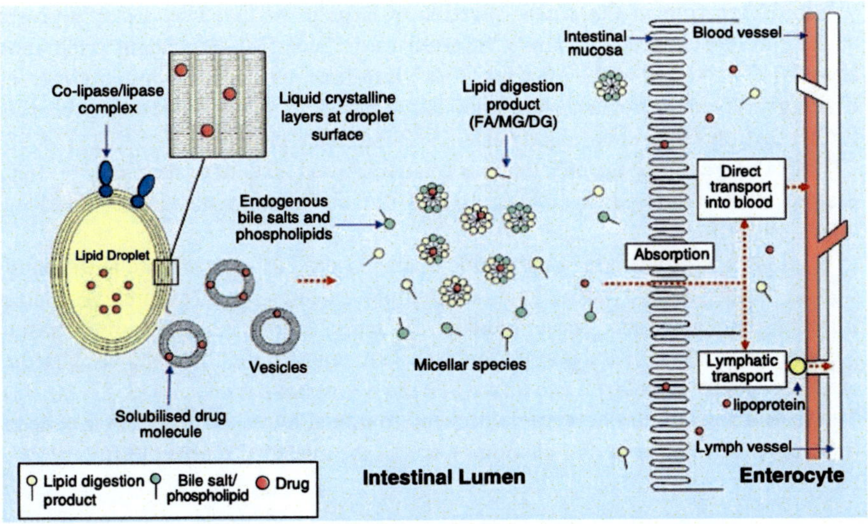

Fig. 15.1 Schematic diagram outlining the process of lipid digestion and absorption. As the lipid droplet is digested, lipid digestion products accumulate at the surface as liquid crystalline structures. The digestion products are solubilized into vesicles and micelles prior to absorption via the portal blood and lymphatics. Coadministered drug would transfer within the solubilized phases prior to absorption

of the lipid droplets occurs through shear forces generated by gastric peristaltic contractions against the closed or partially open pylorus [4]. Gastric shearing reduces the size of crude lipid to droplets 20–40 μm in diameter [5].

The subsequent digestion and absorption of lipids is illustrates schematically in Figure 15.1. The digestion of dietary or formulation lipids, in amounts as low as 2 g may influence processes such as gastric motility [6] and secretion of digestive fluids [6]. Changes to the nature of gastric chyme such as pH [7, 8], osmolality [9–11], and energy content [8, 12, 13] may stimulate digestive processes. Ingested nutrients such as long-chain fatty acids activate duodenal receptors [14], releasing mediators such as apolipoprotein A-IV [15], cholecystokinin (CCK), neurotensin, peptide YY, and proglucagon-derived peptides.

The contribution of gastric digestion to overall lipid digestion is uncertain but has been reported as approximating 30% of total postprandial digestion of dietary triglycerides [4, 16]. Gastric lipolysis results in hydrolysis of triacylglycerols to diacylglycerols and fatty acids [4, 17]. It has been shown, in rats [18, 19], rabbits [20], and humans [21] that gastric lipolysis is more significant for medium-chain than long-chain triglycerides.

The change in pH on transit from stomach to the more neutral duodenum (pH 5.4–7.5 [22, 23]) induces partial ionization of fatty acids, leading to migration of their polar head groups to the surface, lipid droplet size being concurrently reduced to ≤0.5 μm [24]. The presence of triacylglycerides and free fatty acids in the GIT also stimulates release of bile salts, biliary lipids, and phospholipids from the gall bladder and secretion of digestive enzymes from the pancreas [25, 26].

Lipid digestion in the small intestine is largely mediated by triacylglycerol acyl hydrolase (more commonly referred to as "colipase-dependent pancreatic lipase" [27]). Acting at the oil/water (o/w) interface, pancreatic lipase hydrolyses triacylglyceride into diacylglyceride and fatty acid, with the diacylglyceride further hydrolysed to another fatty acid and 2-monoacylglyceride (Fig. 15.1) [24, 25, 27, 28]. Pancreatic lipase efficiency is relatively independent of droplet size; lipid droplet sizes ranging from 0.7 to 10 μm did not significantly affect duodenal digestion [29].

As digestion progresses, amphiphilic digestion products accumulate at the emulsion droplet surface [24] and at liquid crystalline structures, which eventually separate from the droplet surface. These liquid crystalline structures interact with increasing proportions of bile salts and phospholipids to form uni- and multilamellar vesicles that are subsequently solubilized into mixed bile salt–phospholipid micelles [1, 27, 30–33]. The lipid digestion products are solubilized to mixed micelles, providing a concentration gradient that promotes absorption across the intestinal mucosa [30].

15.1.2 Absorption and Postabsorptive Processing

The process of lipid absorption is not critical to understanding the role of lipids in controlled release systems; gastric and intestinal events and/or release from a matrix largely govern the rate of absorption. However, in the case of lipid prodrugs, and access to the intestinal lymphatic system, it is useful to consider the absorption process and postabsorptive intracellular processing of lipophilic molecules for impact on rate of absorption.

Absorption of digested lipid components and lipophilic compounds is complex, involving passive diffusion from the lumen across the apical membrane of the enterocyte [4, 34], active transport and binding to cytosolic transport proteins [35]. Micellar lipid complexes are not absorbed as intact aggregates since micellar bile acids are absorbed via active transport in the ileum.

Once across the intestinal mucosa, lipidic compounds pass to the systemic circulation via the portal vein or undergo lymphatic transport [34]. Colloidal fat droplets comprising lipophilic compounds (e.g. lipophilic drugs, fat soluble vitamins) and triacylglycerides (reassembled from absorbed fatty acids and monoacylglycerides) associated with a surface layer of phospholipid, free cholesterol, apolipoproteins, and glycoproteins from lipoproteins undergo exocytosis from the enterocyte into the lymph [36, 37] for transport to peripheral tissues. Most drugs are transported via the hepatic portal route but some highly lipophilic compounds may be processed via the lymphatics, thereby avoiding first-pass metabolism [25]. Charman et al. proposed that metabolically stable drugs with log $P > 5$ (pH 7.4) and lipid solubility >50 mg/g in LCT are suitable for lymphatic transport, although the triglyceride solubility requirement has been recently questioned for interfacially active drugs [38]. Lymphatic transport is generally highest after coadministration with a high fat meal.

Although such transport may contribute over 50% of the dose appearing in the general circulation, it is difficult to discern any differences in absorption rates between hepatic and lymphatic transported drug. Hence, although important in understanding and optimizing lipid delivery systems, the relevance of lymphatic transport to delayed or extended absorption per se may not be significant.

15.1.3 Classes of Release Kinetics Obtained Using Lipids

Lipids may control drug release and absorption in a number of ways. The primary modalities discussed in this chapter are slow release, delayed release, and pulsatile release.

Slow release: Drug release from the dosage form can be slowed by coating particles or dosage units with a lipid-based coat or by embedding drug in a lipidic matrix.

Delayed release: Ionizable lipids can provide similar pH sensitivity to enteric materials such that release occurs in the small intestine where lipid digestion (and associated matrix/coat erosion) is most efficient.

Relatively small amounts of lipid (~2 g) can reportedly delay gastric emptying ("ileal brake" effect). Absorption is delayed, consequent to delayed passage to the small intestine. This can render formulation with lipids a viable strategy for delaying absorption, in cases where enteric coating may not be suitable due to drug–enteric polymer interactions or in clinical conditions where gastric pH is higher than normal.

Lipids are typically less dense than water so they may have applicability for floating dose forms, consequent gastric retention, and sustained drug absorption.

Pulsatile release is used to deliver "doses" of drug(s) at separate time intervals for reasons related to efficacy, safety, and patient compliance. There are few reports of lipids being used for such delivery but microencapsulated liposomes have been coupled with enzymatic systems to release the liposome drug payload. The Time Clock™ system, comprising dosage forms coated with lipids such as carnauba wax, beeswax, and surfactants provide consistent delivery intervals and absorption profiles [39]. Multipulse systems employing lipids have not been reported. Hence, many reported "pulsatile release" systems are in essence delayed release dose forms.

15.2 Mechanisms of Controlled Release Using Lipids

One or more underlying mechanisms control the behavior of the drug within the formulation to limit its release, or control disposition of the dosage unit in the body prior to release.

15.2.1 *Matrix Controlled Release*

Matrices provide a barrier to slow either the appearance of dissolved drug in gastrointestinal fluids, by inhibiting diffusion out of the matrix to slow dissolution of the drug by providing a physical barrier, or by requiring erosion of the matrix before exposure of undissolved drug particles to the gastrointestinal fluids.

Lipid barriers function in a number of ways. If the matrix is semisolid or liquid at room temperature, diffusion of a hydrophobic drug through the lipid is necessary for release to occur. Complex structures such as liquid crystals can dramatically influence diffusional pathways, slowing release (Sect. 15.3.3). If the lipid is solid or waxy, diffusion may not occur and lipid digestion and/or erosion is required for drug release and dissolution.

Lipids can also stimulate other processes that delay absorption. Thus, the contributions of digestion and absorption can make it difficult to elucidate the contribution of a lipid matrix and complicate attempts to achieve in vitro–in vivo correlations.

15.2.2 *Gastroretention*

Gastroretention strategies aim to retain the dosage form in the stomach, preventing transit, before complete drug release. Prolonged retention coupled with slow drug release can prolong absorption and therapeutic effect. For lipid-based systems, three mechanisms have been evaluated for gastro-retention, namely delayed emptying, floating dose forms, or mucoadhesion.

15.2.2.1 Delayed Gastric Emptying

The relatively low surface area of gastric lining and thick mucosal barrier makes transfer of contents into the small intestine a necessary and often rate-limiting step for absorption. Gastric emptying time is approximately 20–40 min under regular fasted conditions [6]. A fat-rich meal or large quantities of lipids in the dosage form can prolong emptying by hours [40], with long-chain lipids being more effective than medium-chain lipids. Although the mechanisms are not completely understood, interactions between fatty acids with duodenal receptors are believed to be responsible [15, 41].

Section 15.3.1 contains more specific information about the effect of "simple" lipid-based formulations on gastric-emptying rate.

15.2.2.2 Floating Dose Forms

Floating dosage forms are designed to slow transit of the unit from the stomach to the small intestine, their buoyancy in gastric contents delaying gastric emptying. Gelucire™

(polyethylene glycol glycerides) lipids have been evaluated for such purposes using pellets prepared by melt extrusion [42–46]. Diltiazem-containing Gelucire pellets remained in the stomach for 6 h after administration [47]. Riboflavin excretion profiles also revealed that similar pellets were better retained in the fed rather than fasted state, indicating that a physiologically based gastric emptying mechanism may also influence behavior [48, 49]. Glyceryl monooleate (GMO), a component of Gelucire 43/01 has been separately reported to be a useful floatation enhancer, being retained in the upper gastric region for up to 6 h after administration [50].

15.2.2.3 Mucoadhesion

Mucoadhesive-based dosage forms may prolong or boost absorption of poorly permeable drugs such as peptides. Mucoadhesion may result from interaction with glycoproteins in mucus via electrostatic attraction, e.g. cationic particles or polymers such as chitosan with the glycoproteins, by interactions with sugars on the cell surface [51], or following water imbibition and swelling in the vicinity of the mucus layer [52].

While mucoadhesiveness is not a typical target or attribute for lipid-based formulations, there are some reports of novel lipid formulations claiming mucoadhesion as a mechanism for altered drug absorption. A recent report of a lipid-based paclitaxel formulation claims up to 30% increase in oral bioavailability. The formulation can only be gleaned from a patent claiming monoolein and tricaprylin [53]. It is not made clear whether mucoadhesiveness or permeability enhancement by the digestion of tricaprylin to form caprylic acid is responsible for the improved bioavailability [54]. The mucoadhesiveness of swelling lipids such as monoolein (GMO) has been noted in vitro [52] but not in vivo, possibly due to complicating influences of digestion and absorption. Mucoadhesive liposomes have also been reported that incorporate chitosan or other polymers [55], but the mucoadhesion is not ascribed to the lipids per se [56].

15.2.3 Stimulation of Lymphatic Transport

Lymphatic transport of drugs is beneficial where bioavailability is limited by first-pass metabolism. Drugs that are highly lipophilic and have high triglyceride solubility may be transported in relatively high proportion (often up to 30%) via intestinal lymphatics. Drug transport via lymph is increased with increasing amounts of coadministered long-chain lipids swelling the chylomicrons and providing a greater pool into which drug is likely to partition in preference to direct entry to systemic circulation. Such lymphatic transport avoids hepatic first-pass metabolism experienced by drug molecules transported via the portal blood system. Hence for drugs with high hepatic extraction, delivery via lymph may boost systemic levels. For drugs that are not lipophilic, the distribution to the lymphatic system is low, even in the presence of lipids that enhance lymphatic transport. They are transported almost exclusively via the portal blood system, and, if metabolized extensively by the liver on first pass have low bioavailability. One logical strategy for bioavailability enhancement is to prepare a lipophilic prodrug to target lymphatic distribution. Methylnortestosterone undecanoate, administered with food, provided a 50-fold increase in systemic availability compared to methylnortestosterone [57]. Sixty per cent of the absorbed dose of a phospholipid prodrug of valproic acid has also been reported to be lymphatically transported [58]. While not discussed further in this chapter, as the understanding of mechanisms of lymphatic transport increase, the use of formulation approaches to target lipophilic drugs and prodrugs toward lymphatic distribution is certain to become more prevalent.

15.3 Technologies for Controlled Release Using Lipids

15.3.1 Classical Lipid-Based Formulations Used in Oral Drug Delivery

Lipid-based formulations range from simple lipid solutions to more complex systems incorporating triglycerides, partially digested triglycerides, semisynthetic ester glycerides, lipophilic and hydrophilic surfactants and cosolvents [59, 60]. The formulation can influence digestibility, dispersion and solubilization of the lipid vehicle in vivo, in turn influencing drug absorption. Pouton has proposed a lipid classification scheme to provide insight on lipid behavior on oral dosage (Table 15.1) [59].

- Type I formulations comprise drug dissolved in dietary glycerides or vegetable oils and bioavailability is highly dependent on digestion [60]. Lipolysis of the lipid results in dispersion and solubilization of the drug through mixed micellar systems.
- Type II formulations are self-emulsifying systems (SEDDS), combining lipid and water insoluble surfactant(s), which are self-emulsified by gentle agitation [61]. The hydrophobic nature of the surfactant and physicochemical properties of the lipid dictate the type and capacity of the formed solubilized entity.
- Type III formulations contain greater quantities of hydrophilic and/or hydrophobic surfactants to enhance dispersion. They are less dependent on digestion to provide reduced particle size. Type IIIB systems are commonly described as self-microemulsifying drug delivery systems (SMEDDS) [62], which may be capable of promoting drug absorption independent of digestion [61].
- Type IV formulations do not contain lipids but comprise a combination of drug, surfactant, and cosolvent. They often provide highly solubilized drug in the formulation [59]. They are predisposed to drug precipitation in GI fluids, forming fine suspensions or amorphous particles that readily dissolve [60].

Table 15.1 The classification of lipid-based formulations proposed by Pouton [59]

	Type I	Type II	Type IIIA	Type IIIB	Type IV
Hydrophobic composition	High	→	→	→	Low
Lipid: TG, DG, MG (%)	100	40–80	40–80	<20	–
Water insoluble surfactants HLB < 12 (%)	–	20–60	–	–	0–20
Water soluble surfactants HLB > 12 (%)	–	–	20–40	20–50	30–80
Hydrophilic cosolvents (%)	–	–	0–40	20–50	0–50
Particle size on dispersion (nm)	Coarse	100–250	100–250	50–100	<100
Importance of digestion	Crucial	→	→	→	Not recognized as essential but may occur
Effect of aqueous dilution	Not important	→	→	→	Loss of solvent capacity

Lipid-based formulations commonly utilize lipids that are liquids at room temperature; hence, they are not designated as matrix or erosion controlled release systems. Rapid dispersion in the GIT, assisted by digestion or the high interfacial activity in the formulation can engender short diffusional distances and rapid release [63, 64]. Thus in the absence of physiological effects that might delay gastric emptying or absorption, these formulations will not constrain but may even increase the rate of drug availability for absorption. Consequently, there are reports of both rapid and delayed absorption from lipid-based formulations as listed in Table 15.2.

Simple lipid formulations often exhibit a delayed T_{max} relative to other presentations for poorly water soluble drugs. Figure 15.2 illustrates the relative systemic exposure for griseofulvin when administered in lipid emulsion compared to a lipid or aqueous suspension. In addition to the greatly increased exposure from the emulsion, a delayed T_{max} is also evident.

Conversely, plasma profiles may be unchanged compared to a conventional tablet or simple suspension. For example, although the bioavailability of cinnarizine was improved fourfold in beagles when dosed as an oleic acid solution, compared to a tablet [69], there was no delay in reaching T_{max}. Delayed absorption, reflected in greater T_{max} values may reflect the time taken for lipid digestion and drug transport. Such delay is probably not clinically relevant in most cases but the increased exposure (bioavailability) can clearly be beneficial.

15.3.2 *Solid Lipid Systems*

High melting point lipids formed into pellets by melt-solidification or extrusion-spheronization, or filled directly into capsules can be used to slow the release of poorly soluble drugs. Waxy lipids that are crystalline at room temperature, such as Compritol® 888 ATO (glyceryl behenate) [71, 72] or high melting ethoxylated lipids such as Gelucire® 50/02 [73] can slow release by a combination of diffusion and erosion mechanisms [74, 75].

Formulation of piroxicam as Gelucire 50/13 microspheres provided rapid release that was ascribable to liquid crystal formation in an aqueous environment [76]. This may result in earlier onset of action, a desirable attribute with this anti-inflammatory agent.

15.3.2.1 Lipid–Polymer Solid Dispersions

Solid dispersion of poorly water soluble drugs in polymer may improve dissolution characteristics [77]. Strategies include crystallization of lipids within solid polymer matrices such as PEG 6000 [78–80]. This has led to the formulation of solid microemulsion preconcentrates that spontaneously form a microemulsion on polymer dissolution [81]. Fine dispersion of the lipid domains results in rapid dissolution and/or dispersion of drug compared to conventional drug suspensions and may accelerate absorption.

Table 15.2 Examples of lipid-based suspension, solutions, and emulsions used in enhancing the oral bioavailability (BA) of poorly water soluble drugs

Compound	Formulation	Study	T_{max}	Reference
Danazol	Oil-in-water emulsion of monoolein	Oral BA in fed humans	4.9 h for the emulsion vs. 3.1 h for the capsule	[65]
Griseofulvin	Corn oil-in-water emulsion	Oral BA in humans	5.7 h vs. 3.3 h for suspension	[66]
Ontazolast	SBO oil-in-water emulsion	Oral BA in rats	4.6 h vs. 3.3 h for suspension	[67]
Halofantrine	Solutions of peanut oil (LCT)	Oral BA in rats	T_{max} approximately 10 h cf. 6 h for suspension	[68]
Cinnarizine	Oleic acid solution	Oral BA in beagle dogs	T_{max} unchanged cf. tablet	[69]
Griseofulvin	Oil suspensions of peanut oil (LCT), Captex 355 (MCT) and triacetin (SCT)	Oral BA in rats	Oil suspensions the same or slightly reduced T_{max} for lipid formulations compared to suspension	[70]

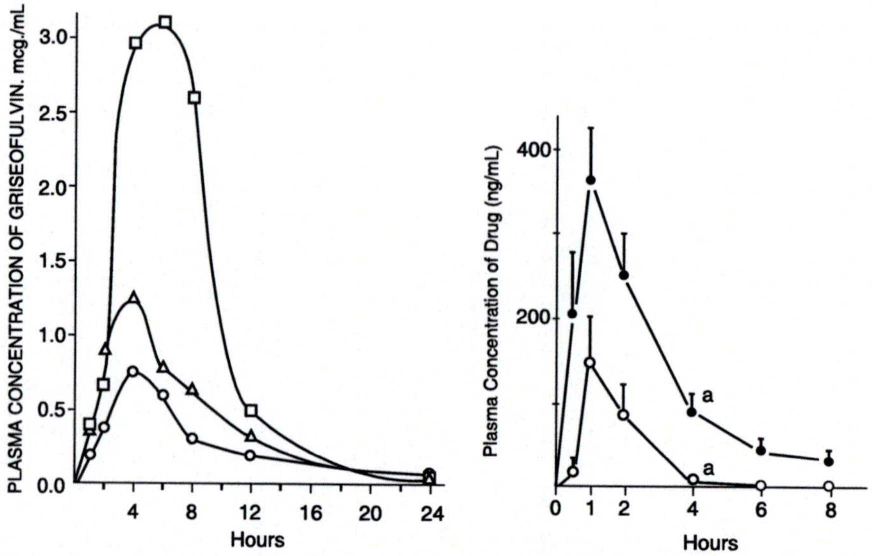

Fig. 15.2 Plasma vs. time profiles after oral administration of poorly water soluble drugs from lipid formulations compared to suspensions. *Left hand panel* is oral administration of griseofulvin to rats as a corn oil in water emulsion (*squares*) versus suspension (*circles*) (Reproduced from [89]). *Right hand panel* is administration of cinnarizine in oleic acid or an aqueous suspension to beagle dogs (*closed* and *open circles*, respectively) (Reproduced from [69])

15.3.2.2 Solid Lipid Nanoparticles

Solid lipid nanoparticles (SLNs) have the potential to encapsulate drugs for oral delivery. Slow erosion of the lipid controls drug release, prolonging plasma residence and "damping down" peak plasma concentrations. In vitro release is slowed, relative to other formulations [82–85].

In vivo studies using SLNs for oral administration (human and animal) are relatively limited. Camptothecin-loaded SLNs altered the drug's oral pharmacokinetics, resulting in two peaks in the plasma profile. The first peak was attributed to rapidly released free drug, the second peak to slower release from the SLN particles. Peak plasma concentration (C_{max}) occurred approximately 6 h after administration [86].

Other reports have shown that:

- An SLN-based cyclosporine product exhibited prolonged plasma presence compared to Neoral™ formulation in rats [87] and pigs (Fig. 15.3) [88].
- T_{max} for quercetin was increased from 5 to 8 h, accompanied by a fourfold increase in bioavailability [90].
- T_{max} was increased by only 20 min with melatonin-loaded SLNs [91].
- T_{max} of buspirone, formulated as SLNs, was not different from a simple solution [92].
- Vinpocetine delivered in six different SLN formulations did not show a difference in T_{max} values compared to drug administered in a simple solution [93].

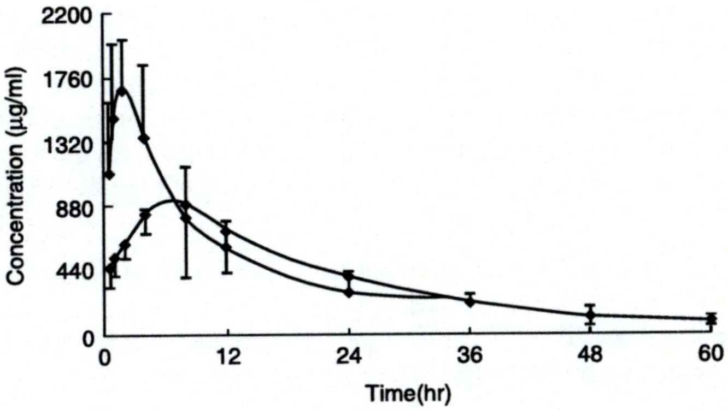

Fig. 15.3 Illustration of plasma profiles for cyclosporine after oral administration in stearic acid SLNs (*lower curve*) vs. a microemulsifying lipid formulation (Neoral) (*upper curve*) (reproduced from [87])

It would seem therefore that, in many cases, the formulation of poorly soluble drugs in SLN form does not affect absorption characteristics, relative to solution or suspension dosage. Further work is indicated to explain such inconsistent behavior.

SLNs pose a number of challenges that may limit more widespread application.

- Drug may be driven to the particle surface during lipid solidification, resulting in either burst or other nonuniform release [94].
- In vivo degradation of SLNs is not well characterized; hence, interpretation of biopharmaceutical behavior may be difficult.
- The crystalline nature of the lipids themselves can be questionable, as supercooling after processing may occur. Size reduction induces supercooling (as with fat emulsions) [95], again complicating the interpretation of in vitro release and in vivo kinetic data.

15.3.3 *Liquid Crystalline Systems*

Unlike micellar systems, which are prone to dissociation on dilution under physiological conditions, some self-assembled liquid crystalline structures retain their form, even on dilution with water. Consequently, they may have potential as sustained release matrices.

15.3.3.1 Lipid-Based Liquid Crystalline Structures

Amphiphilic molecules belonging to the Class II polar lipids self-assemble in water to form condensed structures that combine the order of crystalline materials with the disorder of liquids. Self-assembly occurs due to intermolecular interactions such as hydrogen bonding and van der Waal's forces [96] with the hydrophobic effect minimizing contact of alkyl chains with water. The relationship between lipid structure and LC phase formation can be defined by the dimensionless critical packing parameter *v/al*, where *v* is the effective hydrocarbon volume, *l* is the fully extended hydrocarbon chain length, and *a* is the area of the hydrophilic headgroup (see Fig. 15.4) [97].

- Normal (Type 1) self-assembled structures, with positive curvature toward water, are formed when $v/al < 1$, whereas
- Inverse structures (Type 2) exhibit negative curvature away from water, $v/al > 1$.
- At the tipping point, i.e. $v/al = 1$ L_α bilayers are observed.

The liquid crystalline structure formed depends not only on the concentration and chemical structure of the lipid but also on factors such as temperature, pressure, solvent composition, and concentration [98, 99].

Self-assembled phases can have structures in one, two, or three dimensions (yielding, e.g. lamellar, hexagonal and cubic phase structures, respectively). Lamellar phases consist of stacked bilayers, where lipid molecules are arranged such that the hydrophobic tails meet to form a lipidic domain, and hydrophilic head groups oppose on opposite sides of a hydrophilic domain. Water occupies the hydrophilic domains and interacts with the head groups lining the lamellae.

Cubic liquid crystalline phases, aside from exhibiting both positive and negative curvature, can be further categorized as "discontinuous" and "bicontinuous." Discontinuous cubic phases, usually designated as either I_1 (normal) or I_2 (inverted) are intermediate liquid crystalline phases and reside between hexagonal phases and micelles in the order of progression of LC structures (Fig. 15.5). Discontinuity is attributed to discretely ordered aggregates of micellar structures and thus either the water or hydrocarbon phase is noncontinuous. Bicontinuous cubic LC phases, annotated as V_1 or Q_I (normal) or V_2 or Q_{II} (inverse) are located on either side of the lamellar structure and differ from micellar cubic phases, both lipid bilayer and water domains being continuous but separate domains. The cubic phases can be formed from ionic soaps, nonionic surfactants and glycolipids (a comprehensive list has been compiled by Fontell et al. [100]), and have been noted to have potential significance in cell membranes and fat digestion [101–104]. The bicontinuous structure imparts high viscosity [105] due to the three-dimensional tortuous configuration, which can provide a slow release environment for incorporated drugs.

Hexagonal liquid crystals are positioned between the bicontinuous and discontinuous cubic phases in the order of progression of liquid crystalline structures. The H_{II} phase forms when *v/al* is ~2 [98, 106], and is characterized by water-filled cylindrical micelles of indefinite length packed in hexagonal lattices (Fig. 15.5).

Importantly, for some liquid crystalline structures prepared from poorly soluble lipids, particularly the V_2, H_2, I_2 or L_α phases, addition of water does not result in transition to normal phases. Thus they may coexist with excess water at high dilution. Diacyl phospholipids exemplify such behavior. These form a lamellar phase in

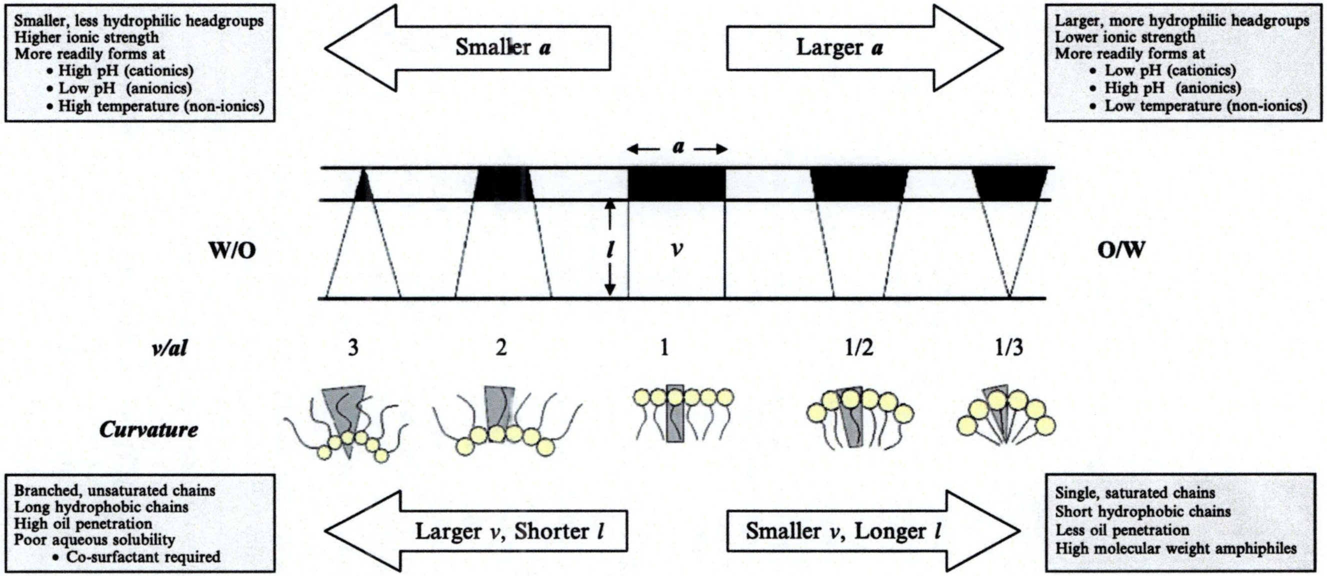

Fig. 15.4 The impact of lipid structure, increasing water concentration and external stimuli on parameters of the dimensionless packing parameter and curvature. The diagram is adapted with permission from Israelachvili et al.

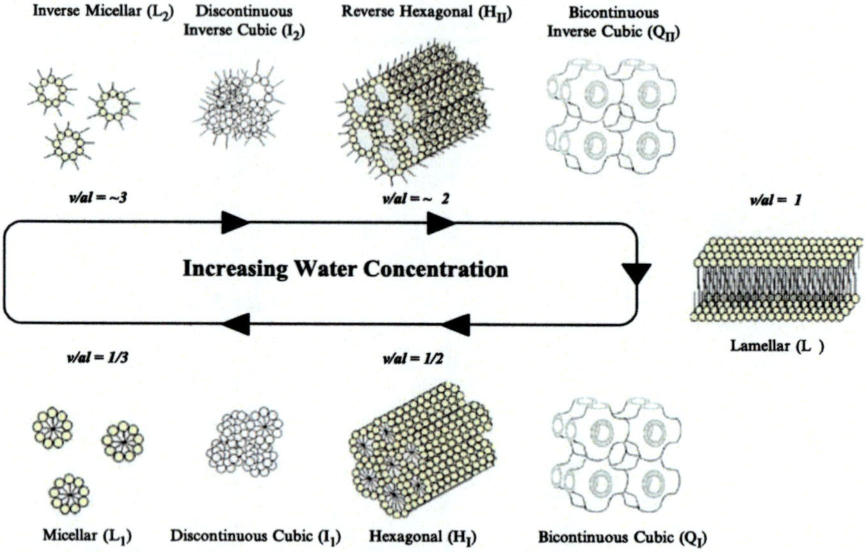

Fig. 15.5 Outline of the lyotropic liquid crystalline phases that can be formed when water is added to anhydrous amphiphilic lipid. Normal (o/w) phases are designated I and inverse (w/o) phases II with decreasing packing parameters as water concentrations increase. Adapted with permission from Israelachvili et al. [106]

excess water (hence their primary role as membrane-forming lipids). A consequence of such behavior is that, when administered to a water-rich environment the liquid crystalline matrix maintains its structure, potentially providing a persistent slow-releasing reservoir.

The stability of some liquid crystals on dilution also means that they can be dispersed to form submicron-sized particles. Dispersed particles of the cubic and hexagonal phases have been termed "cubosomes®" and "hexosomes®," respectively (registered trademarks of Camurus AB Sweden). Cubosomes and hexosomes have been formed using various techniques, with particle diameters ranging from 50 to 250 nm [107]. Cubosomes often present a square profile in microscopy imaging while hexosomes present a hexagonal shape (Fig. 15.6) [107].

The most common methods of preparation of cubosomes and hexosomes involve fragmenting equilibrated bulk liquid crystalline phases, or dispersing molten lipid in aqueous medium via high pressure homogenization or ultrasonication [108–110]. To prevent reaggregation of dispersed particles, a stabilizer is added. Bile salts, amphiphilic proteins, and block copolymers are used to sterically stabilize and maintain dispersion [111]. The block copolymer surfactant Pluronic F127 (PF127) is most commonly utilized [108]. Importantly, the particles most often retain the internal structure of the "parent" bulk phase. They have been considered for sustaining drug release, although their small size and high surface area may preclude applicability for small molecule drugs under sink release conditions [112].

Commonly employed amphiphile structures that form nonlamellar phases in excess water are illustrated in Fig. 15.7. They include glycerate surfactants

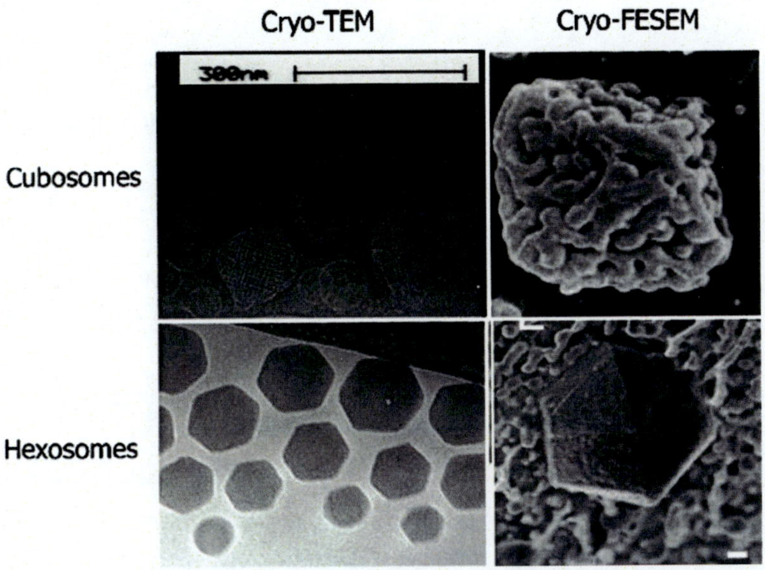

Fig. 15.6 Images of cubosomes and hexosomes using cryo-transmission electron microscopy (cryo-TEM) and cryo-field emission scanning electron microscopy (cryo-FESEM)

[113, 114] (e.g. oleyl glycerate), glycerol ethers [115, 116] (selachyl alcohol) and esters [117] (such as GMO), glycolipids [118, 119] (1-*O*-phytanyl-β-D-xyloside), branched chain alcohols [120] (phytantriol) and nonionic urea surfactants [121, 122] (1-(2-hydroxyethyl)-1-oleyl urea). Many have been investigated, especially GMO as it is widely available and is used in food products [123, 124].

15.3.3.2 Controlled Release from Liquid Crystalline Structures

The tortuous structure of cubic phases and the closed micellar structure of the inverse hexagonal phase have potential for sustaining delivery [125]. The aqueous and lipid domains of the matrix may provide solubilization and sustained release for drugs of widely varying polarity and structure. Wyatt et al. assessed the in vitro release of drugs of varying chemical properties from GMO-based cubic phases [126]. In the presence of simulated gastrointestinal fluids the V_2 matrix provided sustained, diffusion-controlled release of aspirin, with 70% released after 24 h, compared to 100% dissolution of the control tablet after 2 h. Such diffusion control has been subsequently confirmed. Clogston and Caffrey studied release rates of hydrophilic compounds, ranging from small molecules to DNA macromolecules, from cubic phases [127].

Release rate may also depend on matrix nanostructure. Release rates of the hydrophobic anticancer drugs paclitaxel and irinotecan were more rapid from cubic than from hexagonal phases. The same patterns were observed with the hydrophilic compounds octreotide, histidine, and glucose [128]. Glucose release depended on

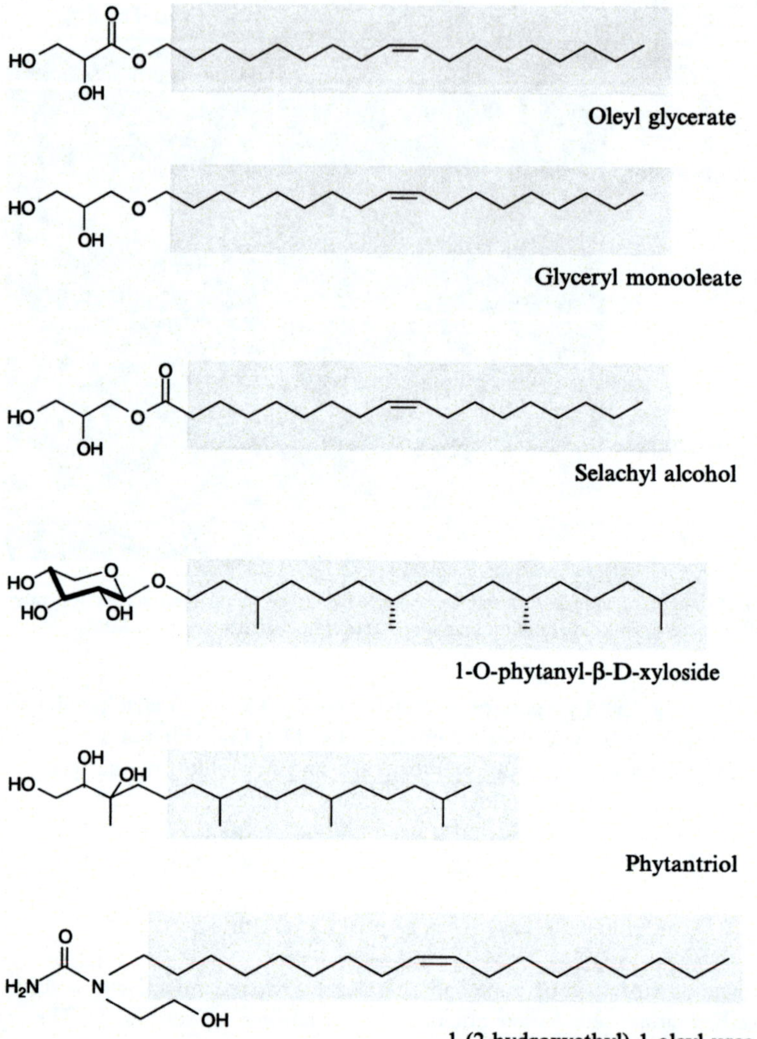

Fig. 15.7 Structures of amphiphilic lipids that form liquid crystal phases in excess water (the hydrophobic region of the molecule is *shaded*)

the lipid in the cubic phase, release kinetics being slower from the smaller channel size in cubic phase assembled from phytantriol than from GMO-assembled cubic phase (see also below) [129]. Chlorpheniramine release from lamellar liquid crystals was faster than from cubic phase material [130].

Inducing transitions between liquid crystalline structures such as cubic and hexagonal phases by temperature changes can reverse the drug release rates from different nanostructures. Figure 15.8 illustrates the reversible change in glucose

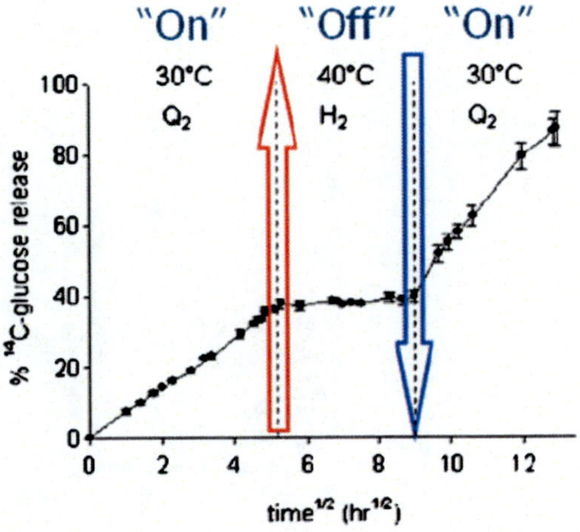

Fig. 15.8 In vitro release of glucose as a model hydrophilic drug from liquid crystalline structures using temperature to reversibly induce transitions between the cubic and hexagonal phase structures. Modified from [131]

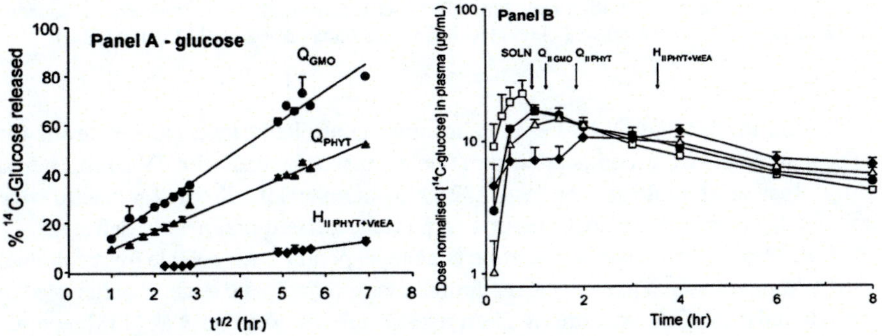

Fig. 15.9 *Panel A* shows in vitro release profiles for glucose as a model hydrophilic drug from different liquid crystalline phases formed using phytantriol (PHYT), glyceryl monooleate (GMO) and vitamin E acetate (VitEA). *Panel B* shows plasma concentrations vs. time after oral administration of liquid crystalline systems. Figures modified from Lee et al. [129]

concentration when switching between the cubic phase at 30°C and the hexagonal phase structure at 40°C [131]. The relevance of such behavior for controlling oral administration is not apparent at this time (while speculating on possibilities at body temperature). However, it illustrates the concept of changing release through changes in other stimuli.

The wealth of in vitro release studies reported for lipid-based liquid crystalline systems contrasts with limited reports on in vivo performance. For hydrophilic drugs, two reports are apparent. Lee et al. used glucose as a model hydrophilic compound [129]. Figure 15.9 illustrates in vitro release from different cubic and hexagonal phase structures and correlation with in vivo oral absorption. Release was slowest from the inverse hexagonal phase, with in vivo T_{max} delayed significantly compared to the cubic phase.

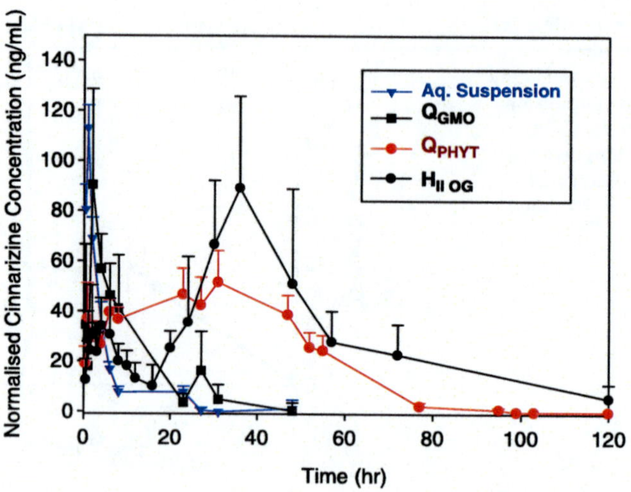

Fig. 15.10 Plasma concentration over time after oral administration of cinnarizine in aqueous suspension, GMO, phytantriol and oleyl glycerate to rats (data taken from [128] and [134])

Insulin loaded into GMO-based cubosomes controlled blood glucose levels for approximately 4 h, a comparable duration to that achieved with IV-administered drug, albeit at a tenfold greater dose [132]. No pharmacokinetic data for insulin were provided, a common feature of reports on particle-assisted oral insulin delivery.

A number of studies used cinnarizine as a model poorly water soluble (hydrophobic) drug administered in liquid crystalline formulations, or in lipidic systems anticipated to form liquid crystalline structures in gastrointestinal fluids. Kossena et al. [133] reported that monolaurin and lauric acid formed a cubic phase in the presence of simulated intestinal fluids. The cubic phase was not stable to dilution but was considered representative of structures formed during digestion of trilaurin, a common component of medium chain triglycerides. Intraduodenal administration of the cubic phase containing cinnarizine resulted in a twofold prolongation of T_{max} compared to a control suspension formulation, indicating the potential of cubic phases for controlling absorption.

GMO and oleyl glycerate, which form cubic and hexagonal phases, respectively, have also been evaluated as vehicles for cinnarizine. When cinnarizine was dissolved in GMO or oleyl glycerate and administered orally to rats, there was a 1.2- and 14-fold extension in T_{max}, respectively, compared to drug in simple suspension (Fig. 15.10) [128]. The plasma profile for GMO was somewhat typical for a lipid-based formulation, with a slight increase in T_{max} being attributable to delayed gastric emptying. However the dramatic effect with the oleyl glycerate vehicle was unexpected, and was accompanied by a fourfold enhancement in bioavailability. Oleyl glycerate has identical molecular structure to GMO (Fig. 15.7), except that the ester is reversed, in which case lipolysis should liberate oleyl alcohol rather than oleic acid. It was hypothesized (and supported by in vitro digestion experiments) that this might inhibit digestion and contribute to the sustained absorption profile, contrasting with GMO which is known to be readily digested in vivo.

To further probe this effect, a subsequent experiment used phytantriol as the vehicle for delivery of cinnarizine [134]. Phytantriol is inherently indigestible due to a lack of an ester link (Fig. 15.7) but forms cubic phases in excess water [110]. Plasma profiles for cinnarizine when using phytantriol and oleyl glycerate as the vehicle were very similar, viz. prolonged T_{max} and greater AUC (Fig. 15.10). Prolonged gastric retention, consequent to poor digestibility is a plausible explanation. In the case of phytantriol, the sustained plasma concentrations were linked to improved gastric retention. Cinnarizine, formulated as an oleic acid emulsion provided profiles similar to those from GMO [135]. An explanation may be that if GMO is readily digested soon after administration, oleic acid would be formed. Oleic acid does not form liquid crystals. Neither did it provide extended plasma profiles, in contrast to oleyl glycerate.

Formulations containing poorly digestible or nondigestible lipids do not typically enhance drug absorption as readily as those comprising digestible components. Yamahira et al. showed that the poorly digestible lipid vehicle *N*-α-methylbenzyl linoleamide reduced absorption of the lipophilic drug SL-512 in rats, compared to administration in readily digested MCT [136]. Myers et al. also reported reduced oral bioavailability of penclomedine, given intraduodenally in mineral oil, compared to soybean oil, although, an improvement in bioavailability was evident relative to a simple aqueous suspension [137]. Oleyl alcohol has also been noted to significantly reduce the absorption of sulfisoxazole acetyl, dicumarol, and griseofulvin when compared to more digestible lipids [138]. Gastroretention effects have not been tendered as explanation for such findings.

To further probe the relative roles of digestibility and liquid crystal formation, selachyl alcohol was used as a vehicle, being nondigestible, but similar in structure to GMO and oleyl glycerate (Fig. 15.6); it also forms liquid crystalline structures. Selachyl alcohol dramatically extended cinnarizine plasma presence [135] (T_{max} approximately 24 h), indicating that liquid crystalline delivery vehicles should incorporate a poorly or nondigestible lipid if sustained absorption is desired. Investigations continue to examine mechanisms of gastric retention. Moreover, it has emerged that GMO possesses mucoadhesive properties [52, 139]. Similarly structured lipids may also possess such properties.

The gastric compartment presents a generally low absorption environment and has limited variability in volume; hence, sink conditions may not pertain with some poorly soluble drugs administered in gastroretentive formulations. The relative roles of slow release and nonsink conditions in determining release kinetics and subsequent absorption from the cinnarizine-containing lipid-based liquid crystalline systems described earlier were not elucidated. Consequently, additional studies were undertaken using dispersed cubosomes loaded with cinnarizine. Drug and lipid load was the same as in the earlier work. The hypothesis was that submicron particles would deliver burst release. Rapid absorption would be expected if drug and/or particles were not retained in the stomach. Figure 15.11 shows that extended cinnarizine plasma profiles were obtained once more with the nondigestible lipid vehicle, while GMO and oleic acid-based systems provided similar pharmacokinetics to bulk GMO [135, 140]. This is consistent with Lai who reported a slightly delayed T_{max} for simvastatin in GMO cubosomes (1.50 h for cubosomes vs. 0.8 h for

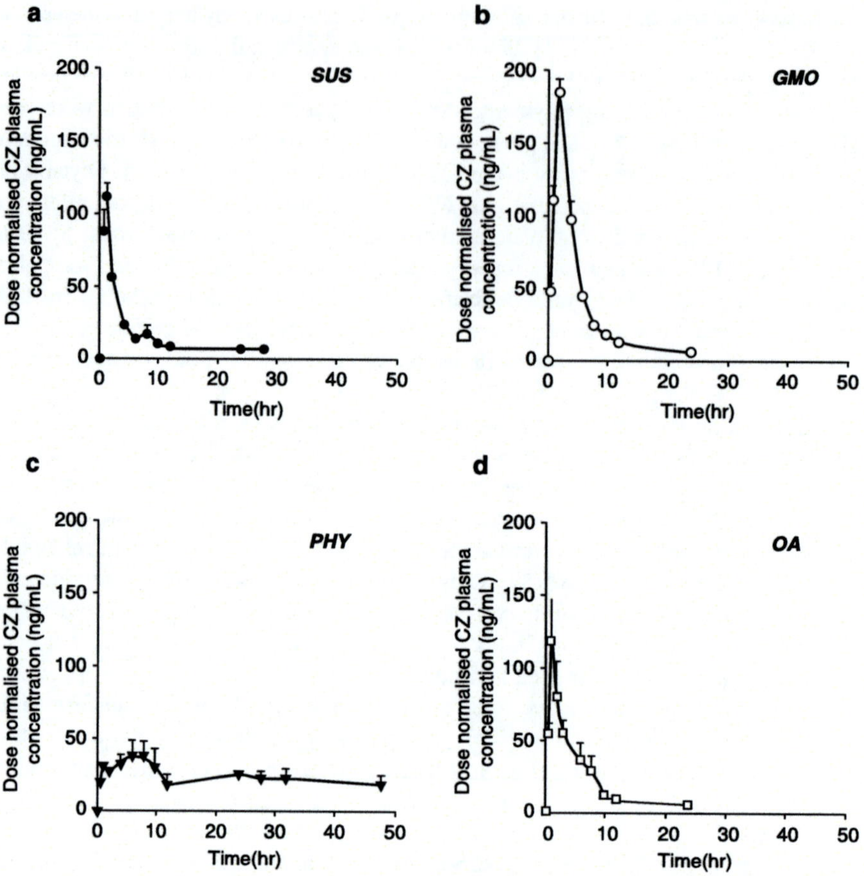

Fig. 15.11 Administration of cinnarizine in different particulate vehicles: (**a**) drug suspension, (**b**) GMO cubosomes (digestible liquid crystalline lipid), (**c**) phytantriol cubosomes (nondigestible lipid), (**d**) oleic acid emulsion (nondigestible, nonliquid crystalline) [140]

a drug suspension) when administered orally to beagles [141]. The findings confirmed that gastroretention, reduced digestibility, and nonsink conditions contribute to the prolonged absorption of cinnarizine from these systems (Fig. 15.11).

15.4 Limitations and Future Developments

The evolution of complicated multicompartmental devices and complex polymer systems for controlling oral delivery is progressing but few technologies are reaching full development due to substantial regulatory hurdles for such systems. This chapter has highlighted the potential application of simple lipid-based formulation approaches for the controlled delivery of compounds with a focus on poorly water soluble compounds. The applications of solid lipid systems and liquid crystal systems, while still in their relative infancy have benefits of biocompatibility and likely regulatory acceptance, and hence are expected to be increasingly important classes of controlled release systems for oral delivery in the future.

References

1. MacGregor K, Embleton J, Lacy J, Perry E, Pouton C (1997) Influence of lipolysis of drug absorption from the gastrointestinal tract. Adv Drug Deliv Rev 25:33–46
2. Fahy E, Subramaniam S, Brown H, Glass C, Merrill A, Murphy R, Raetz C, Russell D, Seyama Y, Shaw W, Shimizu T, Spener F, Meer G, Van Nieuwenhze M, White S, Witztum J, Dennis E (2005) A comprehensive classification system of lipids. J Lipid Res 46:839–861
3. Small D (1968) A classification of biological lipids based upon their interaction in aqueous systems. J Am Oil Chem Soc 45:108–119
4. Carey MC, Small D, Bliss C (1983) Lipid digestion and absorption. Annu Rev Physiol 45:651–677
5. Zanenberg N (2001) A dynamic in vitro digestion model. The Danish School of Pharmacy, Copenhagen
6. Kossena G, Charman WN, Wilson C, O'Mahony B, Lindsay B, Hempenstall J, Davison C, Crowley P, Porter CJH (2007) Low dose lipid formulations: Effects on gastric emptying and biliary secretion. Pharm Res 24:2084–2096
7. Hunt J, Knox M (1961) The regulation of gastric emptying of meals containing citric acid and salts of citric acid. J Physiol 163:34–45
8. Zittel T, Rothenhofer I, Meyer J, Raybould H (1994) Small intestinal capsaicin-sensitive afferents mediate feedback inhibition of gastric emptying in rats. Am J Physiol Gastrointest Liver Physiol 267:G1142–G1145
9. Hunt J, Pathak J (1960) Osmotic effects of some simple molecules and ions on gastric emptying. J Physiol 154:254–269
10. Meeroff J, Go V, Phillips S (1975) Control of gastric emptying by osmolality of duodenal contents in man. Gastroenterology 68:1144–1151
11. Miller J, Kauffman G, Elashoff J, Ohashi H, Carter D, Meyer J (1981) Search for resistances controlling canine gastric emptying of liquid meals. Am J Physiol Gastrointest Liver Physiol 241:G403–G415
12. Hunt J, Stubbs D (1975) The volume and energy content of meals as determinants of gastric emptying. J Physiol 245:209–225
13. McHugh P, Moran T (1979) Calories and gastric emptying: A regulatory capacity with implications for feeding. Am J Physiol Regul Integr Comp Physiol 236:R254–R260
14. Hunt J, Knox M (1968) A relation between the chain length of fatty acids and the slowing of gastric emptying. J Physiol 194:327–336
15. Tso P, Liu M (2004) Apolipoprotein a-iv, food intake, and obesity. Physiol Behav 83:631–643
16. Armand M, Borel P, Pasquier B, Dubois C, Senft M, Andre M, Peyrot J, Salducci J, Lairon D (1996) Physicochemical characteristics of emulsions during fat digestion in human stomach and duodenum. Am J Physiol 271:G172–G183
17. Carriere F, Barrowman J, Verger R, Laugier R (1993) Secretion and contribution to lipolysis of gastric and pancreatic lipases during a test meal in humans. Gastroenterology 105: 876–888
18. Fernando-Warnakulasuriya G, Staggers JE, Frost S, Wells M (1981) Studies of fat digestion, absorption, and transport in the suckling rat. I. Fatty acid composition and concentrations of major lipid components. J Lipid Res 22:668–674

19. Lai H, Ney D (1998) Gastric digestion modifies absorption of butterfat into lymph chylomicrons in rats. J Nutr 128:2403–2410
20. Borel P, Armand M, Ythier P, Dutot G, Senft M, Lanfont H, Lairon D (1994) Hydrolysis of emulsions with different triglycerides and droplet sizes by gastric lipase in vitro. Effect on pancreatic lipase activity. J Nutr Biochem 5:124–133
21. DeNigris S, Hamosh M, Dinkar K, Fink C, Lee T, Hamosh P (1985) Secretion of human gastric lipase from dispersed gastric glands. Biochem Biophys Acta Lipids Lipid Metab 836:67–72
22. Dressman J, Berardi RR, Dermentzoglou LC, Russell TL, Schmaltz SP, Barnett JL, Jarvenpaa KM (1990) Upper gastrointestinal (GI) pH in young, healthy men and women. Pharm Res 7:756–761
23. Evans DF, Pye G, Bramley R, Clark AG, Dyson TJ, Hardcastle JD (1988) Measurement of gastrointestinal ph profiles in normal ambulant human subjects. Gut 29:1035–1041
24. Ros E (2000) Intestinal absorption of triglycerides and cholesterol. Dietary and pharmacological inhibition to reduce cardiovascular risk. Atherosclerosis 151:357–379
25. Humberstone A, Charman WN (1997) Lipid-based vehicles for the oral delivery of poorly water soluble drugs. Adv Drug Deliv Rev 25:103–128
26. Sek L, Porter CJH, Kaukonen A-M, Charman WN (2001) Characterisation and quantification of medium and long chain triglycerides and their in vitro digestion products, by HPTLC coupled with in situ densiometric analysis. J Pharmaceut Biomed Anal 25:651–661
27. Embleton J, Pouton C (1997) Structure and function of gastrointestinal lipases. Adv Drug Deliv Rev 25:15–32
28. Porter CJH, Charman WN (2001) In vitro assessment of oral lipid based formulations. Adv Drug Deliv Rev 50:S127–S147
29. Armand M, Pasquier B, Andre M, Borel P, Senft M, Peyrot J, Salducci J, Portugal H, Jaussan V, Lairon D (1999) Digestion and absorption of 2 fat emulsions with different droplet sizes in the human digestive tract. Am J Clin Nutr 70:1096–1106
30. Hernell O, Staggers JE, Carey MC (1990) Physical-chemical behaviour of dietary and biliary lipids during intestinal digestion and absorption. 2. Phase analysis and aggregation states of luminal lipids during duodenal fat digestion in healthy adult human beings. Biochemistry 29:2041–2056
31. Kossena GA, Boyd BJ, Porter CJH, Charman WN (2003) Separation and characterization of the colloidal phases produced on digestion of common formulation lipids and assessment of their impact on the apparent solubility of selected poorly water soluble drugs. J Pharm Sci 92:634–648
32. Rigler M, Honkanen R, Patton J (1986) Visualization by freeze fracture in-vitro and in-vivo of the products of fat digestion. J Lipid Res 27:836–857
33. Staggers JE, Hernell O, Stafford R, Carey M (1990) Physical-chemical behavior of dietary and biliary lipids during intestinal digestion and absorption. 1. Phase behavior and aggregation states of model lipid systems patterned after aqueous duodenal contents of healthy adult human beings. Biochemistry 29:2028–2040
34. Nordskog B, Phan C, Nutting D, Tso P (2001) An examination of the factors affecting intestinal lymphatic absorption of dietary lipids. Adv Drug Deliv Rev 50:21–44
35. Stremmel W (1988) Uptake of fatty acids by jejunal mucosal cells is mediated by a fatty acid binding membrane protein. J Clin Invest 82:2001–2010
36. Black D (2007) Development and physiological regulation of intestinal lipid absorption. I. Development of intestinal lipid absorption: Cellular events in chylomicron assembly and secretion. Am J Physiol Gastrointest Liver Physiol 293:G519–G524
37. Charman WN, Stella VJ (1991) Transport of lipophilic molecules by the intestinal lymphatic system. Adv Drug Deliv Rev 7:1–14
38. Trevaskis NL, Shanker RM, Charman WN, Porter CJH (2010) The mechanism of lymphatic access of two cholesteryl ester transfer protein inhibitors (CP524,515 and CP532,623) and evaluation of their impact on lymph lipoprotein profiles. Pharm Res 27:1949–1964
39. Pozzi F, Furlani P, Gazzaniga A, Davis SS, Wilding IR (1994) The time clock system - a new oral dosage form for fast and complete release of drug after a predetermined lag time. J Control Release 31:99–108

40. Hunt JN, Knox MT (1968) Control of gastric emptying. Am J Dig Dis 13:372–375
41. Hunt JN, Knox MT (1968) A relation between the chain length of fatty acids and the slowing of gastric emptying. J Physiol (Lond) 194:327–336
42. Chauhan H, Shimpi S, Mahadik KR, Paradkar A (2005) Preparation and evaluation of floating risedronate sodium-Gelucire® 43/01 formulations. Drug Dev Ind Pharm 31:851–860
43. Chansanroj K, Betz G, Leuenberger H, Mitrevej A, Sinchaipanid N (2007) Development of a multi-unit floating drug delivery system by hot melt coating technique with drug-lipid dispersion. J Drug Deliv Sci Technol 17:333–338
44. Siripuram PK, Bandari S, Jukanti R, Veerareddy PR (2010) Formulation and characterization of floating gelucire matrices of metoprolol succinate. Dissol Technol 17:34–39
45. Patel DM, Patel NM, Patel VF, Bhatt DA (2007) Floating granules of ranitidine hydrochloride-gelucire 43/01: Formulation optimization using factorial design. AAPS Pharmscitech 8:Article 30
46. Upadhyay P, Pandit JK, Upadhyay S, Ghosh AK (2010) Studies on formulation and optimization of gastro retentive multi-particulates of glibenclamide and metformin hydrochloride for the treatment of type ii diabetes mellitus using gelucire: A review. J Pharm Sci Res 2:351–354
47. Shimpi S, Chauhan B, Mahadik KR, Paradkar A (2004) Preparation and evaluation of diltiazem hydrochloride-gelucire 43/01 floating granules prepared by melt granulation. AAPS Pharmscitech 5:e43
48. Hamdani J, Goole J, Moes AJ, Amighi K (2006) In vitro and in vivo evaluation of floating riboflavin pellets developed using the melt pelletization process. Int J Pharm 323:86–92
49. Hamdani J, Moes AJ, Amighi K (2006) Development and in vitro evaluation of a novel floating multiple unit dosage form obtained by melt pelletization. Int J Pharm 322:96–103
50. Kumar KM, Shah MH, Ketkar A, Mahadik KR, Paradkar A (2004) Effect of drug solubility and different excipients on floating behaviour and release from glyceryl monooleate matrices. Int J Pharm 272:151–160
51. Woodley JF (2000) Lectins for gastrointestinal targeting - 15 years on. J Drug Target 7:325–333
52. Nielsen LS, Schubert L, Hansen J (1998) Bioadhesive drug delivery systems - i. Characterisation of mucoadhesive properties of systems based on glyceryl mono-oleate and glyceryl monolinoleate. Eur J Pharm Sci 6:231–239
53. Chung H, Jeong SY, Kwon IC, Park Y, Lee IH, Yuk SH, Choi YW, Park JH, Chung JW (2004) Composition for solubilization of paclitaxel and preparation method thereof. KR PCT/KR2003/001427 (18 July 2003)
54. Shin BS, Kim HJ, Hong SH, Lee JB, Hwang SW, Lee MH, Yoo SD (2009) Enhanced absorption and tissue distribution of paclitaxel following oral administration of DHP 107, a novel mucoadhesive lipid dosage form. Cancer Chemother Pharmacol 64:87–94
55. Yamamoto H, Takeuchi H, Hino T, Kawashima Y (2000) Mucoadhesive liposomes: Physicochemical properties and release behavior of water-soluble drugs from chitosan-coated liposomes. STP Pharma Sci 10:63–68
56. Lindmark T, Nikkila T, Artursson P (1995) Mechanisms of absorption enhancement by medium-chain fatty-acids in intestinal epithelial caco-2 cell monolayers. J Pharmacol Exp Ther 275:958–964
57. White KL, Nguyen G, Charman WN, Edwards GA, Faassen WA, Porter CJH (2009) Lymphatic transport of methylnortestosterone undecanoate (MU) and the bioavailability of methylnortestosterone are highly sensitive to the mass of coadministered lipid after oral administration of MU. J Pharmacol Exp Ther 331:700–709
58. Dahan A, Duvdevani R, Shapiro I, Elmann A, Finkelstein E, Hoffman A (2008) The oral absorption of phospholipid prodrugs: In vivo and in vitro mechanistic investigation of trafficking of a lecithin-valproic acid conjugate following oral administration. J Control Release 126:1–9
59. Pouton CW (2006) Formulation of poorly water-soluble drugs for oral administration: Physicochemical and physiological issues and the lipid formulation classification system. Eur J Pharm Sci 29:278–287
60. Pouton C, Porter C (2008) Formulation of lipid-based delivery systems for oral administration: Materials, methods and strategies. Adv Drug Deliv Rev 60:625–637

61. Pouton CW (2000) Lipid formulations for oral administration of drugs: Non-emulsifying, self emulsifying and 'self microemulsifying' drug delivery systems. Eur J Pharm Sci 11(suppl 2):S93–S98
62. Cuine J, McEvoy C, Charman W, Pouton C, Edwards G, Benameur H, Porter C (2007) Evaluation of the impact of surfactant digestion on the bioavailability of danazol after oral administration of lipidic self-emulsifying formulations to dogs. J Pharm Sci 97:995–1012
63. Magenheim B, Levy MY, Benita S (1993) A new in vitro technique for the evaluation of drug release profile from colloidal carriers - ultrafiltration technique at low pressure. Int J Pharm 94:115–123
64. Levy MY, Benita S (1990) Drug release from submicronized o/w emulsion - a new invitro kinetic evaluation model. Int J Pharm 66:29–37
65. Charman W, Rogge M, Boddy A, Berger B (1993) Effect of food and a monoglyceride emulsion formulation on danazol bioavailability. J Clin Pharmacol 33:381–386
66. Bates T, Sequeira J (1975) Bioavailability of micronized griseofulvin from corn oil-in-water emulsion, aqueous suspension, and commercial tablet dosage forms in humans. J Pharm Sci 64:793–797
67. Hauss D, Fogal S, Ficorilli J, Price C, Roy T, Jayaraj A, Keirns J (1998) Lipid based delivery systems for improving the bioavailability and lymphatic transport of a poorly water-soluble ltb4 inhibitor. J Pharm Sci 87:164–169
68. Caliph S, Charman WN, Porter CJH (2000) Effect of short-, medium-, and long-chain fatty acid-based vehicles on the absolute oral bioavailability and intestinal lymphatic transport of halofantrine and assessment of mass balance in lymph-cannulated and non-cannulated rats. J Pharm Sci 89:1073–1084
69. Tokumura T, Tsushima Y, Tatsuishi K, Kayano M, Machida Y, Nagai T (1987) Enhancement of the oral bioavailability of cinnarizine in oleic acid in beagle dogs. J Pharm Sci 76:286–288
70. Dahan A, Hoffman A (2007) The effect of different lipid based formulations on oral absorption of lipophilic drugs: The ability of in vitro lipolysis and consecutive ex vivo intestinal permeability data to predict in vivo bioavailability in rats. Eur J Pharm Biopharm 67:96–107
71. Faham A, Prinderre P, Piccerelle P, Farah N, Joachim J (2000) Hot melt coating technology: Influence of compritol (r) 888 ato and granule size on chloroquine release. Pharmazie 55:444–448
72. Faham A, Prinderre P, Farah N, Eichler KD, Kalantzis G, Joachim J (2000) Hot-melt coating technology. I. Influence of Compritol 888 ato and granule size on theophylline release. Drug Dev Ind Pharm 26:167–176
73. Montousse C, Pruvost M, Rodriguez F, Brossard C (1999) Extrusion-spheronization manufacture of gelucire (r) matrix beads. Drug Dev Ind Pharm 25:75–80
74. Siepmann F, Muschert S, Flament MP, Leterme P, Gayot A, Siepmann J (2006) Controlled drug release from gelucire-based matrix pellets: Experiment and theory. Int J Pharm 317:136–143
75. Zhou F, Vervaet C, Remon JP (1996) Matrix pellets based on the combination of waxes, starches and maltodextrins. Int J Pharm 133:155–160
76. Qi S, Marchaud D, Craig DQM (2010) An investigation into the mechanism of dissolution rate enhancement of poorly water-soluble drugs from spray chilled Gelucire 50/13 microspheres. J Pharm Sci 99:262–274
77. Serajuddin ATM (1999) Solid dispersion of poorly water-soluble drugs: Early promises, subsequent problems, and recent breakthroughs. J Pharm Sci 88:1058–1066
78. Khoo S, Porter CJH, Charman WN (2000) The formulation of halofantrine as either non-solubilising PEG 6000 or solubilising lipid based solid dispersions: Physical stability and absolute bioavailability assessment. Int J Pharm 205:65–78
79. Damian F, Blaton N, Kinget R, Van den Mooter G (2002) Physical stability of solid dispersions of the antiviral agent UC-781 with PEG 6000, Gelucire® 44/14 and PVP K30. Int J Pharm 244:87–98

80. Unga J, Matsson P, Mahlin D (2010) Understanding polymer-lipid solid dispersions-the properties of incorporated lipids govern the crystallisation behaviour of PEG. Int J Pharm 386:61–70
81. Li P, Hynes SR, Haefele TF, Pudipeddi M, Royce AE, Serajuddin ATM (2009) Development of clinical dosage forms for a poorly water-soluble drug ii: Formulation and characterization of a novel solid microemulsion preconcentrate system for oral delivery of a poorly water-soluble drug. J Pharm Sci 98:1750–1764
82. Muller RH, Mader K, Gohla S (2000) Solid lipid nanoparticles (SLN) for controlled drug delivery - a review of the state of the art. Eur J Pharm Biopharm 50:161–177
83. Rai S, Paliwal R, Gupta PN, Khatri K, Goyal AK, Vaidya B, Vyas SP (2008) Solid lipid nanoparticles (SLNs) as a rising tool in drug delivery science: One step up in nanotechnology. Curr Nanosci 4:30–44
84. Uner M, Yener G (2007) Importance of solid lipid nanoparticles (SLN) in various administration routes and future perspectives. Int J Nanomed 2:289–300
85. Souto EB, Doktorovova S (2009) Solid lipid nanoparticle formulations: Pharmacokinetic and biopharmaceutical aspects in drug delivery. Methods Enzymol Liposomes F 464:105–129
86. Yang SC, Zhu JB, Lu Y, Liang BW, Yang CZ (1999) Body distribution of camptothecin solid lipid nanoparticles after oral administration. Pharm Res 16:751–757
87. Zhang QN, Yie GQ, Li Y, Yang QS, Nagai YT (2000) Studies on the cyclosporin a loaded stearic acid nanoparticles. Int J Pharm 200:153–159
88. Penkler L, Muller RH, Runge SA, Ravelli V (1999) Pharmaceutical cyclosporin formulation with improved biopharmaceutical properties, improved physical quality and greater stability, and method for producing said formulation. WO/1999/056733
89. Carrigan PJ, Bates TR (1973) Biopharmaceutics of drugs administered in lipid-containing dosage forms i: GI absorption of griseofulvin from an oil-in-water emulstions in the rat. J Pharm Sci 62:1476–1479
90. Li HL, Zhao XB, Ma YK, Zhai GX, Li LB, Lou HX (2009) Enhancement of gastrointestinal absorption of quercetin by solid lipid nanoparticles. J Control Release 133:238–244
91. Priano L, Esposti D, Esposti R, Castagna G, De Medici C, Fraschini F, Gasco MR, Mauro A (2007) Solid lipid nanoparticles incorporating melatonin as new model for sustained oral and transdermal delivery systems. J Nanosci Nanotechnol 7:3596–3601
92. Varshosaz J, Tabbakhian M, Mohammadi MY (2010) Formulation and optimization of solid lipid nanoparticles of buspirone hcl for enhancement of its oral bioavailability. J Liposome Res 20:286–296
93. Luo Y, Chen DW, Ren LX, Zhao XL, Qin J (2006) Solid lipid nanoparticles for enhancing vinpocetine's oral bioavailability. J Control Release 114:53–59
94. Westesen K, Bunjes H, Koch MHJ (1997) Physicochemical characterization of lipid nanoparticles and evaluation of their drug loading capacity and sustained release potential. J Control Release 48:223–236
95. Westesen K, Bunjes H (1995) Do nanoparticles prepared from lipids solid at room temperature always possess a solid lipid matrix? Int J Pharm 115:129–131
96. Kato T (2002) Self-assembly of phase-segregated liquid crystal structures. Science 295:2414–2418
97. Israelachvili JN, Mitchell DJ, Ninham BW (1976) Theory of self-assembly of hydrocarbon amphiphiles into micelles and bilayers. J Chem Soc Faraday Trans II 72:1525–1568
98. Kaasgaard T, Drummond CJ (2006) Ordered 2-D and 3-D nanostructured amphiphile self-assembly materials stable in excess water. Phys Chem Chem Phys 8:4957–4975
99. Caffrey M, Cheng A (1995) Kinetics of lipid phase changes. Curr Opin Struct Biol 5: 548–555
100. Fontell K (1990) Cubic phases in surfactant and surfactant-like lipid systems. Colloid Polym Sci 268:264–285
101. Landh T (1995) From entangled membranes to eclectic morphologies: Cubic membranes as subcellular space organizers. FEBS Lett 369:13–17

102. Lindblom G, Rilfors L (1988) Cubic phases and isotropic structures formed by membrane lipids- possible biological relevance. Biochem Biophys Acta 988:221–256
103. Luzzati V (1997) Biological significance of lipid polymorphism; cubic phases. Curr Opin Struct Biol 7:661–668
104. Mariani P, Luzzati V, Delacroix H (1988) Cubic phases of lipid containing systems structure analysis and biological implications. J Mol Biol 204:165–189
105. Schwarz US, Gompper G (2001) Bending frustration of lipid-water mesophases based on cubic minimal surfaces. Langmuir 17:2084–2096
106. Israelachvili JN (1994) The science and application of emulsions - an overview Colloid Surf A 91:1–8
107. Boyd B, Rizwan S, Dong Y, Hook S, Rades T (2007) Self-assembled geometric liquid-crystalline nanoparticles imaged in three dimensions: Hexosomes are not necessarily flat hexagonal prisms. Langmuir 23:12461–12464
108. Siekmann B, Bunjes H, Koch M, Westesen K (2002) Preparation and structural investigations of colloidal dispersions prepared from cubic phase monoglyceride/water phases. Int J Pharm 244:33–43
109. Esposito E, Cortesi R, Drechsler M, Paccamiccio L, Mariani P, Contado C, Stellin E, Menegatti E, Bonina F, Puglia C (2005) Cubosome dispersions as delivery systems for percutaneous administration of indomethacin. Pharm Res 22:2163–2173
110. Dong YD, Larson I, Hanley T, Boyd BJ (2006) Bulk and dispersed aqueous phase behavior of phytantriol: Effect of vitamin E acetate and F127 polymer on liquid crystal nanostructure. Langmuir 22:9512–9518
111. Yang D, Armitage B, Marder S (2004) Cubic liquid-crystalline nanoparticles. Angew Chem Int Ed 43:4402–4409
112. Boyd BJ (2003) Characterisation of drug release from cubosomes using the pressure ultrafiltration method. Int J Pharm 260:239–247
113. Boyd BJ, Whittaker DV, Khoo S-M, Davey G (2006) Lyotropic liquid crystalline phases formed from glycerate surfactants as sustained release drug delivery systems. Int J Pharm 309:218–226
114. Fong C, Krodkiewska I, Wells D, Boyd BJ, Booth J, Bhargava S, McDowall A, Hartley P (2005) Submicron dispersions of hexosomes based in novel glycerate surfactants. Aust J Chem 58:683–687
115. Barauskas J, Johnsson M, Nylander T, Tiberg F (2006) Hexagonal liquid crystalline nanoparticles in aqueous mixtures of glyceryl monooleyl ether and Pluronic F127. Chem Lett 35:830–831
116. Popescu G, Barauskas J, Nylander T, Tiberg F (2006) Liquid crystalline phases and their dispersions in aqueous mixtures of glycerol monooleate and glycerol monooleyl ether. Langmuir 23:496–503
117. Clogston J, Rathman J, Tomasko D, Walker H, Caffrey M (2000) Phase behavior of a monoacylglycerol (Myverol 18-99K)/water system. Chem Phys Lipids 107:191–220
118. Hato M, Minamikawa H, Salkar R, Matsutani S (2004) Phase behavior of phytanyl-chained akylglycoside/water systems. Prog Colloid Polym Sci 123:56–60
119. Hato M, Yamashita I, Kato T, Abe Y (2004) Aqueous phase behavior of a 1-o-phytanyl-β-d-xyloside/water system. Glycolipid-based bicontinuous cubic phases of crystallographic spacegroups Pn3m and Ia3d. Langmuir 20:11366–11373
120. Barauskas J, Landh T (2003) Phase behaviour of the phytantriol/water system. Langmuir 19:9562–9565
121. Fong C, Wells D, Krodkiewska I, Weerawardeena A, Booth J, Hartley P, Drummond CJ (2007) Diversifying the solid state and lyotropic phase behavior of nonionic urea-based surfactants. J Phys Chem B 111:10713–10722
122. Fong C, Wells D, Krodkiewska I, Hartley P, Drummond C (2006) New role for urea as a surfactant headgroup promoting self-assembly in water. Chem Mater 18:594–597
123. Spicer PT, Hayden KL (2001) Novel process for producing cubic crystalline nanoparticles (Cubosomes). Langmuir 17:5748–5756

124. Shah JC, Sadhale Y, Chilukuri DM (2001) Cubic phase gels as drug delivery systems. Adv Drug Deliv Rev 47:229–250
125. Drummond CJ, Fong C (1999) Surfactant self-assembly objects as novel drug delivery vehicles. Curr Opin Colloid Interface Sci 4:449–456
126. Wyatt DM, Dorshel D (1992) A cubic-phase delivery system composed of glyceryl monooleate and water for sustained release of water-soluble drugs. Pharm Tech 16:116–130
127. Clogston J, Caffrey M (2005) Controlling release from the lipidic cubic phase. Amino acids, peptides, proteins and nucleic acids. J Control Release 107:97–111
128. Boyd BJ, Khoo S-M, Whittaker DV, Davey G, Porter CJH (2007) A lipid-based liquid crystalline matrix that provides sustained release and enhanced oral bioavailability for a model poorly water soluble drug in rats. Int J Pharm 340:52–60
129. Lee KWY, Nguyen TH, Hanley T, Boyd BJ (2009) Nanostructure of liquid crystalline matrix determines in vitro sustained release and in vivo oral absorption kinetics for hydrophilic model drugs. Int J Pharm 365:190–199
130. Chang C-M, Bodmeier R (1998) Low viscosity monoglyceride based drug delivery systems transforming into a highly viscous cubic phase. Int J Pharm 173:51–60
131. Fong WK, Hanley T, Boyd BJ (2009) Stimuli responsive liquid crystals provide 'on-demand' drug delivery in vitro and in vivo. J Control Release 135:218–226
132. Chung H, Kim J, Um JY, Kwon IC, Jeong SY (2002) Self-assembled "Nanocubicle" as a carrier for peroral insulin delivery. Diabetologia 45:448–451
133. Kossena GA, Charman WN, Boyd BJ, Porter CJH (2004) A novel cubic phase of medium chain lipid origin for the delivery of poorly water soluble drugs. J Control Release 99:217–229
134. Nguyen T, Hanley T, Larson I, Porter CJH, Boyd BJ (2010) Phytantriol and glyceryl monooleate cubic liquid crystalline phases as sustained-release oral drug delivery systems for poorly water soluble drugs ii. In vivo evaluation. J Pharm Pharmacol 62:856–865
135. Nguyen T (2009) Investigation of novel liquid crystalline materials for the sustained oral delivery of poorly water soluble drugs. Monash University, Melbourne
136. Yamahira Y, Noguchi T, Takenaka H, Maeda T (1979) Biopharmaceutical studies of lipid-containing oral dosage forms: Relationship between drug absorption rate and digestibility of vehicles. Int J Pharm 3:23–31
137. Myers R, Stella VJ (1992) Systemic bioavailability of penclomedine (NSC-338720) from oil-in-water emulsions administered intraduodenally to rats. Int J Pharm 78:217–226
138. Bloedow D, Hayton W (1976) Effects of lipids on bioavailability of sulfisoxazole acetyl, dicumarol, and griseofulvin in rats. J Pharm Sci 65:328–334
139. Geraghty P, Attwood D, Collett J, Sharma H, Dandiker Y (1997) An investigation of the parameters influencing the bioadhesive properties of Myverol 18-99/water gels. Biomaterials 18:63–67
140. Nguyen T, Hanley T, Porter CJH, Larson I, Boyd BJ (2010) Nanostructured liquid crystalline particles provide long duration sustained-release effect for a poorly water soluble drug after oral administration. J Control Release 153:180–186
141. Lai J, Chen JM, Lu Y, Sun J, Hu FQ, Yin ZN, Wu W (2009) Glyceryl monooleate/Poloxamer 407 cubic nanoparticles as oral drug delivery systems: I. In vitro evaluation and enhanced oral bioavailability of the poorly water-soluble drug simvastatin. AAPS Pharmscitech 10:960–966

Chapter 16
Buccal Drug Delivery

Michael A. Repka, Li-lan Chen, and Rick S. Chan

Abstract The buccal cavity can be most useful for delivery of drugs for systemic effects. Advantages can include rapid onset of action, avoidance of presystemic elimination, which may include gastrointestinal degradation, first-pass clearance by the liver, or transformation by intestinal mucosal enzymes, e.g., CYP450. Exposure of a bioactive to gastric acid is also avoided. Rate of delivery can also be controlled by formulation approaches.

Drugs must be carefully selected for delivery via the buccal route. For example the dose must be low and the physicochemical properties must be appropriate. Dose must be low and the physicochemical properties must be appropriate. Furthermore, as different regions of the cavity possess differing permeation characteristics, it may be necessary to ensure prolonged contact between the delivery system and the optimal region for consistent and reliable absorption. Such possibilities, together with manufacturing processes for buccal delivery systems, are presented and discussed in this chapter.

16.1 Introduction

16.1.1 Historical Development of Drug Delivery Through the Oral Cavity

Historically, peroral delivery has been favored for the majority of therapeutic agents targeting systemic effect. Oral administration generally leads to "transmucosal" absorption in the GI tract. However, this enteral route of delivery subjects many compounds to extensive pre-systemic elimination, which may include GI degradation,

M.A. Repka (✉)
Department of Pharmaceutics, National Center for Natural Products Research,
School of Pharmacy, University of Mississippi, University, MS, USA
e-mail: marepka@olemiss.edu

C.G. Wilson and P.J. Crowley (eds.), *Controlled Release in Oral Drug Delivery*,
Advances in Delivery Science and Technology, DOI 10.1007/978-1-4614-1004-1_16,

metabolism, or first-pass clearance via the liver. This biotransformation has often resulted in low systemic bioavailability, short duration of therapeutic activity, and/or formation of inactive, and at times toxic metabolites [1]. Parenteral routes, such as intravenous (IV) or intramuscular (IM) administration, unlike oral delivery, permit therapeutic agents to gain direct entry into the systemic circulation and therefore reach the intended site of action more rapidly (with total bioavailability achieved in many IV cases). Unfortunately, such a mode of drug administration entails certain health risks, requires specialized equipment, often requires close medical supervision of the medication or may even necessitate the hospitalization of patients during treatment.

Systemic transmucosal delivery via the epithelial linings of accessible body cavities such as the oral cavity (mouth), nose, rectum, and vagina have received renewed interest within the last decade. These routes have numerous advantages over their peroral counterpart such as bypassing hepatic first-pass clearance and therefore potentially improving systemic bioavailability. In addition, these transmucosal routes may eliminate the disadvantages of IV and IM administration.

16.1.2 Definition of Intraoral Drug Delivery

The oral cavity offers a variety of easily accessible mucosal tissues for use as drug delivery routes. These include the highly permeable sublingual region, richly vascular buccal region, and the soft palate region. An array of controlled and fast release formulations can be administered through the different regions in oral cavity by selecting the right route of administration.

The peroral route has been the most widely used mode of drug administration for centuries. While it retains its preeminence, it has certain limitations. Some drugs degrade in the pH and enzymatic environments of the gastrointestinal tract and/or undergo hepatic first-pass clearance. This reduces bioavailability, duration of therapeutic activity, and may lead to formation of toxic metabolites/breakdown products [1].

Transmucosal absorption of nitroglycerin solution from the oral cavity was demonstrated by Sobrero in the mid-nineteenth century [2]. Subsequently, various delivery systems (conventional and novel) have been proposed and evaluated. Until the mid-1980s, oral transmucosal delivery utilized mainly conventional dosage forms, developed for other routes of administration. However, it became apparent that various factors could decrease the efficiency or exclude the use of systemic absorption from the oral cavity. These included:

- Low residence time at the application site.
- High variability in bioavailability due to salivary secretion.
- Oral musculature activity.

Increased awareness of factors affecting absorption, together with technological advances in biomaterials and techniques has now resulted in delivery systems suited to the oral cavity and aligned with the physicochemical properties of the drug. Strategically, these systems must address one or more of the following requirements:

- Attain rapid release and absorption
- Sustain the release and/or duration of the absorption process
- Develop bidirectional or unidirectional systems
- Fabricate patient-friendly oral mucosal delivery systems

Patient compliance considerations have also shifted towards once-daily regimens and convenience, particularly where the medication is used chronically. This has stimulated investigations of drugs with high potency and sustained effects (e.g., peptide drugs) [3]. Both small and large molecules may be viable candidates for intraoral delivery.

16.2 Structure of the Oral Cavity

16.2.1 The Mucosa

Selection of drug candidates for oral transmucosal drug delivery (TMD) requires cognizance of the physicochemical properties of the drug and likely effect on absorption. The anatomy and physiology of the delivery site also needs to be recognized and evaluated as structural and regional variations in oral mucosa can influence drug permeation characteristics.

All covering and lining tissues of the body consist of a surface epithelium supported by fibrous connective tissue [4]. Comparing the structure of skin and oral mucosa to that of the GI tract, major differences are apparent in the organization of the epithelium. These reflect the different functions of the regions. The linings of the stomach and small and large intestine comprise a simple epithelium. Skin and oral mucosa, in contrast, are covered by a stratified epithelium composed of multiple layers of cells showing various patterns of differentiation between the deepest cell layer and the surface (Fig. 16.1). Such differentiation reflects the functional purposes and demands required of such tissue, such as mobility or rigidity and resistance to mechanical or other damage.

In general, three different types of oral mucosa are recognized (Table 16.1):

- Masticatory mucosa: This covers the gingiva and hard palate, regions that are subjected to the mechanical forces of mastication. It comprises keratinized epithelium resembling the epidermis of the skin and is usually tightly attached to underlying structures by a collagenous connective tissue.
- Specialized mucosa. This has characteristics of both masticatory and lining mucosa and is found on the dorsal region of the tongue. Its surface consists of areas of both keratinized and nonkeratinized epithelium, tightly bound to the underlying muscle of the tongue.
- Lining mucosa.

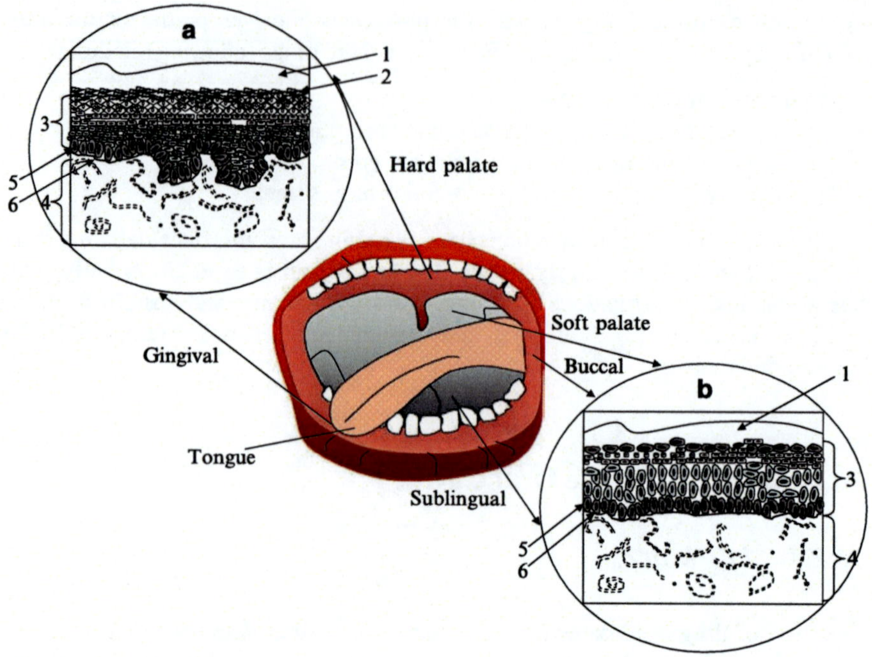

Fig. 16.1 Mucosal regions of the mouth with insets showing the compositions of (**a**) keratinized and (**b**) nonkeratinized mucosal epithelium lining various mucosal regions: 1. Mucus layer, 2. Parakeratinized layer, 3. Epithelium, 4. Lamina propria, 5. Stratum basale, 6. Basal lamina

Table 16.1 Regional variation of mucosal tissue within the oral cavity

Mucosa type	Characteristics
Masticatory	Keratinized epithelium
	Hard palate, gingival
	25% of total surface area of oral cavity
Lining	Nonkeratinized epithelium
	Cheek, sublingual, alveolar
	60% of total surface area
Specialized	Both keratinized and nonkeratinized
	Tongue
	15% of total surface area

16.2.2 *Sites of Administration in the Oral Cavity*

The lining mucosa of the oral cavity is covered by a stratified, nonkeratinized squamous epithelium. This is located within the cheek (buccal), soft palate (palatal), part of the tongue (lingual), the floor of the mouth (sublingual), and the alveolar lining mucosa (labial mucosa) regions. The buccal, labial, and sublingual tissue are primary targets for drug delivery due to their potential permeability. Figure 16.1 illustrates these anatomical locations within the oral cavity as well as mucosal type. Although the surface area of the oral mucosa is relatively small in comparison to the skin and the GI tract, its high vasculature lends itself to potential drug absorption.

The oral cavity may be divided into three sections depending on variations in the thickness and nature of the mucosal lining.

- The sublingual mucosa lines the floor of the mouth and is the thinnest and the most permeable region in the oral cavity. It is supplied with high blood flow and has sufficient surface area to make it a location of choice when rapid absorption/ onset of drug action is necessary. However, its surface is constantly washed by saliva and this plus tongue activity which makes it difficult to keep the dosage form in contact with the mucosa [5].
- The buccal mucosa lines the interiors of the cheek and can be used for systemic as well as local delivery. The surface of buccal mucosa is smooth, relatively immobile, and more permeable than other mucosal tissues makes it a location of choice for controlled release systems that need to stay adhered for an extended period. Buccal mucosa is also more robust and tolerant to irritation and permanent damage from adhesion. However, salivary production and composition may contribute to chemical modification of certain drugs. Moreover, involuntary swallowing can result in drug loss from the site of absorption. Furthermore, constant salivary scavenging within the oral cavity makes it difficult for dosage forms to be retained for an extended period of time to facilitate absorption. The relatively small absorption area and barrier properties can limit this route of delivery [5].
- The soft palate is suspended from the posterior border of the hard palate connecting the oral and nasal parts of the pharynx in the roof of the oral cavity. The palatal mucosa found in the oral cavity is highly vascularized, thin and mostly covered with stratified squamous epithelium.

The term "buccal delivery" will be used hereafter, for reader convenience to describe delivery via any of these intraoral locations.

16.3 Advantages and Challenges of Buccal Drug Delivery

Table 16.2 compares buccal delivery with those for the GI tract and dermal tissues while Table 16.3 lists its advantages and disadvantages [6, 7]. The membranes lining the oral cavity are readily accessible, robust, and exhibit fast cellular recovery following local stress or damage. Indeed, the oral cavity can be considered as designed to handle and withstand, when necessary a wide array of exogenous materials. Properly constructed oral TMD systems are easy and painless to administer and well accepted by the patient. Precise dosage form localization is also possible. The ability to terminate delivery (remove the device) can

Table 16.2 Comparison of routes for systemic drug delivery available to the formulation scientist

Issues	Oral mucosal	Gastrointestinal	Dermal
Accessibility	√√	√	√√√
First-pass clearance	√√√	√	√√√
Acceptability	√√√	√√√	√√√
Surface area	√√	√√√	√√√
Onset of action	√√	√√√	√
Robustness	√√	√√	√√√
Duration	√√√	√	√√
Permeability	√√	√√√	√
Vascular drainage	√√√	√√√	√
Surface environment	√√√	√	√√

√ Not favorable; √√ Intermediate; √√√ Very favorable

Table 16.3 Advantages and disadvantages of transmucosal drug delivery via the oral cavity

Advantages	Disadvantages
Avoids hepatic first-pass metabolism	Expensive
Avoids chemically hostile GI environment	Multilayering – uncomfortable to wear
Avoids GI distress	Processing method reproducibility
Allows use of drugs with short $t_{½}$s	Not applicable for high dose/blood level drugs
Controls plasma levels of potent drugs	Relatively low surface area
Can interrupt drug input quickly (toxicity)	Limited absorption of high MW drugs
Reduces multiple dosing	
Improvement in patient compliance	
Fast cellular recovery following stress	
Useful for pediatric and geriatric patients	

also be a major advantage if side effects become apparent. Patients could also conceivably control the duration of administration to optimize the desired therapeutic effect.

The thin mucin film on the surface of the oral cavity provides the capability to prolong mucosal contact using mucoadhesive formulation components. This provides close contact with the absorbing tissue, thus optimizing concentration gradient with the systemic circulation and reducing the diffusional pathway. Such features offer the potential route for rapid onset, as well as controlled or sustained systemic delivery.

Enzymatic barriers [8] can compromise transmucosal absorption, as can physiological and histological variations of the mucosa. The former can degrade peptides and proteins while the latter hinders transport of large molecules. However, tailoring the formulation to include enzyme inhibitors and/or permeation enhancers may minimize or eliminate these barriers. Thus, the numerous advantages of oral transmucosal delivery continue to be explored for potential delivery of diverse agents for local as well as systemic therapeutic effects.

16.4 Drug Release Mechanisms of Buccal Drug Delivery Systems

The suitability of the oral cavity for systemic absorption and the clinical outcome of such delivery depend on:

- The ability to maintain plasma concentration within the therapeutic range.
- The physicochemical properties of drug.
- Release mechanism of the delivery system.

If the active ingredient has favorable physicochemical properties the release profile must then be designed to deliver the requisite plasma concentrations.

A number of dosage form types have been used for transmucosal delivery, including chewing gums, hollow fibers, bioadhesive tablets, laminated systems, and patches [9–13]. All can provide different drug release characteristics depending on the drug, the excipients, and the manufacturing process.

Mechanism of drug release can be broadly classified as either diffusion controlled, erosion controlled, or combinations thereof (Fig. 16.2). An understanding of the release mechanism provides better understanding for designing systems for optimum effect.

Equation (16.1) can be used to describe the kinetics of drug release (or dissolution) [14]:

$$M_t/M_\infty = kt^n, \tag{16.1}$$

where

M_t/M_∞ is the fraction of drug released, "t" is the release time, "k" is a kinetic constant characteristic of the drug polymer system, "n" is a release exponent indicative of the release mechanism of the drug.

When $n=0.5$, drug is released via Fickian diffusion. For $0.5<n<1$, a non-Fickian solute diffusion is observed. When $n=1$ case II transport (erosion) pertains, with zero order kinetics [15].

In matrix systems, drug release rate is mainly controlled by diffusion; the release exponent is 0.5 for planar surfaces. These systems require that the polymeric carrier be an insoluble but swellable, inert matrix or a polymer of high viscosity grade in which the drug is appreciably soluble or diffusible. For release rate to be diffusion controlled, the rate of dissolution of drug in the solution within the matrix should be much faster than the diffusion rate of the dissolved drug leaving the matrix. Hence, drug solubility is important. Equations for release rates from such matrices are based on Fick's Laws and have been derived by Higuchi [16], indicating that the amount of drug released is a function of the square root of time.

In general, for erosion to be the dominant release mechanism the matrix should contain a hydrophilic polymer of low viscosity grade and the incorporated drug should have low water solubility and/or low diffusivity. Drug dispersed in such a matrix is then released primarily due to erosion of the polymer. There are two

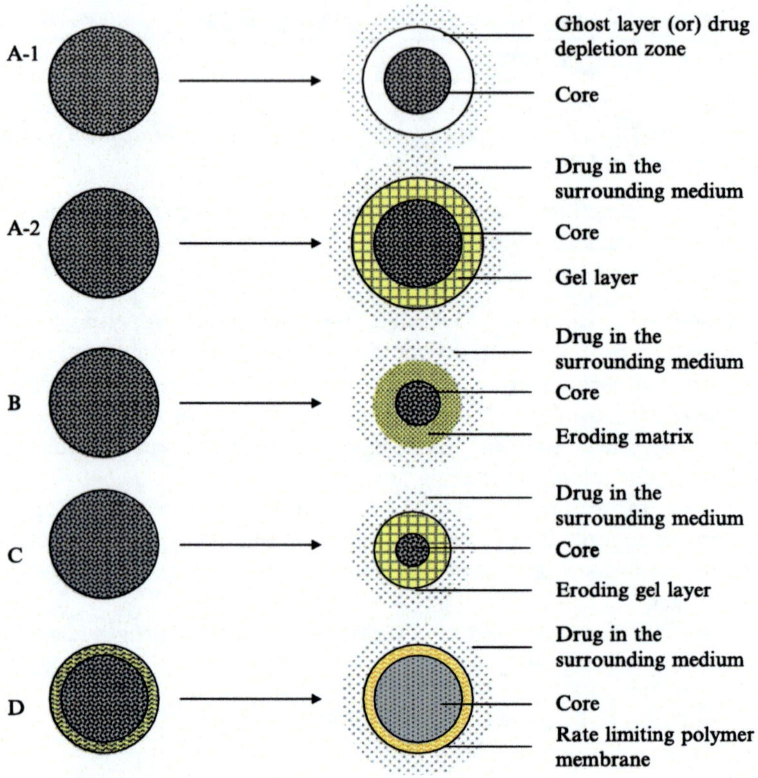

Fig. 16.2 Mechanisms of drug release for transmucoasl drug delivery devices. (**a**) Matric diffusion systems: (**a**-1) swellable matrix (hydrophilic polymers), (**a**-2) nonswellable matrix (lipophilic matrix); (**b**) matrix erosion systems; (**c**) matrix diffusion and erosion systems; (**d**) reservoir systems

types of release from the matrix erosion systems based on the properties of the polymer and the drug:

- If the drug is covalently linked to the polymer backbone directly or via a spacer group, then the drug is released when its bond with the backbone is cleaved by chemical hydrolysis or enzymatic cleavage.
- When the active agent (which is practically insoluble) is homogenously dispersed in an erodible polymer, drug release occurs by dissolution or erosion of the gel layer formed, or by polymer disintegration. This mechanism may exhibit zero-order release kinetics with a release exponent equal to 1.0, as described earlier.

A combination of two hydrophilic gel formers, which influence each other's swelling process, may also give a zero-order release profile. Researchers have formulated a hot-melt extruded matrix erodible film system containing low molecular weight hydroxypropyl cellulose (HPC) and a low molecular weight polyethylene oxide (PEO) incorporated with a poorly soluble drug (clotrimazole) [17].

Zero-order release was exhibited by the matrix [17]. Such surface erodible systems, though difficult to achieve, have several advantages, including the ability to control the rate of drug delivery and the ability to deliver a variety of therapeutic agents. This may be accomplished by varying the drug loading within the matrix, thereby controlling the functioning life of the matrix, by varying its thickness and other physical attributes. Such a "platform technology" may have wide applicability.

Matrix diffusion and erosion systems are formed by polymers that do not possess highly resistant gel structures. A combination of two different low viscosity polymers having desired characteristics can be used to achieve this type of release. Release rate follows anomalous transport and does not obey Fick's laws, with the release exponent between 0.5 and 1.0 for planar surfaces. Release from such matrices is controlled both by diffusion of the drug and erosion of the gelled polymer. They may be employed to sustain release of both water soluble and insoluble drugs and can release 100% of the incorporated drug, unlike its matrix diffusion-only counterpart. Zero-order release may also be possible with such systems if care is taken in polymer selection [17].

16.5 Factors Affecting Drug Absorption from the Oral Mucosa

Until the mid-1980s, only a handful of drugs were regarded to be suitable, or even clinically necessary, for delivery via the oral mucosal route. However, new and more potent drugs and improved biomaterials have lead to a renaissance of this mode of delivery. In addition, research in the transdermal area has identified certain penetration enhancers that may improve the permeability of oral mucosal tissue when coupled with appropriate delivery systems (buccal patches, films) [18]. At the same time, there has been an explosion in knowledge of factors affecting drug release and its absorption across the oral mucosa.

The main factors affecting peroral drug absorption are the physicochemical properties of the drug, biological processes in the oral cavity, and the composition (formulation) of the delivery system. The therapeutic dose of the drug is also important, as is its dose response. Variability in drug absorption can also be important if drug that is swallowed is subject to extensive first-pass metabolism.

16.5.1 Physicochemical Factors

Passive diffusion is the primary mechanism of drug permeation across the oral mucosa [19]. Facilitated diffusion has also been shown to occur, but primarily with nutrients. The phenomenon of "buccal partitioning" has been used to describe the membrane reservoir effect noted in some studies [20, 21]. However, this drug binding effect has not been characterized very well to date. Thus, this review focuses on passive diffusion.

Parameters such as partition coefficient (K_p), degree of ionization, and molecular mass of the drug can greatly influence delivery across the oral mucosa membrane [22]. The extent of absorption is generally proportional to the drug's lipophilicity or oil-in-water partition coefficient. However, there is a delicate balance between K_p and solubility for a drug to be a successful candidate for oral mucosal absorption [23]. If a drug is too lipophilic, it cannot dissolve appreciably in the aqueous mucin/saliva and thus may not be available for significant absorption.

Generally, the unionized form of a drug is the most lipid soluble and thus most diffusible across biological membranes. Therefore, the pK_a of the drug must be addressed when diffusivity and ultimate bioavailability are being considered. The average pH of human saliva is approximately 6.4 and remains relatively constant. The effect of pH on drug absorption via the oral mucosa has been studied extensively [22]. Most highlight the importance of the state of drug ionization.

In general, small molecules penetrate the mucosa more rapidly than larger molecules. However, penetration enhancers can dramatically improve the permeability of some high molecular weight mucopolysaccharides [24].

The area of application is also a limiting factor for oral mucosal drug delivery. It has been estimated that the total amount of drug that could be systemically delivered across the buccal mucosa from a 2-cm^2 system in 1 day is 10–20 mg [18]. Mucosal area available for a matrix or reservoir system is also a limiting factor. Drug potency becomes of paramount importance in the systems' formulation and design.

16.5.2 Biological Factors

Saliva may be considered a positive or a negative factor for oral transmucosal drug absorption. Adult males have an average volume of approximately 0.9 mL and females a slightly smaller average volume of about 0.8 mL [25]. Salivary secretions include mucins, proline-rich proteins, and other proteins such as histatins, cystatins, statherins, and α-amylases. These protect the integrity of the hard and soft oral tissues. Changes in salivary volume, flow rate, and pH can alter drug absorption which occurs more readily across moist mucous membranes than those that lack mucous [22].

A drug must dissolve before absorption can occur – and in TMD systems the dissolution medium is aqueous mucin/saliva. Excess saliva has been shown to adversely affect absorption for some systems [26]. Disease states or co-administered medications may cause xerostomia and inhibit drug permeation due to decreased salivary flow. These factors may in turn affect the condition of the mucosa, causing increased or decreased drug penetration.

It is generally agreed that the oral mucosa tissue is more permeable to drugs than the skin but differences are widely debated. Estimates range from a 4-fold to a 4,000-fold greater permeability for the oral mucosa [6, 27]. Such wide differences may reflect the fact that different mucosal regions have different structures and thus different permeability. It is also generally accepted that a permeability "barrier" exists in the outermost one-fourth to one-third of the mucosal epithelium, resulting from intercellular material that is derived from membrane coating granules (MCGs) [28]. In keratinized epithelia, these intercellular lipid substances include sphingomyelin, glucosylceramides, ceramides, and other nonpolar lipids. Unkeratinized epithelia contain primarily cholesterol esters, cholesterol, and glycosphingolipids [29]. Studies have demonstrated that a decrease in compound permeability may not be due to keratinization alone in that MCGs may play a significant role in the permeation [30, 31]. Due to other factors discussed previously, a primary strategy for TMD should be to target the oral cavity's lining mucosa.

Oral mucosal anatomy suggests that there are two permeability barriers. Intercellular spaces and cytoplasm are essentially hydrophilic in nature and function as a transport barrier to lipophilic compounds. The second barrier, the cell membrane is lipophilic and impedes diffusion of hydrophilic compounds due to low partition coefficient. The presence of both regions in oral mucosa suggests two possible routes for drug transport, viz., paracellular (intercellular) and transcellular (intracellular). Drug transport by these two routes most likely occurs simultaneously, although one route usually dominates depending on the physicochemical properties of the diffusant. Some researchers maintain that the paracellular route is the primary route for hydrophilic compounds, and thus, the intercellular space is the preferred route for hydrophilic drug transport [18]. Limited surface area of the intercellular space and the labyrinthal pathways within this area are the main limitations for this route. Intercellular flux (J_H) may be quantitated by (16.2) [18]:

$$J_{\mathrm{H}} = D_{\mathrm{H}}\varepsilon C_{\mathrm{D}}/h_{\mathrm{H}}. \tag{16.2}$$

where ε is the fraction of surface area of the paracellular route, D_H is the diffusion coefficient of the intercellular spaces, h_H is the path length of the paracellular route, and C_D is the donor side drug concentration.

The cell membrane is relatively lipophilic in nature and thus for lipophilic compounds partition propensity is high. Surface area for the transcellular route is large and the pathways for transcellular movement are relatively short. Thus, it is believed that the permeability of lipophilic compounds across the epithelial cell membrane is typically high. The flux via the transcellular route (J_L) can be expressed as:

$$J_{\mathrm{L}} = (1-\varepsilon)D_{\mathrm{L}}K_{\mathrm{P}}C_{\mathrm{D}}/h_{\mathrm{L}}, \tag{16.3}$$

where K_P is the partition coefficient between the lipophilic region (cell membrane) and the hydrophilic region, and h_L is the path length of the transcellular route. ε is the area fraction of the paracellular route and D_L is the diffusion coefficient of the drug in the transcellular spaces.

The primary route of transport is that which offers the least resistance to penetration of drug.

16.5.3 Formulation Factors

16.5.3.1 Bioadhesion and Bioadhesive Polymers

Bioadhesion is a phenomenon related to the ability of biological or synthetic material to adhere to a biological substrate. Oral mucosal drug delivery necessitates the use of mucoadhesive polymers as dosage forms should ideally adhere to the mucosa and withstand salivation, tongue movement, and swallowing for a predetermined period of time. Mucoadhesion has received considerable attention due its importance in drug delivery.

A widely used approach to characterize adhesivity involves consideration of interatomic or intermolecular forces at the adhesive:substrate interface. Numerous mechanisms of adhesion or mucoadhesion have been established and proposed. These include:

- Hydrogen bonding
- Surface energy and contact angle
- Polymer chain interpenetration
- Swelling rate of a polymer interacting with mucin [11, 32–34]

Adequate bioadhesion is essential for the success of a bioadhesive drug delivery system such that release is maintained over time to elicit the desired therapeutic response. Examples of mucoadhesive polymers that have been investigated previously are listed in Table 16.4 [33, 34, 36–67].

Several properties and associated techniques for in vitro determination of bioadhesion have been reported [68]. They include tensile strength testing, shear stress testing, adhesion weight method, fluorescent probe method, flow channel techniques, and a colloidal gold staining method. More recently, Texture Analyzer® (TA.XT2i) has been used for mucoadhesive studies [17]. Texture analyzer variables include contact force, contact time, and speed of probe withdrawal from the chosen substrate [17].

16.5.3.2 Penetration Enhancement of the Oral Mucosa

Though the oral cavity has been used as a localized site for systemic drug delivery, the amount of drug that can be absorbed perorally has historically been too little or too slow in many cases for the attainment of a desired therapeutic response. As discussed previously, this is partly due to the relatively small surface area of the oral mucosa and low permeability to various therapeutic agents. Unlike surface area, which is relatively constant, permeability of the oral mucosa may be temporarily altered to allow a greater drug flux and thus increased bioavailability [69]. Permeation enhancers have been used successfully to increase epithelial permeability and bioavailability in some studies [18]. Various classes of the transmucosal permeation enhancers are presented in Table 16.5 [70–82].

Table 16.4 Bioadhesives studied for transmucosal drug delivery systems [35]

Bioadhesive
Hydroxypropyl cellulose
Carbopol
Polyvinylpyrrolidone
Carboxymethyl cellulose
Hydroxypropyl methylcellulose
Hydroxyethyl cellulose
Poly(vinyl alcohol)
Poly(isobutylene)
Xanthan gum
Locust bean gum
Chitosan
Pectin
Polycarbophil
Hyaluronic acid benzyl esters
Poly(acrylic acid)
Poly(methacrylic acid)
Poly(acrylic acid-co-acrylamide)
Poly(acrylic acid-co-methylmethacrylate)
Poly(acrylic acid-co-butylacrylate)
Polytetramethylene glycol
Cydot®[polymeric blend of carbomer and poly(isobutylene)]
Poly(acrylic acid-co-polyethylene glycol)
Polyethylene glycol
Drum dried waxy maize starch

Table 16.5 Penetration enhancers for potential use in oral transmucosal

Chemical class	Example(s)
Amides	*m,m*-dimethyl-*m*-toluamide
Anionic surfactants	Sodium lauryl sulfate
Benzoic acids	Sodium salicylate, methoxy salicylate
Cationic surfactants	*N,N*-bis(2 hydroxyl ethyl) oleylamine, benzalkonium chloride, cetylpyridinium chloride, cetyltrimethyl ammonium bromide
Fatty acids	Lauric acid, oleic acid, undecanoic acid, methyl oleate
Fatty alcohols	Octanol, nonanol
Lactam	Laurocapram (Azone®)
Nonionic surfactants	Polyoxyethylene (20) sorbitan mono oleate, 23-lauryl ether
Polyols	Propylene glycol, polyethylene glycol
Sugars	Cyclodextrins
Sulfoxides	Dimethyl sulfoxide, dodecyl methyl sulfoxide
Terpenes	Menthol, thymol, limonene
Trypsin Inhibitors	Aprotinin
Ureas	Urea
Zwitterionic – surfactants	Dodecyl dimethyl ammoniopropane sulfate
Others	Selected synthetic polymers

An ideal penetration enhancer possesses the following properties:

- Is non-toxic, non-irritating, and non-allergenic.
- Provides rapid onset of increased permeability.
- Allows rapid recovery of normal barrier properties when removed.
- Physically and chemically compatible with a wide range of drugs.

Such an ideal enhancer has not been identified to date. An improved understanding of enhancer structure and mechanism of action is vital for formulation of transmucosal delivery systems. In general, penetration enhancers are thought to function in one or more of the following ways [83]:

- Fluidize intercellular lipids.
- Alter protein conformation.
- Modify drug solubility parameters.

Judicious use of a combination of enhancers may provide a synergistic impact on membrane permeability while minimizing individual enhancer concentration, thus decreasing tissue toxicity. A study of current penetration agents coupled with the discovery of novel agents may lead to a new era in oral mucosal drug delivery systems. Sustained and intensified effort is also needed to thoroughly explore the enhancer agents currently available.

16.5.3.3 Enzyme Inhibitors

Resident proteases and peptidases (e.g., pepsin, chymotrypsin, elastase, carboxypeptidase, aminopeptidase) can cause protein and peptide degradation in the GI tract [84]. Although buccal mucosa is relatively low in enzymatic activity the presence of the same above enzymes in saliva can cause degradation. Such enzymatic activity may limit mucosal absorption. Walker et al. [85] evaluated the peptidase activity on the surface of porcine buccal mucosa in vitro and reported that, while aminopeptidase N activity was detected using Leu-*p*-nitroanilide, no carboxypeptidase or dipeptidyl peptidase IV activity was evident. Such enzymatic activity could be minimized by co-administration of enzyme inhibitors (e.g., bacitracin, aprotinin, diprotin A) to improve the bioavailability. Langoth [86], Yang [87], Bird [88], and Garren [89] have demonstrated the role of enzyme inhibitors in their respective studies. Peptide and protein delivery will be discussed in a subsequent section.

16.5.3.4 System Types and Production Methods

In addition to material selection for oral transmucosal systems (e.g., bioadhesives, polymeric carriers), consideration should also be given to the design of the dosage system and its commercial manufacture. Oral transmucosal fentanyl citrate (Actiq®) is a good example of such a system. It comprises a fentanyl-containing matrix that dissolves when the patient rubs it against the buccal mucosa. A simple, but unique solid-molded lozenge presentation was transformed into a transmucosal "lollipop-type" drug delivery system. When the molded matrix dissolves, approximately 25% of the contained drug is absorbed very rapidly through the buccal mucosa and enters the bloodstream with no first-pass metabolism [90]. The transmucosal fraction provides rapid release of pain (within 5–10 min) but duration of effect is short [91]. Non-absorbed drug is swallowed, absorbed from the GI tract, and undergoes first-pass metabolism. About one-third of this amount is bioavailable, achieving a total bioavailability of about 50% [90]. The transmucosal fraction provides rapid relief of pain (within 5–10 min) but a short duration of effect [91]. These characteristics make transmucosal fentanyl appropriate for treating breakthrough pain in certain cancer conditions [92].

Some researchers have studied bilayer oral transmucosal systems, viz., a fast release layer bonded to a sustained release layer, combining rapid and sustained effects. A system was reported by Nozaki et al. [93] for such controlled systemic delivery of the anti-anginal, isosorbide dinitrate. This drug is widely used for the treatment of angina pectoris. However, when administered orally, it has a short duration of action and poor bioavailability. There are several conventional dosage forms of isosorbide dinitrate for application to the oral cavity, such as sprays and sublingual preparations. These permit rapid onset but have short duration of efficacy, due to inadequate residence time of the dosage form in the oral cavity. Mucoadhesive dosage forms have accordingly been developed that prolong residence time at the application site and minimize swallowing of drug dissolved in the saliva. Transmucosal Therapeutic System (TmTs) consists of fast and sustained release layers to provide rapid onset and sustained duration of action. A combination of polyvinylpyrrolidone and D-mannitol or polyacrylic acid was used as the matrix. The system was applied to the gingival mucosa of beagle dogs for 12 h and provided a rapid rise in plasma level of isosorbide dinitrate while maintaining the minimum effective concentration over an adequate period of time [93].

A bilayer nicotine-containing mucoadhesive tablet was evaluated by Park and Munday [94] for nicotine replacement to aid smoking cessation. The bilayered tablet was inserted in the buccal sulcus of human volunteers, with the controlled release layer in contact with the buccal membrane (lining of the cheek). HPC concentrations of 20–30% w/w and Carbopol® 934 20% w/w provided suitable adhesion and nicotine release that approached zero-order kinetics over 4 h. A fast releasing layer provided an initial "burst" release of drug. Thus, this bilayer delivery system provides early and sustained nicotine levels.

Lidocaine is used in the oral cavity for local pain relief or anesthesia. Okamoto et al. [95] developed polymeric dosage forms of lidocaine and studied the effect of the preparation method on performance. Lidocaine-containing films were prepared by direct compression of a mixture of drug and polymer(s) (DCPM method), direct compression of the spray dried powder containing drug and polymer mix (DCSD method), or solvent evaporation (SE method). Lidocaine release rate and buccal penetration rate were affected by the mode of preparation. The SE method provided the highest drug release and permeation rate followed by the DCSD method and lastly the DCPM technique (i.e., SE > DCSD > DCPM).

Table 16.6 Examples of commercially available drug delivery systems for systemic delivery by the oral mucosal route

Drug	Product name	Dosage form
Sublingual		
Nitroglycerine	Suscard	Tablet
	Nitrostat	Tablet
Erythrityl tetranitrate	Cardilate	Tablet
Isosorbide dinitrate	Sorbitrate	Tablet
Isosorbide mononitrate	Imdur	Tablet
	Isordil	Tablet
Apomorphine	Uprima	Tablet
Buprenorphine + Naloxone	Suboxone	Tablet
Buprenorphine	Subutex	Tablet
Testosterone	Androspray	Spray
	Testatropinol	Tablet
	SubDiol Z	Tablet
Androstenediol (androdiol)	Cyclodiol	Tablet
Estrogen	DHEA	Liquid drops
Buccal		
Nicotine	Nicorette	Tablet, chewing gum
	Nicotinell	Lozenge
Prochlorperazine	Buccastem	Tablet
Androgen (testosterone)	Striant	Buccal patch
Miconazole	Tibozole	Mucoadhesive buccal tablet
Fentanyl citrate	Actiq	Lozenge

Most patch or film delivery systems utilize solvent casting or solvent evaporation methods. These have been moderately successful but improved methods of film production for oral TMD are warranted.

As discussed earlier, hot-melt extrusion (HME) has been shown to be viable for producing films for TMD. Although not suitable for all heat labile drugs, this technology is feasible for many drugs and applications. Clotrimazole has been incorporated into HME films, attaining good drug stability and zero-order release profiles [17]. Hydrocortisone was not only shown to be chemically stable when incorporated into HPC films, but also acted as a plasticizer for the extrudate [96].

16.6 Buccal Drug Delivery Systems

Table 16.6 lists commercially available products for oral TMD. Most are conventional dosage forms adapted for oral mucosal delivery (e.g., tablets, sprays) but there is an increasing presence of novel systems for delivery of traditional drugs, proteins, peptides, and vaccines. Combinations of unifunctional and multifunctional materials may decrease previous limitations for product design. Indeed, new excipients, including a new polymer specifically designed for melt extrusion, Soluplus®, may also expand the opportunities for developing a variety of TMD systems. The utility of newer production methods, such as HME techniques may also increase feasibility while providing environmentally friendly processes. More choices and opportunities are also becoming available for drug selection for novel oral TMD systems. The last decade has stimulated a renaissance for oral transmucosal or buccal systems. The result is that the pharmaceutical scientist may more efficiently tailor new systems for drug delivery to the oral cavity. This section will focus on three examples, viz.

- Medicated chewing gum.
- Dissolvable film strip.
- Hot melt extrusion.

16.6.1 Medicated Chewing Gum

16.6.1.1 General Aspects and Definition

Chewing gum has existed for several centuries but was not used for drug delivery until 1924 when the first medicated chewing gum, Aspergum® (aspirin chewing gum), was introduced in the USA [97]. This mode of delivery of medications was not generally well accepted until nicotine chewing gum was commercialized as a smoking cessation product in 1978. In 1991, Commission of European Communities approved chewing gum as a pharmaceutical dosage form.

The European Pharmacopeia defines chewing gum as a "*solid, single-dose preparation with a base consisting mainly of gums that are intended to be chewed but not swallowed. They contain one or more active substances which are released by chewing. After dissolution or dispersion of the active substances in saliva, chewing gums are intended to be used for local treatment of mouth diseases or systemic delivery after absorption through the buccal mucosa or from the gastrointestinal tract*" [98].

16.6.1.2 Gum Composition and Influence on Drug Absorption

In addition to the active, medicated chewing gums consist mainly of water insoluble taste-free gum bases and water soluble ingredients such as fillers, antioxidants, buffers, sweeteners, colors, and flavoring agents [99, 100]. They comprise elastomers, resins, waxes, fats, and emulsifiers. The elastomers are inert synthetic rubbers, such as styrene–butadiene copolymers blended with polyisobutene. Polyvinyl acetate and waxes are added to reduce adhesion to teeth. Partially hydrogenated fatty acids esters are used to soften the gum [99] and improve texture and chewability. Emulsifiers, such as glycerol monostearate and lecithin, promote saliva uptake by the gum to facilitate the release of actives and/or flavors.

Table 16.7 The therapeutic applications of medicated chewing gum for systemic absorption

Therapeutic category	Active ingredient	Indication
Analgesic	Aspirin	Pain, fever, and/or inflammation reduction
	Salicylamide	Pain reliever
Vitamin	Ascorbic acid	Vitamin C deficiency
Stimulant	Caffeine	Alertness, fatigue stopper, and tiredness reducer
Anti-allergy	Loratadine	Relieve allergy symptoms
Anti-emetics	Dimenhydrinate	Motion sickness treatment
Anti-hypertension	Verapamil	Reduce high blood pressure
Anti-tussive	Noscapine	Cough suppression
Anti-addictive	Methadone	Opioids dependence
	Nicotine	Nicotine withdrawal symptoms

The composition (lipophilic/hydrophilic ingredient ratio) and amount of gum base determines the amount and rate of drug release. Gum texture also affects release. A reasonable chewing time (acceptable duration is around 30 min [101]) is required to prolong the contact time in the oral cavity and allow more absorption.

The systemic therapeutic effects for drugs delivered by medicated chewing gums reflect the relative amounts of drug absorbed via oral mucosal tissues as well as in the GI tract. Chewing exposes more surface area, thus allowing more drugs to be released for absorption. However, the mastication process also facilitates salivation which dilutes the drug concentration in saliva and increases swallowing [100], reducing availability for buccal absorption. Individual chewing habits, such as biting frequency and chewing intensity further contribute to variation in the amount of drug absorbed as typified by nicotine chewing gums.

Table 16.7 lists some therapeutic applications of medicated chewing gum [102–111].

16.6.1.3 Methods for Manufacture

Manufacturing processes for medicated chewing gum can be classified as conventional melting, kneading, and direct compression methods.

Conventional Method

A typical process involves softening or melting the gum base (between 50 and 70°C). Other excipients, such as sweeteners, emulsifiers, and fillers, are successively blended with the base [99]. Flavors are usually added later due to their volatility or thermal lability. The mixture is cooled, rolled into sheets, cut into strips, chopped and packaged as unit doses. A light coating of fine sugar may be added during rolling to prevent sticking and to enhance any added flavor [112].

Table 16.8 Typical formulations of medicated chewing gum

Ingredient	Conventional method	Direct compression
API	Nicotine polacrilex	Nicotine polacrilex
Gum base	Synthetic/natural chewing gum	Directly compressible gum powder
Powdered sugars	Sorbitol	Sorbitol
Buffer	Calcium bicarbonate	Sodium biocarbonate
Glidant	n/a	Magnesium stearate
Softener	Glycerin/70% sorbitol	n/a

Such a process has several limitations that can constrain its use with thermally labile drugs. It is also difficult to control dose uniformity. Conversion of manufacturing equipment, normally used for confectionery manufacture to meet pharmaceutical requirements can also be a challenge.

Direct Compression Method

Use of gums that are readily compressible may facilitate direct compression. The gum, in powder form is blended with binders, lubricants, and sweeteners and compressed using a conventional compaction press.

16.6.1.4 Chewing Gum Products

Nicotine chewing gums for smoking cessation are the most prominent medicated gums. Table 16.8 lists other examples, made by conventional melting and kneading as well as by direct compression [113, 114].

16.6.2 Fast Dissolving Film Strip

A dissolvable oral strip is a pliable film, applied lingually that quickly disintegrates/dissolves to release drug into the oral cavity. Good wettability, large surface, and low thickness (less than 100 μm) facilitate fast release. The physicochemical properties of drugs and polymers used to fabricate the film can influence whether the drug is absorbed via the oral mucosa or is swallowed, which would lead to absorption of the bioactive from the GI tract.

Advantages of the thin film strip format, such as discreetness, convenience, portability, dose accuracy, and tolerability, can benefit healthcare professionals and consumer alike. Table 16.9 summarizes currently commercial pharmaceutical/OTC thin film products [115, 116].

Table 16.9 Commercial available pharmaceutical/OTC thin film products

Distributor	Brand	API	Strength	Packaging
Del	Orajel	Menthol/pectin	2 mg/30 mg	Multiple strips/cassette dispenser
InnoZen	Suppress	Menthol	2.5 mg	Multiple strips/cassette dispenser
Novartis	Gas-X	Simethicone	62.5 mg	Foil–Foil peelable non-CR pouch
Novartis	Theraflu	Phenylephrine HCl/Dextromethorpah HBr	10 mg/20 mg	Foil–Foil CR pouch
Novartis	Theraflu	Phenylephrine HCl/Diphenylhydramine HCl	10 mg/25 mg	Foil–Foil CR pouch
Novartis	Theraflu	Dextromethorphan HBr	15 mg	Foil–Foil CR pouch
Novartis	Theraflu	Diphenylhydramine HCl	25 mg	Foil–Foil CR pouch
Novartis	Triaminic	Phenylephrine HCl	2.5 mg	Foil–Foil CR pouch
Novartis	Triaminic	Phenylephrine HCl/Dextromethorphan HBr	2.5 mg/5 mg	Foil–Foil CR pouch
Novartis	Triaminic	Dextromethorphan HBr	7.5 mg	Foil–Foil CR pouch
Novartis	Triaminic	Phenylephrine HCl/Diphenylhydramine HCl	5 mg/7.5 mg	Foil–Foil CR pouch
Pfizer	Benadryl	Diphenylhydramine HCl	12.5 mg	Foil–Foil CR pouch
Pfizer	Benadryl	Diphenylhydramine HCl	25 mg	Foil–Foil CR pouch
Pfizer	Sudafed	Phenylephrine HCl	10 mg	Foil–Foil CR pouch
Prestige	Chloraseptic	Benzocaine/menthol	3 mg/3 mg	Multiple strips/cassette

16.6.2.1 Composition

Principal components of dissolvable thin film strip are polymer(s) with the capability to form a continuous film and plasticizer(s) to reduce the glass transition temperature of the polymer forming a pliable, soft film. Such polymers are invariably hydrophilic and water soluble. They can comprise natural gums, hydrocolloids, starches, and cellulose derivatives. In some cases, water insoluble enteric polymers, with pH-dependent solubility provide technical novelty and advantages for some active ingredients [117].

In addition to the active ingredient (API), excipients such as polymers, plasticizers neutralizing agents (depending on the choice of polymers), sweeteners, flavors, and/or taste masking agents may be required to provide palatable dissolvable films. Aversive taste of the API can be a major challenge. Large amounts of sweeteners and flavors, and in some instances, microencapsulated API (to mask bitterness) are used to counteract such properties. Latterly, biological approaches have been used that involve blocking the bitterness receptor [118].

It is desirable to have film thickness of less than 125 μm to maintain good wettability and achieve rapid dissolution on contact with saliva. However, this restricts the drug payload per unit to less than 50% weight. This confines application to potent drugs.

16.6.2.2 Manufacture

Manufacture comprises three unit operations:

- Prepare the coating solution.
- Coat and dry the film.
- Cut into film strips and insert in pouches.

Prepare the Coating Solution

Water and ethanol (Class III solvent) are the most commonly used solvents to dissolve the API and excipients. Air entrapment must be avoided by optimizing the processing parameters or having a deaeration step, as bubble formation and persistence can compromise film integrity. The process needs to take account of and obviate possibilities for loss of volatile ingredients (such as flavors and solvents). Typically, viscosity, pH, solids content, and homogeneity of drug content are key quality attributes for coating solution.

Coating and Drying

This usually consists of three stages, viz., casting, drying, and rewinding the film as shown in Fig. 16.3.

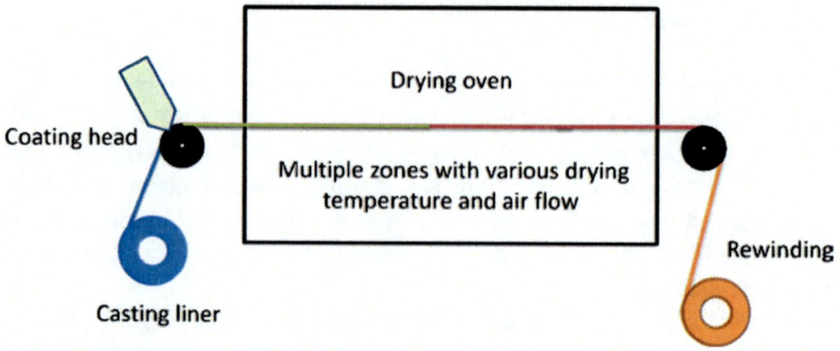

Fig. 16.3 Solvent coating process of making thin oral film

The coating solution is deposited evenly over the web on an inert substrate with comparable surface tension to the coating solution. There are many techniques to ensure that a predefined amount of solution is cast on the liner to give the requisite film thickness. These include coating heads such as forward rolls, reverse rolls, gravure rolls, knife-over-roll, metering rods, slot die, curtains, dip, and air knife. Slot die and forward rolls are the most common. The line/web speed needs to be maintained at an appropriate rate to produce dry film with a predefined and uniform thickness, acceptable residual solvent, at acceptable cost. Solvents are removed by convectional drying. Air velocity, flow pattern, and distribution in the drying oven are critical for effective solvent removal and to provide a uniform cast film, creating the final appearance of the film strip. Typically, 85–90% of solvent is evaporated in the first zone. In-process controls during film casting include coat weight (dry), drug content, and residual solvents. The most common processing problems concern deposits of dried coated mass from the coating heads onto the molten cast film, blisters, gels, streaks, foreign particles, and cross-web variability.

Dried cast film is rewound after coating and drying. It is preferable and highly recommended that the film be cooled or conditioned to room temperature before rewinding. The winding tension, the width of the web, and the storage condition need to be defined to match downstream packaging requirements.

The final unit operation converts bulk film into a unit dosage with desired dimensions using rotary knives or steel rule dies with custom-made equipment. Each film strip is placed into a single foil pouch that is sealed by applying heat or ultrasound technology. The pouching process can be performed in a continuous (high speed) or intermittent (low speed) manner. The most commonly used foil pouch materials are PET/foil/PET and PET/foil/Barex, and the choice depends on the stability and other requirements of the formulation.

A cast film must have sufficient physical strength to ensure it can withstand slitting, cutting, and pouching without breaking or elongating. Likewise, the film strip must maintain its integrity when removed from the pouch by the user. These goals can only be achieved through clearly defined specifications for attributes such as tensile strength, Young's Modulus, and elongation that are determined during product development.

16.6.3 Hot-Melt Extrusion Technology

16.6.3.1 Historical Development

HME technology has been widely used in the plastics industry for more than half of a century. It is now being adopted in the pharmaceutical industry and being categorized as a "continuous process." Pharmaceutical companies have gradually focused on HME techniques for many drug delivery applications. Such interest is reflected by over 200 research papers and several reviews on the topic [119–123]. HME patents issued worldwide for pharmaceutical systems have also increased since the early 1980s.

16.6.3.2 Advantages

HME offers the following advantages over traditional processing techniques for pharmaceutical applications.

- The process is anhydrous
- It is largely a continuous operation with fewer processing steps
- It requires no compression of the actives
- It can improve bioavailability due to dispersion of the drug at the molecular level in the dosage form [124–127]

The components in the formulation must be thermally stable at the processing conditions utilized. This may limit the extrusion of thermosensitive drugs. However, the advent of new techniques such as a combination of HME with nanotechnology [128], powder coating [129, 130], and complexation [131] over the last several years continuously adds active pharmaceutical ingredients to the HME candidate list. Advances in types of extruders have also expanded the number of drugs available for processing. Twin-screw extruders with flexible screw designs and inclusion of injection ports at different stages within the barrel permit the pharmaceutical scientist more freedom to employ wider range of drugs or more complex dosage forms.

16.6.3.3 Composition and Processing

The extrusion process involves conversion of the API and excipients into a product of uniform shape (such as a fast-dissolving film or tablet) and density by forcing materials through a die under controlled conditions. In HME, the meltable substances may be polymeric [96, 132–137] or low melting point waxes [138, 139]. Hot-melt extruded films may be produced separately and layered, postextrusion, or a multilayered system may be coextruded. The onset of action and/or bioavailability of the drug may be improved if dispersed at the molecular level. New chemical entities, isolated natural compounds, and currently marketed drugs with low bioavailability due to solubility are prime candidates for this technology. Thus, hot-melt oral transmucosal dosage forms present the pharmaceutical scientist with new opportunities, including the attainment of zero-order drug release if desired.

Polymeric films for oral or transmucosal delivery should be flexible, elastic, and soft, yet sufficiently bioadhesive to withstand the mechanical stress of the oral cavity. Prodduturi et al. developed clotrimazole (CT) polymeric films utilizing different molecular weight grades of HPC (HPC-JF, GF, and MF) [140] and PEO (PEO N-80 and PEO N-750) [141]. Both systems exhibited zero-order drug release and release rate was dependent on the molecular weight of the polymer. Release rate constant and release mechanism were independent of % drug loading. Thus, size of the dosage form and/or release of the drug from extruded film could be tailored by altering drug load without affecting the release mechanism. A 55:35 ratio of polymers (HPC/PEO) produced optimal stability of the API. Such findings are germane to the development of successful transmucosal systems, whether for immediate or controlled release.

Lidocaine and chlorpheniramine maleate have also been successfully incorporated into HPC/PEO matrices for transmucosal delivery [96, 142].

Depending on the materials and the processing conditions, the drug incorporated within hot melt extrusion-based systems need only be subjected to high temperatures for brief periods (30–120 s) [143]. Thermal degradation may not therefore be an issue. Further exploration of the technique may expand its capability for local and systemic delivery of therapeutics.

16.7 Conclusions

This chapter has attempted to provide an overview of the various formulation, processing, and physiological factors that can influence drug absorption via the oral cavity. A thorough understanding of the "biology" of the drug (mode of action, dose response, metabolism, and pharmacokinetics) as well as its physicochemical characteristics, along with appropriate choice of ingredients to target absorption sites in the oral cavity may identify opportunities to enhance onset of action, bioavailability, and therapeutic efficacy, whether locally or systemically. The emergence of new polymeric carriers and manufacturing processes will continue to make buccal delivery systems more feasible as a suitable and attractive alternative to peroral administration.

References

1. Chien YW (1995) Biopharmaceutics basis for transmucosal delivery. STP Pharma Sci 5(4):257–275
2. Sobrero A (1847) Sur plusieurs composes detonnants produits avec l'acide nitrique et le sucre, la dextrine, la lactine, la mannite et la glycerine. C R Hebd Seances Acad Sci 224:247

3. Frokjaer S, Hovgaard L (2000) In: Frokjaer S, Hovgaard L (eds) Pharmaceutical formulation development of peptide and proteins. Taylor & Francis, London, UK, p 238
4. Squier CA, Nanny D (1985) Measurement of blood flow in the oral mucosa and skin of the rhesus monkey using radiolabeled microspheres. Arch Oral Biol 30(4):313–318
5. Madhav NVS et al (2009) Orotransmucosal drug delivery systems: a review. J Control Release 140(1):2–11
6. Shojaei AH (1998) Buccal mucosa as a route for the systemic drug delivery: a review. J Pharm Pharm sci 1(1):15–30
7. Welling PG (2002) Absorption of drugs. Encyclopedia of pharmaceutical technology. Marcel Dekker, New York, pp 8–22
8. Sayani AP, Chien YW (1996) Systemic delivery of peptides and proteins across absorptive mucosae. Crit Rev Ther Drug Carrier Syst 13(1–2):85–184
9. Jay S et al (2002) Transmucosal delivery of testosterone in rabbits using novel bi-layer mucoadhesive wax-film composite disks. J Pharm Sci 91(9):2016–2025
10. Krishnaiah YSR et al (2002) Three-layer guar gum matrix tablet formulation for oral controlled delivery of highly soluble trimetazidine dihydrochloride. J Control Release 81(1–2):45–56
11. Repka MA, McGinity JW (2001) Bioadhesive properties of hydroxypropylcellulose topical films produced by hot-melt extrusion. J Control Release 70(3):341–351
12. Roller NW, Ship II (1975) Lidocaine topical film strip for oral mucosal biopsies. J Oral Med 30(2):55–58
13. Burgalassi S et al (1996) Development and invitro/invivo testing of mucoadhesive buccal patches releasing benzydamine and lidocaine. Int J Pharm 133(1–2):1–7
14. Korsmeyer RW et al (1983) Mechanism of solute release from porous hydrophilic polymers. Int J Pharm 15:25–35
15. Langer RS, Peppas NA (1981) Present and future application of biomaterials in controlled drug delivery systems. Biomaterials 2(10):201–213
16. Higuchi T (1961) Rate of release of medicaments from ointment bases containing drugs in suspension. J Pharm Pharm Sci 50:873–874
17. Repka MA, Prodduturi S, Stodghill SP (2003) Production and characterization of hot-melt extruded films containing clotrimazole. Drug Dev Ind Pharm 29(7):757–765
18. Zhang H, Robinson JR (1996) Routes of drug transport across oral mucosa. In: Rathbone MJ (ed) Oral mucosal drug delivery. Marcel Dekker, New York, pp 51–63
19. Siegel IA (1984) Permeability of the oral mucosa. In: Siegel IA et al (eds) The structure and function of oral mucosa. Pergamon, New York, pp 95–108
20. Ho NFH (1993) Biophysical kinetic modeling of buccal absorption. Adv Drug Deliv Rev 12(1–2):61–97
21. Henry JA et al (1980) Drug recovery following buccal absorption of propranolol. Br J Clin Pharmacol 10(1):61–65
22. McElnay JC, Hughes CM (2002) Drug delivery-buccal route. In: Encyclopedia of pharmaceutical technology. Marcel Dekker, New York, pp 800–810
23. Zhang H, Zhang J, Streisand JB (2002) Oral mucosal drug delivery: clinical pharmacokinetics and therapeutic applications. Clin Pharmacokinet 41(9):661–680
24. Merkle HP, Wolany G (1992) Buccal delivery for peptide drugs. J Control Release 21:155–164
25. Lagerlof F, Dawes C (1984) The volume of saliva in the mouth before and after swallowing. J Dental Res 63:618
26. Rathbone MJ, Ponchel G, Ghazali FA (1996) Systemic oral mucosal drug delivery and delivery systems. In: Rathbone MJ (ed) oral mucosal drug delivery. Marcel Dekker, New York

27. Galey WR, Lonsdale HK, Nacht S (1976) The in vitro permeability of skin and buccal mucosa to selected drugs and tritiated water. J Invest Dermatol 67(6):713–717
28. Gandhi RB, Robinson JR (1994) Oral cavity as a site for bioadhesive drug delivery. Adv Drug Deliv Rev 13(1–2):43–74
29. Wertz PW, Squier CA (1991) Cellular and molecular basis of barrier function in oral epithelium. Crit Rev Ther Drug Carrier Syst 8(3):237–269
30. Squier CA, Hall BK (1984) The permeability of mammalian non-keratinized oral epithelia to horseraddish peroxidase applied in vivo and in vitro. Arch Oral Biol 29(3):45–50
31. Hill MW, Squier CA (1979) The permeability of oral palatal mucosa maintained in organ culture. J Anatomy 128(1):169–178
32. Repka MA, McGinity JW (2000) Physical-mechanical, moisture absorption and bioadhesive properties of hydroxypropylcellulose hot-melt extruded films. Biomaterials 21(14):1509–1517
33. Park H, Robinson JR (1987) Mechanisms of mucoadhesion of poly(acrylic acid) hydrogels. Pharm Res 4(6):457–464
34. Park K, Robinson JR (1984) Bioadhesive polymers as platforms for oral-controlled drug delivery: method to study bioadhesion. Int J Pharm 19(2):107–127
35. Repka MA et al (2004) Matrix and reservoir-based oral transmucosal systems: tailoring delivery solutions. Am J Drug Deliv 2(3):173–192
36. Alur HH et al (1999) Transmucosal sustained-delivery of chlorpheniramine maleate in rabbits using a novel, natural mucoadhesive gum as an excipient in buccal tablets. Int J Pharm 188(1):1–10
37. Ishida M, Machida Y, Nambu N, Nagai T (1981) New mucosal dosage form of insulin. Chem Pharm Bull 29(3):810–816
38. Satoh K et al (1989) Factors affecting the bioadhesive property of tablets consisting of hydroxypropyl cellulose and carboxyvinyl polymer. Chem Pharm Bull 37(5):1366–1368
39. Gupta A, Garg S, Khar RK (1994) Interpolymer complexation and its effect on bioadhesion strength and dissolution characteristics of buccal drug delivery systems. Drug Dev Ind Pharm 20(3):315–325
40. Anlar S et al (1994) Formulation and in-vitro-in vivo evaluation of buccoadhesive morphine sulfate tablets. Pharm Res 11(2):231–236
41. Anders R, Merkle HP (1989) Evaluation of laminated muco-adhesive patches for buccal drug delivery. Int J Pharm 49(3):231–240
42. Anders R et al (1983) Buccal absorption of protirelin: an effective way to stimulate thyrotropin and prolactin. J Pharm Sci 72(1–2):1481–1483
43. Ishida M, Nambu N, Nagai T (1982) Mucosal dosage form of lidocaine for toothache using hydroxypropyl cellulose and carbopol. Chem Pharm Bull 30(3):980–984
44. Guo JH (1994) Investigating the surface properties and bioadhesion of buccal patches. J Pharm Pharmacol 46(8):647–650
45. Lehr CM et al (1992) In vitro evaluation of mucoadhesive properties of chitosan and some other natural polymers. Int J Pharm 78(1):43–48
46. Sanzgiri YD et al (1994) Evaluation of mucoadhesive properties of hyaluronic acid benzyl esters. Int Pharm 107(2):91–97
47. Watanabe K et al (1991) Drug release behaviors from hydrogel prepared with water soluble dietary fibers. J Pharm Sci Technol 51(1):29–35
48. Veillard MM et al (1987) Preliminary studies of oral mucosal delivery of peptide drugs. J Control Release 6:123–131
49. Ch'ng HS et al (1985) Bioadhesive polymers as platforms for oral controlled drug delivery II: Synthesis and evaluation of some swelling, water-insoluble bioadhesive polymers. J Pharm Sci 74(4):399–405

50. Garcia-Gonzalez N et al (1993) Design and evaluation of buccoadhesive metoclopramide hydrogels composed of poly(acrylic acid) crosslinked with sucrose. Int J Pharm 100(1–3):65–70
51. Leung S, Robinson JR (1988) The contribution of anionic polymer structural features to mucoadhesion. J Control Release 5(3):223–231
52. Leung SS, Robinson JR (1990) Polymer structure features contributing to mucoadhesion: 2. J Control Release 12:187–194
53. Bodde HE, De Vries ME, Junginger HE (1990) Mucoadhesive polymers for the buccal delivery of peptides, structure-adhesiveness relationships. J Control Release 13(2–3):225–231
54. Benes L et al (1997) Transmucosal, oral controlled-release, and transdermal drug administration in human subjects: a crossover study with melatonin. J Pharm Sci 86(10):1115–1119
55. Cassidy JP, Landzert NM, Quadros E (1993) Controlled buccal delivery of buprenorphine. J Control Release 25(1–2):21–29
56. Nakane S et al (1996) Oramucosal delivery of LHRH: Pharmacokinetic studies of controlled and enhanced transmucosal permeation. Pharm Dev Technol 1(3):251–259
57. NguyenXuan T et al (1996) Mucoadhesive semi-solid formulations for intraoral use containing sucralfate. Eur J Pharm Biopharm 42(2):133–137
58. Nair M, Chien YW (1996) Development of anticandidal delivery systems: (II) Mucoadhesive devices for prolonged drug delivery in the oral cavity. Drug Dev Ind Pharm 22(3):243–253
59. Nozaki Y, Ohta M, Chien YW (1997) Transmucosal controlled systemic delivery of isosorbide dinitrate: in vivo/in vitro correlation. J Control Release 43(2–3):105–114
60. Li X, Shojaei AH (1996) Novel copolymers of acrylic acid and poly ethylene glycol monomethylether monomethacrylate for buccoadhesion: In vitro assessment of buccoadhesion and analysis of surface free energy. Proc Int Symp Control Release Bioact Mater 23:165–166
61. Shojaei AH, Li X (1995) Novel copolymers of acrylic acid and poly ethylene glycol monomethylether monomethacrylate for buccal mucoadhesion: preparation and surface characterization. Pharm Res 12:S1–S210
62. Shojaei AH, Li X (1997) Mechanisms of buccal mucoadhesion of novel copolymers of acrylic acid and polyethylene glycol monomethyl ether monomethacrylate. J Control Release 47(2):151–161
63. Taylan B et al (1996) Design and evaluation of sustained-release and buccal adhesive propranolol hydrochloride tablets. J Control Release 38(1):11–20
64. Yukimatsu K et al (1994) Development of a trans-mucosal controlled-release device for systemic of antianginal drugs pharmacokinetics and pharmacodynamics. Drug Dev Ind Pharm 20(4):503–534
65. Voorspoels J et al (1997) Buccal absorption of progesterone in the dog. Proc Int Symp Control Release Bioact Mater 24:185–186
66. DeGrande G et al (1996) Specialized oral mucosal drug delivery systems: Patches. In: Rathbone MJ (ed) Oral mucosal drug delivery. Marcel Dekker, New York, pp 285–317
67. Choi HK et al (1997) Interpolymer complex prepared by template polymerization for transmucosal drug delivery. Int Symp Control Release Bioact Mater 24:415–416
68. Wong CF, Yuen KH, Peh KK (1999) An in-vitro method for buccal adhesion studies: importance of instrument variables. Int J Pharm 180(1):47–57
69. Zhang H, Robinson JR (1996) In vitro methods for measuring permeability of the oral mucosa. In: Rathbone MJ (ed) Oral mucosal drug delivery. Marcel Dekker, New York, pp 85–100
70. Shin SC, Kim JY (2001) Enhanced permeation of triamcinolone acetonide through the buccal mucosa. Eur J Pharm Biopharm 50(2):217–220
71. Aungst BJ, Rogers NJ (1988) Site dependence of absorption-promoting actions of Laureth-9, Na salicylate, Na2EDTA, and Aprotinin on rectal, nasal, and buccal insulin delivery. Pharm Res 5(5):305–308

72. Aungst BJ, Rogers NJ (1989) Comparison of the effects of various transmucosal absorption promoters on buccal insulin delivery. Int J Pharm 53(3):227–235
73. Gandhi R, Robinson JR (1992) Mechanisms of penetration enhancement for transbuccal delivery of salicylic acid. Int J Pharm 85(1–3):129–140
74. Siegel IA, Gordon HP (1986) Surfactant-induced alterations of permeability of rabbit oral mucosa in vitro. Exp Mol Pathol 44(2):132–137
75. Siegel IA, Gordon HP (1985) Effects of surfactants on the permeability of canine oral mucosa in vitro. Toxicol Lett 26(2–3):153–158
76. Steward A, Bayley DL, Howes C (1994) The effect of enhancers on the buccal absorption of hybrid (BDBB) alpha-interferom. Int J Pharm 104(2):145–149
77. Oh CK, Ritschel WA (1990) Biopharmaceutic aspects of buccal absorption of insulin. Methods Find Exp Clin Pharmacol 12(3):205–212
78. Kurosaki Y et al (1988) Effects of surfactants on the absorption of salicylic acid from hamster cheek pouch as a model of keratinized oral mucosa. Int J Pharm 47:13–19
79. Kurosaki Y et al (1987) Enhanced permeability of keratinized oral-mucosa to salicylic acid with 1-dodecylazacycloheptan-2-one (Azone). In vitro studies in hamster cheek pouch. Int J Pharm 49(1):47–55
80. Manganaro AM, Wertz PW (1996) The effects of permeabilizers on the in vitro penetration of propranolol through porcine buccal epithelium. Mil Med 161:669–672
81. Wolany GJM et al (1990) Buccal absorption of Sandostatin (octreotide) in conscious beagle dogs. Proc Int Symp Control Release Bioact Mater 17:224–225
82. Coutel-Egros A et al (1992) Combined effects of pH, cosolvent and penetration enhancers on the in vitro buccal absorption of propranolol through excised hamster cheek pouch. Int J Pharm 84:117–128
83. Ganem-Quintanar A et al (1997) Mechanisms of oral permeation enhancement. Int J Pharm 156(2):127–142
84. Vyas SP et al (1997) Some approaches to improve bioavailability of peptides and proteins through oral and other mucosal routes. Pharmazie 52(5):339–345
85. Walker GF, Langoth N, Bernkop-Schnurch A (2002) Peptidase activity on the surface of the porcine buccal mucosa. Int J Pharm 233:141–147
86. Langoth N, Kalbe J, Bernkop-Schnurch A (2003) Development of buccal drug delivery systems based on a thiolated polymer. Int J Pharm 252(1–2):141–148
87. Yang F et al (1986) Factors affecting water absorption of the epoxy tetraglycidyl-4,4′-diaminodiphenylmethane cured with diaminodiphenyl sulfone. J Polym Sci A Polym Chem 24(10):2609–2618
88. Bird AP, Faltinek JR, Shojaei AH (2001) Transbuccal peptide delivery: stability and in vitro permeation studies on endomorphin-1. J Control Release 73(1):31–36
89. Garren KW, Topp EM, Repta AJ (1989) Buccal absorption. III. Simultaneous diffusion and metabolism of an aminopeptidase substrate in the hamster cheek pouch. Pharm Res 6(11):966–970
90. Cleary JF (1997) Pharmacokinetics and pharmacodynamic issues in the treatment of breakthrough pain. Semin Oncol 24(suppl 16):S16–S139
91. Fine PG et al (1991) An open label study of oral transmucosal fentanyl citrate (OTFC) for the treatment of breakthrough cancer pain. Pain 45(2):149–153
92. Fine PG (1997) Fentanyl in the treatment of cancer pain. Semin Oncol 24(suppl 16):S16–S207
93. Nozaki Y, Yukimatsu K, Mayumi T (1996) A new transmucosal therapeutic system for the systemic delivery of isosorbide dinitrate: in vitro and in vivo evaluation in beagle dogs. STP Pharma Sci 6(2):134–141
94. Park CR, Munday DL (2002) Development and evaluation of a biphasic buccal adhesive tablet for nicotine replacement therapy. Int J Pharm 237(1–2):215–226

95. Okamoto H et al (2002) Development of polymer film dosage forms of lidocaine for buccal administration: II. Comparison of preparation methods. J Pharm Sci 91(11):2424–2432
96. Repka MA et al (1999) Influence of plasticizers and drugs on the physical-mechanical properties of hydroxypropylcellulose films prepared by hot-melt extrusion. Drug Dev Ind Pharm 25(5):625–633
97. Growth through research: a history of Schering-Plough Corporation (brochure) (1992:3)
98. Council of Europe (2010) Europe pharmacopoeia, 6th edn. The European Directorate for The Quality of Medicine & Health Care, Strasbourg, France
99. Rassing MR (1994) Chewing gum as a drug delivery system. Adv Drug Deliv Rev 13(1–3):89–121
100. Jacobsen J, Christrup L, Jensen N-H (2004) Medicated chewing gum: pros and cons. Am J Drug Deliv 2(2):75–88
101. Barabolak R et al (1991) Gum chewing profiles in the U.S. population. Commun Dent Oral Epidemiol 19(2):125–126
102. Woodford DW, Lesko LJ (1981) Relative bioavailability of aspirin gum. J Pharm Sci 70(12):1341–1343
103. Bousquet E et al (1992) Bioavailability of two formulations of acetylsalicylic acid gums. Pharmazie 47(8):607–609
104. Christrup LL, Bonde J, Eriksen H (1988) Chewing gum as a drug delivery system: III. Bioavailability of salicylamide administered in tablets and chewing gum. Farmaci Sci Ed 16:6–14
105. Christrup L, Rasnmussen S, Rapping M (1988) Chewing gum as a delivery system: IV. Excretion of ascorbic acid in urine after administration of chewing gum and chewing tablets. Farmaci Sci Ed 16:44–47
106. Kamimori GH et al (2002) The rate of absorption and relative bioavailability of caffeine administered in chewing gum versus capsules to normal healthy volunteers. Int J Pharm 234(1–2):159–167
107. Noehr-Jensen L et al (2006) The relative bioavailability of loratadine administered as a chewing gum formulation in healthy volunteers. Eur J Clin Pharmacol 62(6):437–445
108. Valoti M et al (2003) Pharmacokinetics of diphenhydramine in healthy volunteers with a dimenhydrinate 25 mg chewing gum formulation. Methods Find Exp Clin Pharmacol 25(5):377–381
109. Christrup LL et al (1990) Relative bioavailability of (+/–)-verapamil hydrochloride administered in tablets and chewing gum. Acta Pharm Nord 2(6):371–376
110. Jensen LN et al (1991) Chewing gum and lozenges as delivery systems for noscapine. Acta Pharm Nord 3(4):219–222
111. Russell MA, Raw M, Jarvis MJ (1980) Clinical use of nicotine chewing-gum. Br Med J 280(6231):1599–1602
112. Bumrela SB, Kane RN, Dhat SP (2005) Medicated chewing gum: new reformulation technique. http://www.pharmainfo.net
113. Lichtnecketr S, Lundgren C, Ferno O (1975) Chewable smoking substitute composition. US Patent 3901248
114. Morjaria Y et al (2004) In vitro release of nicotine from chewing gum formulations. Dissolution Technol 11(2):12–15
115. Hariharan M, Bogue A (2009) Thin film technology. Drug Deliv Technol 9(2):24–29
116. Vondrak B, Barnhart S (2008) Dissolvable films: dissolvable films for flexible product format in drug delivery. Pharm Tech. http://www.adhesivesresearch.com/Files/Pharm%20Tech%20April%2008%20final%20df%20article.pdf
117. Chan SY et al (2005) Orally dissolving films. Patent WO/2005/009386
118. Devantier HR et al (2008) Quantitative assessment of TRPM5-dependent oral aversiveness of pharmaceuticals using a mouse brief-access taste aversion assay. Behav Pharmacol 19(7):673–682

119. McGinity JW et al (2001) Hot-melt extrusion as a pharmaceutical process. Am Pharm Rev 4:25–37
120. Breitenbach J (2002) Melt extrusion: from process to drug delivery technology. Eur J Pharm Biopharm 54(2):107–117
121. Chokshi R, Hossein Z (2004) Hot–melt extrusion technique: a review. Iran J Pharm Res 3:3–16
122. Crowley MM et al (2007) Pharmaceutical applications of hot-melt extrusion: Part I. Drug Dev Ind Pharm 33(9):909–926
123. Repka MA et al (2007) Pharmaceutical applications of hot-melt extrusion: Part II. Drug Dev Ind Pharm 33(10):1043–1057
124. Forster A, Hempenstall J, Rades T (2001) Characterization of glass solutions of poorly water-soluble drugs produced by melt extrusion with hydrophilic amorphous polymers. J Pharm Pharmacol 53(3):303–315
125. Ndindayino F et al (2002) Bioavailability of hydrochlorothiazide from isomalt-based moulded tablets. Int J Pharm 246(1–2):199–202
126. Kinoshita M et al (2002) Improvement of solubility and oral bioavailability of a poorly water-soluble drug, TAS-301, by its melt-adsorption on a porous calcium silicate. J Pharm Sci 91(2):362–370
127. Breitenbach J, Magerlein M (2003) Melt extruded molecular dispersions. In: Ghebre-Selassie I, Martin C (eds) Pharmaceutical extrusion technology, vol. 133. Informa Healthcare, London
128. Miller DA et al (2007) Hot-melt extrusion for enhanced delivery of drug particles. J Pharm Sci 96(2):361–376
129. Zheng W et al (2004) Properties of theophylline tablets powder-coated with methacrylate ester copolymers. STP Pharma Sci 14(4):319–325
130. Sauer D et al (2007) Influence of processing parameters and formulation factors on the drug release from tablets powder-coated with Eudragit L 100–55. Eur J Pharm Biopharm 67(2):464–475
131. Fukuda M et al (2008) Influence of sulfobutyl ether beta-cyclodextrin (Captisol) on the dissolution properties of a poorly soluble drug from extrudates prepared by hot-melt extrusion. Int J Pharm 350(1–2):188–196
132. Follonier N, Doelker E, Cole ET (1995) Various ways of modulating the release of diltiazem hydrochloride from hot-melt extruded sustained release pellets prepared using polymeric materials. J Control Release 36(3):243–250
133. Zhang F, McGinity JW (1999) Properties of sustained-release tablets prepared by hot-melt extrusion. Pharm Dev Technol 4(2):241–250
134. Grunhagen HH, Muller O (1995) Melt extrusion technology. Pharmaceutical Manufacturing International, pp 166–170
135. Follonier N, Doelker E, Cole ET (1994) Evaluation of hot-melt extrusion as a new technique for the production of polymer-based pellets for sustained release capsules containing high loadings of freely soluble drugs. Drug Dev Ind Pharm 20(8):1323–1339
136. Aitken-Nichol C, Zhang F, McGinity JW (1996) Hot melt extrusion of acrylic films. Pharm Res 13(5):804–808
137. Cuff G, Raouf F (1998) A preliminary evaluation of injection molding as a technology to produce tablets. Pharm Technol 22:96–106
138. Miyagawa Y et al (1996) Controlled release of diclofenac sodium from wax matrix granule. Int J Pharm 138(2):215–224
139. Sato H, Miyagawa Y (1997) Dissolution mechanism of Diclofenac sodium wax matrix granules. J Pharm Sci 86(8):929–934

140. Prodduturi S et al (2004) Water vapor sorption of hot-melt extruded hydroxypropyl cellulose films: effect on physico-mechanical properties, release characteristics, and stability. J Pharm Sci 93(12):3047–3056
141. Prodduturi S et al (2005) Solid-state stability and characterization of hot-melt extruded poly(ethylene oxide) films. J Pharm Sci 94(10):2232–2245
142. Repka MA, McGinity JW (2001) Influence of chlorpheniramine maleate on topical films produced by hot-melt extrusion. Pharm Dev Technol 6(3):295–302
143. Repka MA et al (2001) Hot-melt extrusion technology. In: Swarbrick J, Boylan J (eds) Encyclopedia of pharmaceutical technology. Marcel Dekker, New York

Chapter 17
Controlling Release by Gastroretention

Mark D. Coffin and Matthew D. Burke

Abstract A perpetual challenge in the pharmaceutical industry is the development of drug products that overcome inherent biopharmaceutical challenges while maximizing the therapeutic benefit and minimizing the potential for adverse events. The concept of a gastric retentive formulation is often very enticing as it has the potential to overcome many common challenges. In this chapter, we provide guidance on key preclinical and clinical concepts to consider when developing a potential gastric retentive formulation to ensure it can survive the "valley of death." In addition, we discuss alternative approaches including bio-inspired concepts and provide "real-world" examples of attempts to develop novel gastric retentive formulations.

17.1 Introduction

The development of gastric retentive formulations (GRFs) has been actively pursued by researchers for more than 30 years [1–3]. The primary motivation behind this effort has been to enable sustained delivery of drugs to the stomach and/or proximal small intestine. This would enable once daily dosing of drugs with short half-lives, and whose site of action or absorption is in the upper gastrointestinal tract. In addition, a GRF may increase the extent and/or duration of absorption for drugs that have saturable or site-specific transport, very low solubility at intestinal pH values, or are chemically unstable in the small or large intestine.

M.D. Burke (✉)
GlaxoSmithKline R&D, Harlow, Essex, UK
e-mail: matthew.d.burke@gsk.com

C.G. Wilson and P.J. Crowley (eds.), *Controlled Release in Oral Drug Delivery*, Advances in Delivery Science and Technology, DOI 10.1007/978-1-4614-1004-1_17,

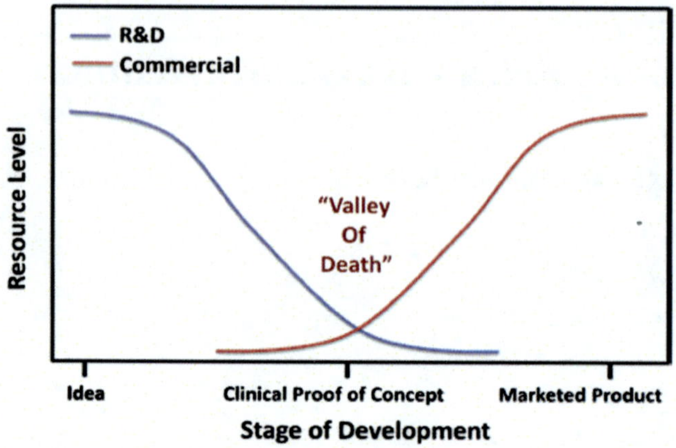

Fig. 17.1 Illustration of when gastric retentive formulations fail in development

Lastly, GRFs may be useful in reducing the intra- and intersubject variability of the absorption of the aforementioned drugs by normalizing gastrointestinal transit time.

A wealth of information is available on past attempts, common strategies, and challenges; however, a true gastric retentive dosage form has yet to emerge from the plethora of attempts. We define a true gastric retentive dosage form as one whose performance is not explicitly dependent upon consumption of a large meal, regular snacks, or intake of liquids. Some dosage forms are claimed to be gastric retentive, but appear to offer no advantage over a large, nondisintegrating tablet taken with a meal. We concur with Waterman's comment in his review article on gastric retention [4]: "Although the goal remains valuable, the promise of gastric retentive systems remains unfulfilled at this time."

As gastric retentive dosage forms enter into the key clinical proof of concept stage, resource requirements begin to shift. When they transition from preclinical evaluation, to clinical and commercial feasibility, the highest rate of failure is observed. This has been referred to as the "Valley of Death" (Fig. 17.1).

To our knowledge, no truly gastric retentive dosage form has survived the "valley of death." In this chapter, we will focus on the key challenges during the clinical proof of concept stage to assist drug delivery scientists to propel their dosage forms past this stage to be deemed worthy of the title "gastric retentive dosage form." Specifically, in this chapter we will:

- Offer recommendations on how to assess if a GRF will improve the performance of a drug
- Review physiologic challenges and strategies to achieve gastric retention
- Provide guidance on how to develop and evaluate a GRF
- Present a case study using previously unpublished results of original research on a proposed GRF

17.2 Will a Gastric Retentive Dosage Form Improve Drug Performance?

The first consideration in developing a gastroretentive formulation for a given compound is to assess the potential benefit of sustained drug delivery to the stomach. Optimization of the drug's pharmacokinetic (PK) profile could lead to enhancement of a drug's safety and/or efficacy. This may not be possible with more conventional techniques due to the properties of the drug, such as a short PK and/or pharmacodynamic (PD) half-life, significant pH-dependent solubility or localized transporter absorption mechanism. This challenge can be further magnified if the particular receptor/target is located in the upper gastrointestinal tract itself. In these cases, PK may not be an appropriate tool to assess improvement and optimization of drug release and a surrogate marker would need to be identified. However, in most cases the theoretical benefits of a gastroretentive formulation are based more on "best guesses" or modeling PK/PD relationships and may not be grounded in reality. Therefore, it is highly recommended to obtain the appropriate data before launching into the development abyss of a gastric retentive dosage form. The two key proof of concept areas that are highly valuable to pursue are:

1. Mimicking the performance of a GRF in a human clinical study
2. Investigating the drug properties in the colon

17.2.1 Proof of Concept: Mimicking the Performance of a Gastric Retentive Formulation

It is relatively simple to determine the potential benefit of a GRF without actually designing a dosage form. An excellent literature example of this type of clinical study was performed using Acyclovir by Lewis et al. [5] as shown in Fig. 17.2. Lewis found that use of a simple "sipping" method was extremely beneficial. These investigators dosed 400 mg of acyclovir via an oral immediate release tablet (2×200 mg tablets), a duodenal infusion (500 mL of a 5% dextrose solution, constant infusion rate over 4 h) and a "sipped solution" (10.4 mL of the 500 mL 5% dextrose solution every 5 min over 4 h). Both the duodenal infusion and "sipping" methods provided AUCs almost twice that of the tablets. The sipping technique is an extremely simple and cost-effective approach to determine if saturable transport is limiting bioavailability.

In addition to investigating a specific mechanism that is resulting in reduced exposure, this technique can be used to generate drug release rates similar to what would be expected from a GRF. In Fig. 17.3, a single IR tablet was compared to eight sequential doses of 1/8 the total IR dose over a 4-h period [6]. In this case, the AUC increased by 27% indicating a benefit in pursuing a gastroretentive formulation. In both cases, an appropriate clinical study design generated clear data to determine if a gastric retentive dosage form would be worth pursuing.

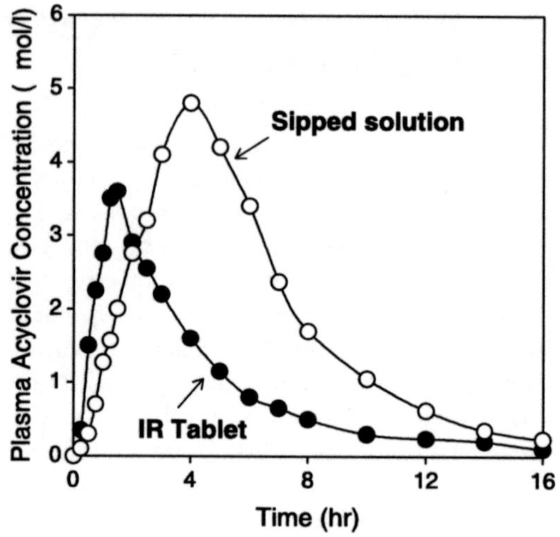

Fig. 17.2 Effect of administration rate of acyclovir on its plasma concentration profile in healthy volunteers. 400 mg dose administered either as 2 × 200 mg tablets at one time, or sipped as a solution every 5 min over a 4-h period [5]

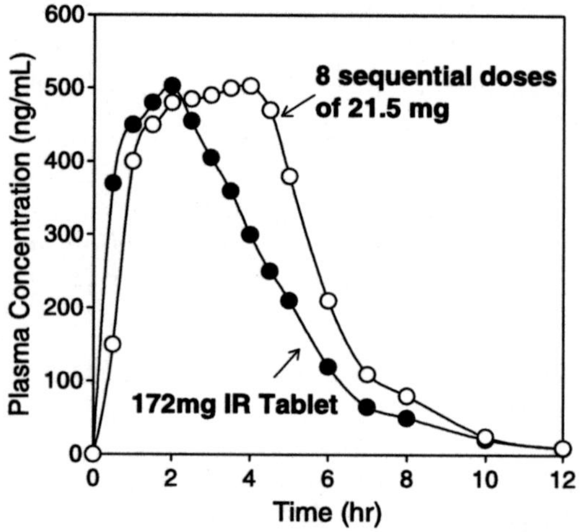

Fig. 17.3 Effect of providing eight sequential doses of 12.5 mg each over 4 h versus a single dose of 172 mg on the plasma profile of a GSK Compound [6]

17.3 Investigating Drug Properties in the Colon

Sustained release of a drug from an oral dosage form can be achieved by several methods as detailed elsewhere in this book (i.e. matrix tablets, osmotic pumps, multiparticulates, etc). However, due to the gastrointestinal transit time in the fasted state (from mouth to colon) being roughly 4 h, almost all nongastroretentive, sustained release dosage forms will require substantial absorption of the drug from the colon. The colonic environment is substantially different than the upper GI tract, with a limited amount of unbound water, complex transit through the ascending colon (colonic sieving), compaction of fecal material, high microbial content, and a thicker mucus layer lining the intestinal wall. All of these factors can contribute to

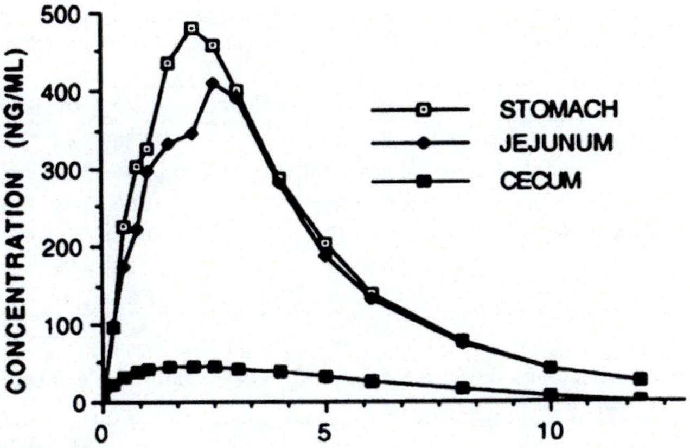

Fig. 17.4 Ranitidine plasma concentration profiles when 150 mg dose administered to different regions of GI tract via nasoenteric tubes [7]

poor absorption of drugs from the colonic region. Proper assessment of the challenges that can be faced in the colonic environment can be evaluated using the following techniques:

- Regional drug absorption studies
- Stability studies in simulated colonic media

Human regional drug absorption studies, utilizing nasointestinal intubation or remote drug delivery capsules, have proven extremely useful in explicitly determining a drug's extent of absorption from various regions of the gastrointestinal tract. While cell membrane transport studies have grown increasingly predictive of human intestinal permeability, they remain an indirect measure of human intestinal absorption. Human regional absorption studies remain the gold standard in assessing the extent of absorption from different regions of the gastrointestinal tract.

Human regional absorption studies first began using a nasoenteric tube to administer the drug solution. Williams et al. [7] showed that when a ranitidine solution is delivered to the cecum, its extent of absorption is approximately 15% that of the same solution administered to the stomach. Figure 17.4 provides the plasma concentration profiles as a function of time when these solutions were administered via a nasoenteric tube to the stomach, jejunum, or cecum.

Using modern technology comprising remote drug delivery capsules, one can use a less invasive technique to obtain similar information. Pithavala et al. [8] used such remote drug delivery capsules to evaluate the regional absorption of ranitidine and obtained similar results to Williams [7]. One primary advantage of using intubation tubes is that one can study the impact of various infusion rates on absorption. However, the technique has the disadvantage of being invasive and does not enable the impact of dosing a powder, which was shown to be useful by Menon and coworkers [9].

Table 17.1 Relative bioavailability of Acipimox when administered as a powder to the distal small bowel or colon as the free acid or sodium salt, using the oral dose as the reference treatment (Adapted from [8])

Dosing location	Free acid or Na salt	$AUC_{0\text{-}inf}$ (μg h/ml)	F_{rel} (%)
Stomach (immediate release capsule)	Free acid	12.88±3.03	–
Distal small bowel	Free acid	3.94±3.08	29.6
Colon	Free acid	1.66±1.39	12.9
Stomach (immediate release capsule)	Na salt	14.1±1.81	–
Distal small bowel	Na salt	7.34±4.98	52.0
Colon	Na salt	1.80±1.49	12.6

Table 17.2 Extent of absorption from the colon for GlaxoSmithKline compounds and other compounds referenced in the literature

Colonic absorption (%)	GSK (% of compounds)	Literature (non-GSK) (% of compounds)
0–30	50	38
30–60	5	7
60+	45	55

Menon et al. [9] demonstrated the value of human regional drug absorption studies to guide the appropriate formulation strategy. These investigators compared the extent of absorption of the hypolipidemic agent, Acipimox when administered as a powder to the stomach, distal small bowel (DSB) and colon. The free acid and sodium salt forms of the drug were dosed to determine the role of drug solubility. Their results are shown in Table 17.1, and demonstrate that Acipimox is poorly absorbed from the colon (<15%) as both the salt and free acid, but that the more soluble sodium salt is better absorbed than the free acid from the DSB (~50% vs. 30%). These data demonstrate that Acipimox could benefit from a GRF.

In the past, colonic absorption was used to predict the likelihood to achieve a successful modified release product with conventional techniques. It was suggested that a colonic absorption of 60% or higher was very likely to be successful, between 30 and 60% was difficult but achievable, and less than 30% was not likely to be successful with conventional techniques [10]. Therefore, the group of compounds which had a colonic absorption of less than 30% would be good candidates for a GRF. Table 17.2 provides a breakdown of GlaxoSmithKline (GSK) compounds and compounds from the literature in regard to their extent of absorption from the colon.

Based on these data it is clear that a high percentage of compounds suffer from poor colonic absorption and would be good candidates for a GRF.

Another factor that can steer product development toward a gastric retentive dosage form is the gastrointestinal stability of a drug candidate. In the fasted state, the typical transit time from mouth to the colon is 4 h. Therefore, if the duration of drug release required to obtain the target pharmacokinetic profile is more than 4 h, the drug could spend a significant amount of time in the colon. Basit et al. [11] demonstrated the susceptibility of drugs to degradation in the colon. Using a batch culture fermenter, they showed that essentially no nizatidine was intact after 12 h of incubation. Conversely, the H_2-receptor antagonists cimetidine and famotidine showed no significant degradation after 24 h. These data highlight another typical challenge for successful sustained release of a drug using a conventional dosage form, while a gastric retentive dosage form enables this challenge to be avoided.

17.4 Guidance on Development and Evaluation of Gastric Retentive Formulations

Given the fact that, at this time, there is no evidence to date to suggest that a truly GRF is available, the authors would like to provide their recommendations on the steps to consider in the development and evaluation of gastroretentive formulations:

1. Define what success is for the dosage form (based on its clinical and commercial needs).
2. Select an imaging modality (and marker when appropriate) that enables explicit determination of in vivo performance.
3. Test the proposed GRF for success in human subjects, not in animal models.
4. Carefully control the type of meals being given, and their time of consumption.
5. Use cross-over study designs to ensure potential outliers are accounted for.

It is important to note that these fundamental steps do not account for the obvious necessities of a GRF to have appropriate drug release rate along with a commercially viable manufacturing method. However, given the fact that no GRF has yet been successful in providing "value-added" gastric retention, we are limiting the scope of our guidance to clinical evaluation.

1. *Definition of success*
   This may vary from drug to drug, depending on its particular pharmacokinetic requirements. For example, one drug may only require dosing at breakfast with a gastric residence time of approximately 6 h to ensure it will be retained until lunch. Another compound may require 12 h of gastric residence time to provide constant drug levels throughout the entire day, regardless of meal size and frequency.
2. *Select an imaging modality (and marker when appropriate) that enables explicit determination of in vivo performance.*
   Gamma scintigraphy remains the method of choice for in vivo evaluation of gastric residence time. While magnetic marker monitoring, magnetic residence imaging, X-ray and swallowable cameras are viable options, gamma scintigraphy is the method of choice. It offers the advantages of being noninvasive, highly sensitive, and able to provide real-time, dynamic information. The primary disadvantages of this imaging modality are cost and identification of a suitable radiolabeling method. It is critical to note that in gamma scintigraphy, the camera only tracks a gamma emitting radionuclide that is incorporated into the dosage form – not the dosage form itself.

3. *Test the proposed GRF for success in human subjects, not animal models.* While animal models are useful for assessing the safety of a proposed GRF, or for verifying suitability of a radiolabel, they have not proven to be useful predictors of success in man [3, 12, 13].
4. *Carefully control the type of meals being given and their time of consumption.* Several investigators have described the critical impact of food on the gastric emptying of large, nondisintegrating dosage forms. Wilson et al. [14] showed that when a large 800 mg strength coated ibuprofen tablet was given after a high fat breakfast (3,327 kJ) average gastric emptying time was 8.8 h, with two subjects having gastric residence times of more than 15 h. It is important to note that the subjects in this study were given a snack (362 kJ) 2.5 h after dosing, and lunch (4,160 kJ) 4.5 h after dosing. The frequent consumption of food most likely reset the migrating motor complex before a housekeeper wave was initiated for those subjects who retained the tablet for 10 h or more. Conversely, when the tablets were given in the fasted state or after a light breakfast (646 kJ), mean gastric emptying times were less than 2 h. Borin et al. [15] reported similar results when 800 mg strength ibuprofen tablets were given after a heavy breakfast (750 kcal), with a mean gastric residence time of 8 h.
5. *Use cross-over study designs to ensure potential outliers are accounted for.* Whenever possible, cross-over study designs should be used. This enables each subject to serve as their own control, and enables the investigator to account for the possibility of some subjects having very rapid or slow gastric emptying. Our experience over a number of studies is that there are subjects who tend to consistently have fast or slow gastric emptying.

17.5 Physiological Challenges and Gastric Retentive Strategies

17.5.1 The Housekeeper Wave: The Number One Challenge To Achieve Gastric Retention

The most significant physiological event which impacts all potential GRFs is the Phase III period of the migrating motor complex or the so-called "housekeeper" wave. The housekeeper waves expel all remaining material (including dosage forms) from the stomach. In order to truly be classified as truly gastroretentive, the dosage form must be retained in the stomach during the housekeeper wave. One key function of the stomach is to deliver nutrients to the small intestine at a rate suitable for absorption. The rule of thumb for this rate of nutrient delivery is 3–4 kcal/min. However, in reality delivery is not truly linear but highly dependent upon meal composition. Phase III contractions of the housekeeper wave are only initiated after no further nutrient delivery is detected by the small intestine. Therefore, by using our rule of thumb and the known quantity of food ingested we can predict when the housekeeper wave will occur in the fed state. After this initial housekeeper wave occurs, it will continue to occur in the fasted state every 1–2 h, unless more food is ingested and the migrating motor complex is reset. More specific anatomical and physiological aspects to specific GRF strategies will be highlighted in the following sections but a fundamental question that remains unresolved is: "Has technology advanced enough to *safely* overcome gastrointestinal physiology (i.e. the housekeeper wave)?"

Several comprehensive reviews have been published in the past such as those by Waterman [4] or Streubel [16]. Waterman correctly concluded that a GRF has yet to have been successfully commercialized. He pointed out that while some companies may claim gastric retention, the current reality is that these so-called GRFs offer no significant advantage over most conventional modified release tablets taken with food. At present, the best commercially available means of prolonging gastric residence time is to take a large, nondisintegrating tablet with food at bedtime. Dosing as bedtime has been shown to delay the onset of a housekeeper wave. As will be described in the case study at the close of this chapter, dosing a nondisintegrating tablet with even a small meal at bedtime is enough to yield 6 h of gastric retention. This highlights the significant impact that the normal anatomy and physiology of the stomach can have on the transit of dosage forms. A clear understanding of the gastric environment and its dynamics, including postural variables, can guide which potential gastric retentive techniques are most likely to succeed.

17.5.2 Gastric Retentive Strategies: Mucoadhesion (Bioadhesives)

The concept of mucoadhesion is based on the formulation first making good contact with the mucus layer of the stomach, followed by its adhesion to the mucus layer. Thus the formulation remains in the stomach until the mucus layer sloughs off or the formulation no longer adheres to the mucus. There are several challenges with this approach: the first is the challenge to make good contact with the mucus layer. In the fed state, nonspecific adhesion could occur with food. In both the fed and fasted state, free mucus may also adhere to the formulation. Once good contact is made, sufficient adhesion is required to hold the formulation to the mucus.

Selection of the proper mucoadhesive polymer is often performed in-vitro prior to in-vivo evaluation. However, Laulicht et al. [17] reveal a lack of IVIVC for predicting suitable performance. Laulicht et al. also concluded that polymers that produce strong bioadhesive bonds may not achieve prolonged gastric retention in-vivo. This brings up the third challenge, even if the formulation successfully avoids nonspecific adhesion and correctly attaches to the mucus layer of the stomach, the turnover rate of the mucus will dictate the residence time in the stomach. The turnover rate for the mucus layer has not been definitively determined, but in some cases is reported to be 5 h and in others 24 or 48 h. However, it is clear that if adhesion to the uppermost layer of the mucus occurs, that mucus will be the first to slough off, limiting the duration of mucoadhesion. Based on all of the challenges encountered with this approach, there is a very low probability of success and to our knowledge no mucoadhesion techniques have been proven to be clinically successful.

17.5.2.1 Polymers for Mucoadhesion

"Mucoadhesive" materials are commonly divided into three categories: anionic, cationic, and neutral. Though many materials are claimed as mucoadhesive, only a few, such as Carbopol 934P/974P have been tested in-vivo. More recent strategies include targeting specific cross-links in the mucus layer, such as disulfide bonds using thiol-based polymers "thiomers" [18] and novel polymers "spheromers" (homo- and copolymers of fumaric, sebacic, and adipic anhydride with or without metal oxides) [19].

17.5.2.2 Maximizing the Surface for Mucoadhesion: A Potential Opportunity

One physical aspect of a mucoadhesive polymer that can influence performance is the surface area capable of interacting with the mucus layer. This is analogous to reducing drug substance particle size to increase surface area in order to have more contact with liquid to increase the rate of dissolution. To maximize the surface area for adhesion to the mucus layer, the polymer should have a high surface area to volume ratio.

Electrospinning techniques have been used to create nanofibers of polymers. The high surface area of such nanofibers and the rougher texture of a nonwoven nanofiber matrix may provide improved mucoadhesion compared to a typical cast film or tablet coat. Takeuchi et al. [20] mention that a particle less than 1 μm in size has the ability to penetrate deeper into the mucus layer, thus allowing more potential for interaction. Interactions with the deeper portions of the mucus layer would also reduce the turnover rate associated with the upper, loosely adherent layer that limits gastric residence time.

Using the Gantrez polymer ES225 (normally used for denture adhesives), an electrospun nanofiber mat was manufactured and compared to a cast film. An SEM of the nanofiber mat is shown in Fig. 17.5. The diameters of the nanofibers are estimated to be 350 nm.

When the nanofiber mat was compared to a cast film of the same material, it demonstrated improved mucoadhesion versus cast films of the same formulation during ex-vivo mucoadhesion testing with rat stomach tissue. The test was performed with freshly excised rat stomach tissue which was held in place using a studded Perspex block with a top plate. A texture analyzer (Model TAXT2i) with a 14 mm diameter stainless steel plate probe was brought in contact with the sample for a period of 5 s at a force of 0.05 N and the force and time required to detach the probe were measured. The area under the detachment force vs. time curve was

Fig. 17.5 SEM image of the Gantrez ES-225 nanofiber matrix which was electrospun from a 20% w/w polymer solution in an 85:15 mixture of ethanol and water. The technique to electrospin the polymer is described by Burke [21]

Table 17.3 Comparison of Gantrez ES225 cast films and an electrospun nanofiber matrix for mucoadhesion to excised rat stomach tissue as measured with a texture analyzer [21]

	Mucoadhesiveness (ns)	"Stringiness" (mm)
Gantrez ES225 Cast Film	0.03	0.79
Gantrez ES225 Electrospun Nanofiber matrix	0.07	4.16

calculated as the mucoadhesiveness of the sample and the time to achieve complete detachment was calculated as the "stringiness." A summary of the findings are shown in Table 17.3.

Based on the significant improvement in the performance of the nanofiber presentation versus a cast film, the Gantrez electrospun nanofiber mat was dosed in mongrel dogs to determine in-vivo gastric emptying performance. The average gastric residence time was evaluated using gamma scintigraphy by tracking a polyvinylacetate nanofiber with entrapped nanoparticles of samarium oxide. The polyvinylacetate nanofibers were used at a very low level and were intertwined with the Gantrez nanofiber matrix polymers. The average gastric emptying time using three mongrel dogs was an impressive 19.3 h.

This case study highlights a key parameter that formulation scientists have to improve mucoadhesive performance. However, the key challenges mentioned earlier in this section still limit the true possibility to develop a successful GRF based on mucoadhesion.

Table 17.4 Gastric residence time of floating calcium alginate beads in the fed [25] and fasted state [26] in healthy volunteers

Fed or fasted	Onset of gastric emptying	Time to completion of gastric emptying
Fed ($n=7$)	Did not begin (still in stomach after at least 5.5 h)	Unknown, all subjects had retained the beads when imaging was stopped
Fasted ($n=5$)	20–82 min	5–55 min

17.5.3 Gastric Retentive Strategies: Buoyancy (Floating and High Density)

The concept of buoyancy has been studied extensively and reported in the literature [22–24]. Floating products that have been marketed include Madopar HBS, Valrelease HBS, Gaviscon, and TUMS Lasting Effects. Floating systems have received considerably more attention than high-density formulations, possibly because high-density products have not reported consistent success at any stage of development (except possibly in some veterinary applications). Floating formulations have proven to be retained longer in the stomach than nonfloating dosage forms when the size of the product is kept small. However, as the size of the formulation is increased, the difference in gastroretention time of the floating versus nonfloating formulations becomes negligible.

Floating systems can be effective at gastroretention provided that sufficient gastric contents are present. Table 17.4 compares the performance of floating beads in the fed and fasted state.

Whitehead et al. [25] showed that when given with a large meal, gastric retention of floating beads was maintained for at least 5.5 h (imaging had stopped at that time for one subject). Comparatively, Stops et al. [26] studied calcium alginate beads in healthy volunteers who had fasted overnight for 10 h and then swallowed the beads with 100 mL of water. It is noteworthy that these subjects did not have anything to eat or drink, nor did they lie down until the beads had emptied from the stomach. Given the limited amount of liquids administered, it was not surprising that the longest gastric residence time in the fasted state was only 90 min. These studies demonstrate that floating systems may enhance gastric residence time in general, but are dependent upon a subject consuming an adequate amount of food or liquid to keep the dosage form in the stomach.

High-density dosage forms have been investigated for prolonging gastric residence in the fed or fasted states. Results from two studies are shown in Table 17.5.

Whitehead et al. [25] studied high-density beads in the fed state, while Clarke et al. [27] evaluated high-density beads in the fasted state. Healthy volunteers were used in both studies. These data showed that there was no advantage to using high-density beads to extend residence time. It is worth noting that one of the subjects in the fasted state had immediate gastric emptying. This was likely due to the ingestion of the dosage form being inadvertently synchronized with the onset of a housekeeper wave, expelling the GRF from the stomach immediately after it entered. Extremely rapid gastric emptying is an inherent problem when dosing in the fasted state. The authors envision that any truly gastric retentive dosage form will have to at least be taken with a small snack to reset the migrating motor complex. It is difficult to imagine

Table 17.5 Gastric residence time of high-density beads in the fed [25] and fasted state [27] in healthy volunteers

Fed/fasted	Density (g/mL)	Size (mm)	Onset of gastric emptying (range, min)	Time to completion or 50% of gastric emptying (range, min)
Fed	1.8	2.1	32–102	9–255[a]
Fasted	1.5	1.3	21–175	87–477[b]
Fasted	2.4	1.3	0–187	1–279[b]

[a]Time to completion
[b]50% of gastric emptying

that a dosage form could be designed to freely move from mouth to stomach, yet still be able to instantaneously become gastric retentive upon entering the stomach.

An anatomical challenge for floating systems concerns the relative positions of the pyloric sphincter and the formulation. Scintigraphy studies on Gaviscon, an antacid which forms a floating raft when coming into contact with the gastric contents, have shown that posture has a significant effect on gastric residence time of the raft. If the subject was lying in a position such that the raft was near the pyloric sphincter (i.e. lying on their left side), the raft emptied from the stomach much more quickly than when lying on their right side (approximately 30 min versus 70 min for 50% of the radioactivity to leave the stomach) [28].

Floating dosage forms offer the potential for longer gastric residence time in the general population but successful retention is completely dependent on a patient maintaining the appropriate posture and food or liquid intake. The reliability and robustness of the approach must accordingly be a concern in practice.

17.5.4 Gastric Retentive Strategies: Coadministration of Gastric Motility Agent

Gastric motility-affecting agents have been studied, albeit to a lesser extent than the other techniques, yet may provide an effective way to control the transit of dosage forms. Cholecystikinin, citric acid, endogenous opioid peptides, curcumin, amylin analogs, GLP-1 agonists, and fatty acids are some of the agents shown to prolong gastric emptying. The main challenges facing this approach are:

- Determining if the effective dose is commercially feasible (cost of goods would be too high for some agents)
- Dosage form size may be too large where the amount of agent is high
- Safety: agents may have pharmacological effects as well as impacting motility
- Onset of action must be faster than gastric emptying of the formulation
- Having sufficient magnitude and duration of action

The diabetic epidemic in conjunction with new therapies may offer some opportunities for prolonging gastric residence time in this patient population. DeFronzo et al. [29] showed that exenatide (a GLP-1 analog) significantly decreases gastric emptying versus sitagliptin. Kong et al. [30] showed in Type 1 diabetic patients that pramlintide (an amylin analog) delayed the onset of liquid emptying from 7 to 69 min. In addition, the lag time in solid material emptying was extended from 44 to 150 min. Very low levels of such mediators might retain the effect on motility with little other pharmacological impact.

17.5.5 Gastric Retentive Strategies: Bioinspired GRFs

The focus on novel, high technology solutions to drug delivery challenges can overlook situations where nature has already developed a solution. For gastric retention, *H. pylori* is able to stay in the stomach for extended periods of time. While these bacteria have a detrimental impact on our stomach, perhaps there are key learnings that can be used to guide a novel bio-inspired GRF. Other organisms capable of intestinal retention are nematodes (or tapeworms).

In certain tropical countries, up to 93% of the population is infected with tapeworms. Even in more developed countries, such as the United States, infections are common [31] (Fig. 17.6).

One investigator of tapeworms concluded that the tapeworm uses the hosts enteric nervous system to alter the intestinal myoelectrical patterns and reduce frequency of the migrating motor complex Phase III contractions (i.e. the housekeeper wave) [33]. To our knowledge, bioadhesion by organisms to epithelial surfaces of other organisms has not received much attention and is an untapped area of potential research for the future. Conceptually, a genetically engineered tapeworm might be delivered/implanted in the GI tract to secrete a therapeutic protein or siRNA continuously into the intestine providing the therapy needed to treat a chronic disease.

17.5.6 Gastric Retentive Strategies: Expandable Dosage Forms

Expandable GRFs are designed to expand to a sufficient physical size and strength to prevent being compressed and pushed through the pylorus. After being retained in the stomach for the desired length of time, the GRF should weaken and/or decrease in size to enable it to be cleared from the stomach. Size and strength attributes of such a GRF providing gastric retention are not explicitly known at this time. Evidence from the endoscopic literature regarding the ingestion of large objects, or formation of gastric bezoars, suggests that a GRF must be fairly rigid and of a size larger than 5 cm long and 2 cm in diameter [34–36]. These physical attributes of size and strength are necessary for a GRF to withstand being expelled from the stomach during a Phase III contraction (i.e. housekeeper wave) of the migrating motor complex.

Several different expansion strategies have been explored, with some being marketed as GRFs. However, to our knowledge, no expandable GRF has offered a significant advantage over a large, nondisintegrating tablet taken with food. The case study which concludes this chapter offers a new addition to the literature in regard to expandable dosage forms failing to cross the "valley of death."

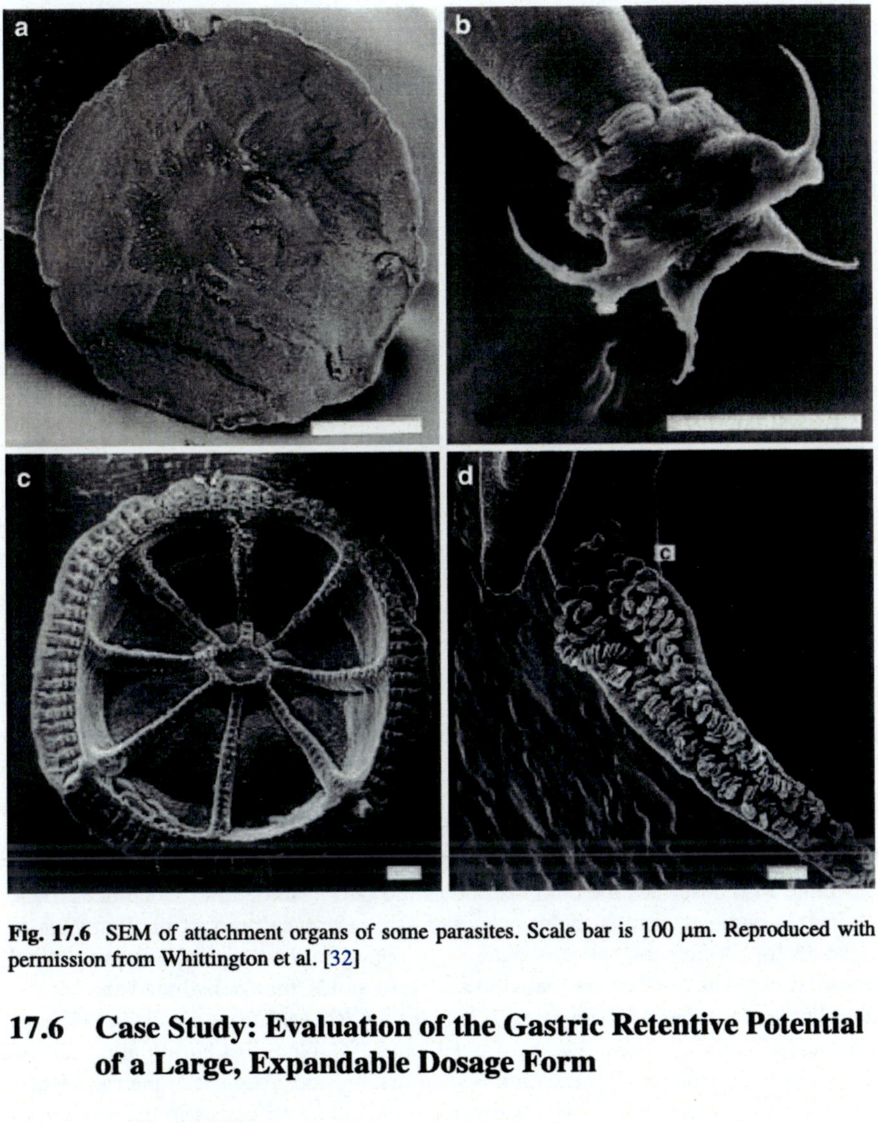

Fig. 17.6 SEM of attachment organs of some parasites. Scale bar is 100 μm. Reproduced with permission from Whittington et al. [32]

17.6 Case Study: Evaluation of the Gastric Retentive Potential of a Large, Expandable Dosage Form

17.6.1 Introduction

Ahmed and Ayres recently published encouraging results on the development of an expandable GRF [37]. Expandable GRFs made from naturally occurring carbohydrate polymers were prepared in various sizes in the shape of a rectangular prism. The GRFs were tested in fasted, healthy volunteers using riboflavin as a model compound. Deconvolution of riboflavin pharmacokinetics, a drug known to have saturable, site-specific absorption in the upper small intestine, was used to estimate the gastric residence time of the various formulations. The largest GRF tested, with dimensions of 7 cm × 1.5 cm × 1 cm before drying (these researchers estimated that the GRFs

Table 17.6 Composition of gastric retentive formulations tested in healthy volunteers

	Grams per 100 g of water (water removed after forming the gel)		
Ingredient	Formulation A (used in Study 1)	Formulation B (used in Study 2)	Formulation C (used in Study 2)
Locust bean gum	0.75	1	0.75
Xanthan gum	0.75	1	0.75
Polyethylene glycol 400	3	3	3
Cellulose acetate[a]	0.3	0.15	0.15
Activated charcoal[a]	0.05	0.02	0.02
Indium chloride[a]	Trace[b]	Trace[b]	Trace[b]

[a]Used to radiolabel the dosage form
[b]Each GRF contained a maximum of 0.5 MBq of Indium-111 at time of dosing

recovered approximately 75% of their original size in about 45 min based upon in vitro and in vivo data), was estimated to have a gastric residence time of approximately 15 h in fasted subjects. However, the authors noted that their results were preliminary, and further investigation was needed; with particular emphasis on studying the influence of the feeding regimen on the in vivo performance of the GRF.

The GRFs evaluated in this case study (previously unpublished clinical results) were of very similar composition to those studied by Ahmed and Ayres, but a variety of shapes were investigated. In addition, no drug was incorporated into the dosage forms as the initial goal was solely to determine whether the formulation was gastroretentive. Burke and coinvestigators [38] described how they developed a robust radiolabeling technique to allow scintigraphic evaluation of gastric residence time in healthy volunteers. The method consisted of adsorbing $^{111}InCl_2$ onto activated charcoal, dispersing the radiolabeled charcoal into molten cellulose acetate, cooling this mixture and grinding it up into small particles, and then incorporating it into the GRF during its manufacture.

Table 17.6 describes the composition of the GRFs studied in two clinical trials. A large, nondisintegrating tablet was used as a reference product in all studies to account for differences between patients in regard to their typical gastrointestinal transit times. The GRFs were prepared as gels in molds, then dehydrated and packed into hard gelatin capsules. Figure 17.7 shows a GRF packed into a clear, 000 capsule and after 0.5, 1, 2 and 3 h in 0.1 N HCl at 37°C. Note that Formulations B and C were developed after Formulation A, and that an improvement in the radiolabeling procedure was made. This improvement enabled a reduction in the amount of cellulose acetate and activated charcoal that was needed. This reduction in nonhydrating solids improved the gel strength and hydration rate of the proposed GRFs.

The first clinical study evaluated GRFs of the same size and shape (7 cm × 1.5 cm × 1 cm) as the best performing GRF reported by Ahmed and Ayres, while the second study evaluated a shorter/wider rectangular prism (5 cm × 3 cm × 1.5 cm), a sphere (3 cm in diameter), a ring (4.7 mm outer diameter, 1.5 cm inner diameter, 1.5 cm thick) and a "pillow" (4 cm × 4 cm in the longest dimension, 1.5 cm thick). All sizes reported for the GRFs are those for freshly prepared hydrated gels prior to drying. Gamma scintigraphy was used to determine the gastric residence time of the GRFs.

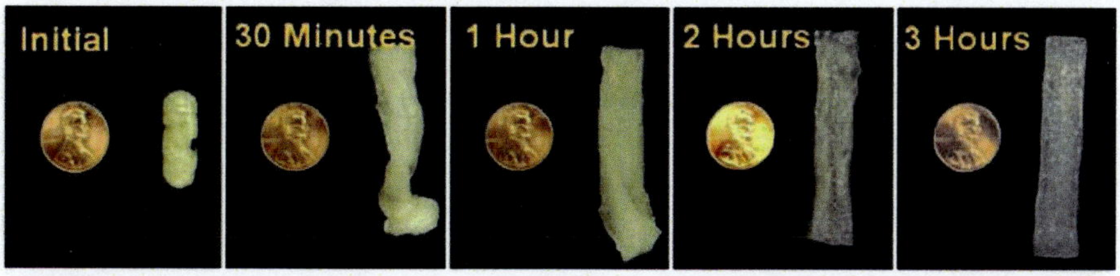

Fig. 17.7 Hydration of the gastric retentive formulation used in Study 1 as a function of time in 0.1 N HCl

Table 17.7 Overview of clinical studies and objectives

Study	Periods	Primary objectives
1	1–4	A. To determine gastric residence time and in vivo performance of Formulation A in the shape of rectangular prism when taken after a low calorie breakfast. B. To determine the effect of nighttime dosing on gastric residence time of Formulation A in the shape of rectangular prism when taken after a low calorie meal.
2	1–8	To determine effect of gel shape and composition on gastric residence time when taken after a low calorie breakfast.
2	9–12	To determine the effect of meal size and frequency on gastric residence time of Formulation B in the shape of rectangular prism.

Table 17.7 provides an overview of the clinical studies, and the primary objectives of each study. In all of these studies, a meal was consumed immediately prior to dosing. This enabled the proposed GRFs to hydrate in the stomach prior to the housekeeper wave (or Phase 3 contraction) of the migrating myoelectric complex (MMC) attempting to empty the dosage form out of the stomach.

Study 1 investigated the effect of morning and nighttime dosing after a low calorie meal, on the gastric residence time of Formulation A. This formulation was prepared in the shape of large, rectangular prism (7 cm × 1.5 cm × 1 cm), identical to the dimensions of the best performing GRF described by Ahmed and Ayres [37]. The results of Study 1 are fully described in Table 17.8. The median gastric emptying time for morning dosing of the GRF was 4 h (range 2.5–24 h), compared to 2.5 h (range 0.5–5 h) for the tablet. The median emptying time of the GRF and the tablet for evening dosing was 6 h (range for the GRF following evening dosing was 4–8 h, while the range for the tablet was from 4 to 7 h). The authors do not consider this small extension in gastric residence time for the GRF after morning dosing to be clinically relevant, but it was interesting to note a trend that the GRF generally provided a small increase in gastric residence time in most subjects.

A potential mechanism for this trend in increased gastric residence time is that the GRF reduced the rate of emptying of the meal, which would subsequently delay the onset of the housekeeper wave. Table 17.9 shows that the GRF delayed gastric emptying of the meal by approximately 1.5 h as estimated by the time for 50% (T_{50}) and 90% (T_{90}) of the meal to empty from the stomach.

As can be seen in Table 17.10, Study 2 explored a variety of gel shapes, to further understand the potential of these polysaccharide gels to enhance gastric residence time.

Table 17.8 Gastric emptying times of Formulation A, prepared in the shape of a rectangular prism (7 cm × 1.5 cm × 1 cm prior to dehydration) in Study 1

	Gastric emptying times following morning dosing (h)		Gastric emptying times following evening dosing (h)	
Subject no.	GRF	Tablet	GRF	Tablet
001	4.5	0.5	Drop out	Drop out
002	4.0	4.0	8.0	6.0
003	4.0	2.5	4.0	4.0
004	2.5	2.5	5.0	6.0
005	4.0	2.5	6.0 and 14.5[a]	6.0
006	24.0 and 25.0[b]	5.5	8.0	7.0
Median	4.0	2.5	6.0	6.0
Min	2.5	0.5	4.0	4.0
Max	24.0	5.5	8.0	7.0

[a]For subject 005, the GRF broke up into two pieces on emptying. One piece emptying at 6 h the other piece was retained and emptied at 14.5 h postdose. Initial emptying was used for all calculations

[b]For subject 006, the GRF broke up into two pieces on emptying from the stomach. One piece emptying at 24 h ,the other piece was retained and emptied at 25 h postdose. Initial emptying was used for all calculations

Table 17.9 Effect of a gastroretentive dosage form (GRF) and night time dosing on the time for 50% (T_{50}), and 90% (T_{90}) of the meal to empty from the stomach in Study 1

Subject	Morning dosing (h)				Evening dosing (h)			
no.	GRF T_{50}	GRF T_{90}	Tablet T_{50}	Tablet T_{90}	GRF T_{50}	GRF T_{90}	Tablet T_{50}	Tablet T_{90}
001	2.9	4.4	1.0	1.7	Drop out		Drop out	
002	2.6	3.7	1.3	3.6	3.3	6.9	2.3	5.4
003	2.8	3.9	0.7	2.0	3.2	3.9	1.3	3.3
004	2.0	2.4	0.7	1.8	2.2	4.6	1.1	3.5
005	1.8	2.9	1.5	2.3	1.1	4.9	0.5	2.0
006	1.3	5.4	1.3	4.7	5.0	7.1	1.3	2.5
Median	2.3	3.8	1.2	2.2	3.2	4.9	1.3	3.3
Min	1.3	2.4	0.7	1.7	1.1	3.9	0.5	2.0
Max	2.9	5.4	1.5	4.7	5.0	7.1	2.3	5.4

The rationale was to identify shapes that would still swell to a size larger than the pylorus, and also be more resistant to compression during a Phase 3 contraction. The shapes evaluated included a rectangular prism, a sphere, a pillow and a ring using a higher concentration of the polysaccharides (Formulation B). The rectangular prism and pillow shapes were further evaluated using a lower concentration of the polysaccharides (Formulation C) to provide a more fundamental understanding of the system. All dosage forms were administered after a low calorie breakfast.

The data in Tables 17.11 and 17.12 show that the modified GRF shapes and higher concentrations of locust bean gum and xanthan gum (Formulation B) did not significantly and consistently extend gastric residence time over the nondisintegrating tablet. In addition, in Study 2 the GRFs had no effect on the rate of gastric

Table 17.10 Description and picture of gastric retentive shapes used in Study 2

Shape	Outer dimensions prior to removal of water	Photograph
Prism (hydrated)	5×3×1.5 cm	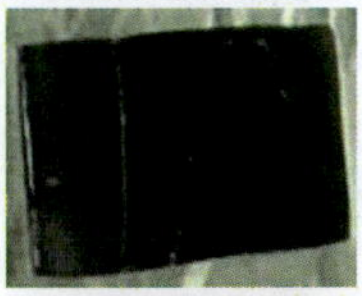
Ring (dehydrated)	4.7 mm outer diameter, 1.5 cm inner diameter, 1.5 cm thick	
Sphere (Hydrated)	Diameter = 3 cm	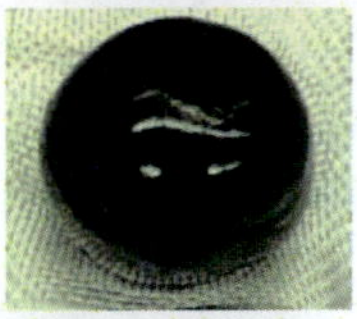
Pillow (Hydrated)	4×4 cm, tapered to 3×3 cm at edges, 1.5 cm thick	

Table 17.11 Gastric emptying times (hours) in Study 2, periods 1–4. All dosage forms given after a low calorie (~180 kcal) breakfast

		Formulation B		
	Reference tablet	Rectangular prism	Ring	Pillow
Subject no.	Gastric emptying time (hours)			
001	2.8	24.4	24.5	3.5
002	3.3	4.8	4.8	3.7
003	2.7	3.3	3.3	2.8
004	Removed from study	Removed from study	Removed from study	62.7[a]
005	2.3	3.3	3.3	2.8
006	2.3	3.2	2.3	2.8
Median	2.7	3.3	3.3	2.8
Min	2.3	3.2	2.3	2.8
Max	3.3	24.4	24.5	3.7

[a]Not used in calculation of mean, median, and maximum because subject removed from study after this dosing period

emptying of the meal. Interestingly, in these first eight dosing periods of Study 2, whenever Formulation B was administered, there was always one subject who retained the GRF in the stomach for more than 20 h. In contrast, neither the reference tablet nor Formulation C provided a single observation of a gastric retention time of more than 5 h. This observation could simply be due to random chance, but

Table 17.12 Gastric emptying times (hours) in Study 2, periods 5–8. All dosage forms given after a low calorie (~180 kcal) breakfast

		Formulation B	Formulation C	
	Reference tablet	Sphere	Pillow	Rectangular prism
Subject no.	Gastric emptying time (hours)			
007	2.8	21.0	3.8	3.8
008	1.8	2.8	3.3	2.8
009	3.2	3.8	3.3	2.8
010	2.7	2.8	3.3	3.3
011	4.7	2.8	2.3	3.3
012	2.3	3.3	2.3	4.3
Median	2.8	3.1	3.3	3.3
Min	1.8	2.8	2.3	2.8
Max	4.7	21.0	3.8	4.3

Table 17.13 Gastric emptying times (hours) in Study 2, periods 9–12

	Low calorie breakfast followed by snack 2.5 h later		High calorie breakfast	
Subject no.	Reference tablet	Formulation B Rectangular prism	Reference tablet	Formulation B Rectangular prism
	Gastric emptying time (hours)			
013	21.0	21.0	9.8	9.8
014	2.3	25.5	21.0	16.0
015	2.3	2.4	16.0	21.0
016	21.0	16.0	26.2	16.0
017	–	–	–	21.0[a]
018	21.1	21.0	9.8	9.8
Median	21.0	21.0	16.0	16.0
Min	2.3	2.4	9.8	9.8
Max	21.0	25.5	26.2	21.0

Dosage forms were given either after a low calorie breakfast (~180 kcal) followed by a snack 2.5 h later; or after a high calorie breakfast (~800 kcal) followed by lunch 5 h later

[a]Not used in calculation of mean, median, and maximum because subject removed from study after this dosing period

it may suggest that these GRFs are nearing the physical attributes necessary for clinically relevant gastric retention. It should be pointed out, that in general, the GRFs left the stomach intact. This observation suggests that the formulations were either compressed to a small enough size to pass through the pylorus, or that there was inadequate rehydration of the gels in vivo.

The last four periods of Study 2 examined the effect of meal size and meal frequency on the gastric residence time of Formulation B, prepared in the shape of a rectangular prism. The results in Table 17.13 show that when either the tablet or GRF were given after a high calorie breakfast, and followed by lunch 5 h later, every subject retained the tablet and the GRF for more than 9 h. In addition, when the dosage forms were given after a low calorie meal, followed by a low calorie snack 2.5 h later, more than half the subjects had gastric emptying times of at least 16 h for both the tablet and the GRF.

These data in Tables 17.11–17.13 clearly demonstrate that meal composition and frequency of meals were the factors that most significantly influenced gastric residence time. As such, it is essential that meals be carefully controlled when evaluating the performance of GRFs in clinical trials. The strategy of dosing after a low calorie breakfast followed by lunch 5 h later appears to be a useful means of screening the potential of a dosage form to provide meaningful gastric retention. It has been shown that the Phase 3 contractions after breakfast are the strongest of the day, and thus likely represent the most challenging physiological environment for a GRF to overcome.

The proposed GRFs in this study were well tolerated by the volunteers, but failed to provide a clinically relevant increase in gastric residence time over a large, non-disintegrating tablet. The reason for this is most likely due to inadequate gel strength of the GRFs to resist compression to a size small enough to pass through the pylorus. However, there are insufficient data to rule out that this failure was due to inadequate swelling of the dehydrated gels.

17.7 Conclusions

While a gastroretentive dosage form could enhance the performance of several drugs, invention of such a dosage form continues to elude scientists. Perhaps a novel strategy will be developed in the future or a combination of the various techniques such as high density and mucoadhesion would enable gastroretention to be achieved. However, the fundamental physiological challenges of gastroretention still remain and as pharmaceutical scientist's progress novel products into the "valley of death," the authors highly recommend following the suggested clinical guidance to rapidly and definitely prove the gastric retentive potential of their new inventions.

Acknowledgments The authors would like to thank Scott Staton, Erin Peters, Ann Vickers, Professor Jim Ayres, Professor Clive Wilson and the research staff at the Bio-Images Research Facility including Professor Clive Wilson, Professor Howard Stevens and Bridget O'Mahoney for their assistance with the clinical evaluation of GRFs discussed in this book chapter.

References

1. Watanabe S, Kayano M, Ishino Y, Miyao K (1976) US Patent 3,976,764
2. Mamajek RC, Moyer ES (1980) US Patent 4,207,890
3. Cargill R, Engle K, Gardner CR, Fix JA, Porter P, Sparer RV (1989) Controlled gastric emptying. II. In vitro erosion and gastric residence times of an erodible device in beagle dogs. Pharm Res 6(6):506–509
4. Waterman KC (2007) A critical review of gastric retentive controlled drug delivery. Pharm Dev Technol 12:1–10

5. Lewis LD, Fowle ASE, Bittiner SB (1986) Human gastrointestinal absorption of acyclovir from tablet duodenal infusion and sipped solution. Br J Clin Pharmacol 21(4):459–462
6. Burke MD, Wilson CG (2006) Gastroretentive dosage forms. Drug Deliv Technol 6(8):26–31
7. Williams MF, Dukes GE, Heizer W, Han Y-H, Hermann DJ, Lampkin T, Hak LJ (1992) Influence of gastrointestinal site of drug delivery on the absorption characteristics of ranitidine. Pharm Res 9(9):1190–1194
8. Pithavala YK, Heizer WD, Parr AF, O'Connor-Semmes RL, Brouwer KLR (1998) Use of the InteliSite Capsule to Study Ranitidine Absorption from various sites with the human intestinal tract. Pharm Res 15(12):1869–1875
9. Menon R, Cefali E, Wilding I, Wray H, Connor A (2009) The assessment of human regional drug absorption of free acid and sodium salt forms of acipimox, in healthy volunteers, to direct modified release formulation strategy. Biopharm Drug Dispos 30:508–516
10. Wilding I (2002) Absorption of drugs released in the colonic environment. In: AAPS Conference: Pharmaceutics and Drug Delivery Conference, Arlington, VA, April 2002
11. Basit AW, Newton JM, Lacey LF (2002) Susceptibility of the H2-receptor antagonists cimetidine, famotidine and nizatidine, to metabolism by the gastrointestinal microflora. Int J Pharm 237(1–2):23–33
12. Klausner EA, Lavy E, Stepensky D, Friedman M, Hoffman A (2002) Novel Gastroretentive dosage forms: evaluation of gastroretentivity and its effect on riboflavin absorption in dogs. Pharm Res 10:1516–1523
13. Klausner EA, Eyal S, Lavy E, Friedman M, Hoffman A (2003) Novel levodopa gastroretentive dosage form: in vivo evaluation in dogs. J Control Release 88:117–126
14. Wilson CG, Washington N, Greaves JL, Kamali F, Rees JA, Sempik AK, Lampard JF (1989) Bimodal release of ibuprofen in a sustained-release formulation: A scintigraphic and pharmacokinetic open study in healthy volunteers under different conditions of food intake. Int J Pharm 50(2):155–161
15. Borin MT, Khare S, Beihn RM, Jay M (1990) The effect of food on gastrointestinal (GI) transit of sustained-release Ibuprofen tablets as evaluated by gamma scintigraphy. Pharm Res 7(3):304–307
16. Streubel A, Siepmann J, Bodmeier R (2006) Drug delivery to the upper small intestine window using gastroretentive technologies. Curr Opin Pharmacol 6:501–508
17. Laulicht B, Cheifetz P, Tripathi A, Mathiowitz E (2009) Are in vivo gastric bioadhesive forces accurately reflected by in vitro experiments? J Control Release 134:103–110
18. Bernkop-Schnurch A, Guggi D, Pinter Y (2004) Thiolated chitosans: development and in-vitro evaluation of a mucoadhesive permeation enhancing oral drug delivery system. J Control Release 94(1):177–186
19. A. Nangia (2008) BIOROD – Bioadhesive based oral delivery system. In: 35th Annual Meeting of the Controlled Release Society, New York, New York, USA
20. Takeuchi H, Yamamoto H, Kawashima Y (2001) Mucoadhesive nanoparticulate systems for peptide drug delivery. Adv Drug Deliv Rev 47(1):39–54
21. Burke MD, Luzhansky D (2007) Nanofiber based drug delivery. In: Deepak T, Michel D, Yashwant P (eds) Nanoparticulate drug delivery systems: recent trends and emerging technologies, Chap. 4. Informa Healthcare, London
22. Arora S, Ali J, Ahuja A, Khar R, Baboota S (2005) Floating drug delivery systems: a review. AAPS Pharm Sci Tech 6(3):372–390
23. Stops F, Fell J, Collett J, Martini L, Sharma H, Smith A-M (2004) Floating dosage forms to prolong gastro-retention: an in vivo study in the fasted state. J Pharm Pharmacol 56:S-67
24. Davis S, Stockwell A, Taylor M, Hardy J, Whalley D, Wilson C, Bechgard H, Christensen F (1986) The effect of density on the gastric emptying of single- and multiple-unit dosage forms. Pharm Res 3(4):208–213
25. Whitehead L, Fell J, Collett JH, Sharma HL, Smith A-M (1998) Floating dosage forms: an in vivo study demonstrating prolonged gastric retention. J Control Release 55:3–12
26. Stops F, Fell J, Collett J, Martini LG, Sharma H, Smith A (2006) The use of citric acid to prolong the in-vivo gastro-retention of a floating dosage form in the fasted state. Int J Pharm 308:8–13

27. Clarke G, Newton J, Short M (1993) Gastrointestinal transit of pellets of differing size and density. Int J Pharm 100:81–92
28. Bennett CE, Hardy JG, Wilson CG (1984) The influence of posture on the gastric emptying of antacids. Int J Pharm 21:341–347
29. DeFronzo RA, Okerson T, Viswanathan P, Guan X, Holcombe JH, MacConell L (2008) Effects of exenatide versus sitagliptin on postprandial glucose, insulin and glucagon secretion, gastric emptying and caloric intake: A randomized cross-over study. Curr Med Res Opin 24(10): 2943–2952
30. Kong MF, King P, Macdonald IA, Stubbs TA, Perkins AC, Blackshaw PE, Moyses C, Tattersall RB (1997) Infusion of pramlintide, a human amylin analogue, delays gastric emptying in men with IDDM. Diabetologia 40:82–88
31. Bagi MEAE, Sammak BM, Mohamed AE, Al Karawi MA, Al SM, Al Thagafi MA et al (2004) Gastrointestinal parasite infestation. Eur Radiol 14:E116–E131
32. Whittington ID, Cribb BW (2001) Adhesive secretions in the platyhelminthes. Adv Parasitol 48:102
33. Dwinell KL, Bass P, Zou F, Oaks JA (2002) Small intestinal transactions decrease the occurrence of tapeworm-induced myoelectric patterns in the rat. Neurogastroenterol Mot 14:349–356
34. Hashemy A, Seleem M, Khadrah A (2002) An extraordinary swallowed foreign body: a dessert spoon in the stomach. Kuwait Med J 34(1):53–55
35. Webb WA (1995) Management of foreign bodies of the upper gastrointestinal tract: Update. Gastrointest Endosc 41(1):39–51
36. Koch H (1977) Operative endoscopy. Gastrointest Endosc 24(2):65–68
37. Ahmed IS, Ayres JW (2007) Bioavailability of riboflavin from a gastric retention formulation. Int J Pharm 330:146–154
38. Burke MD, Staton JS, Vickers AW, Peters EE, Coffin MD (2007) A novel method to radiolabel gastric retentive formulations for gamma scintigraphy assessment. Pharm Res 24(4):695–704

Chapter 18
Drug Delivery to the Colon

Abdul W. Basit and Emma L. McConnell

Abstract The colon is a challenging target for drug delivery, as reaching the distal regions of the gastrointestinal tract presents significant physiological challenges and environmental barriers. Many approaches have been used to surmount these, with mixed success rates. Colonic delivery has historically been limited to treatment of local conditions such as inflammatory bowel disease. Latterly, efforts have also concerned delivery for treating colon cancer and for systemic delivery of selected compounds. Such approaches have concerned use of enteric coatings, sustained release systems, bacterially triggered treatments, or combinations of these. Possibilities are discussed in this chapter, along with historical experiences with systems for treating ulcerative colitis and Crohn's disease.

18.1 Introductory Remarks and Historical Development

Traditionally, the clinical applications of oral colonic drug delivery have been limited to the local treatment of inflammatory bowel disease (IBD). Enteric coatings, sustained release systems, and bacterially triggered treatments have all been used to deliver anti-inflammatory molecules to the colon to treat this debilitating condition. However, for many years the treatment of colonic cancer has been postulated as an ideal candidate for colonic drug delivery but little has been delivered in this field although there are some potential avenues which are starting to be explored. Also, there are other local diseases of the large intestine which could benefit from topical delivery to the colonic mucosa, and the potential of the colon for systemic

A.W. Basit (✉)
The School of Pharmacy, University of London, London, UK
e-mail: Abdul.Basit@pharmacy.ac.uk

C.G. Wilson and P.J. Crowley (eds.), *Controlled Release in Oral Drug Delivery*, Advances in Delivery Science and Technology, DOI 10.1007/978-1-4614-1004-1_18,

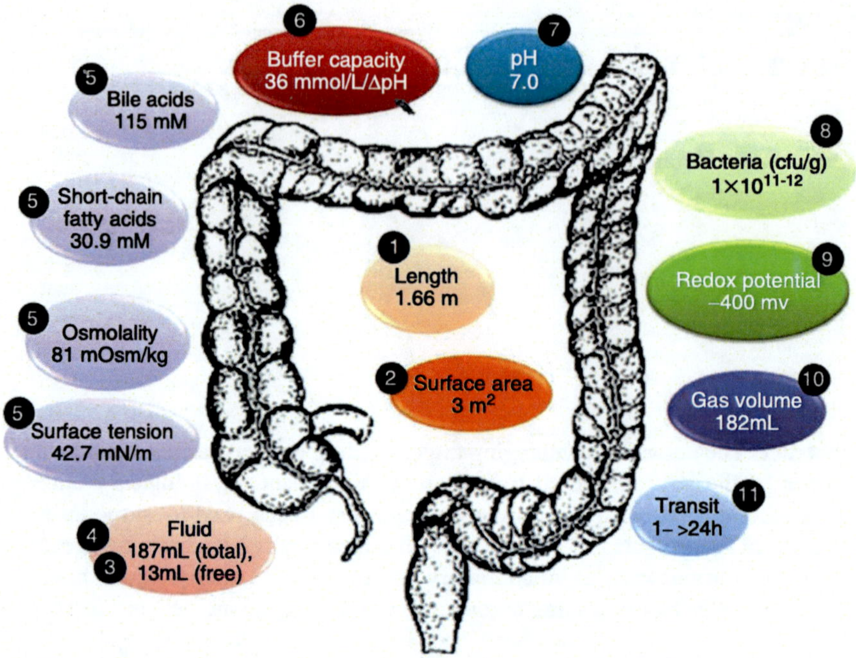

Fig. 18.1 The colonic environment: some physiological features of the colon

drug delivery should not be ignored. Applications for colonic delivery, old and new, are constrained by the physiological difficulties of targeting this distal site and the environmental barriers presented by the colon. There are, however, some overlooked factors which could prove beneficial for colonic drug delivery. This book chapter considers how the colonic physiology and local environment can affect the success or failure of traditional and novel strategies for delivery to the large intestine, local and systemic, and examines some of the up-to-date issues in colonic drug delivery.

The biggest issue affecting the success or failure of colonic drug delivery is the colonic environment (Fig. 18.1), and the difficulties it presents to dosage form design for such delivery. For example, there is a distinct lack of fluid, and that which is present is heterogeneously distributed and very little of it free to solubilize a drug. On the colonic mucosa and mixed in with the colonic fluid and solids are the colonic microbiota. These microorganisms (bacteria, yeasts, and fungi) number 10^{11}–10^{12} cfu/g of material in the colon, and there may be as many as 3,000 different species residing here [1]. They digest polysaccharides, proteins, and even drugs. To date, over 30 drugs have been identified and published as substrates for colonic bacteria [2] and many more are expected to be uncovered. This can have significant consequences for drugs exposed to the intestinal bacteria; they could potentially be activated, inactivated, or made toxic.

Drug absorption in the colon can be influenced by colonic residence time. The colon shows variations in transit; the residence of a dosage form in the colon can be from around 1 h up to several days [3] and this can affect drug bioavailability [4]. Dosage form factors influence colonic transit; tablets (25×9 mm) move ahead of pellets (0.5–1.8 mm) in the ascending colon [5] in a process known as streaming (due to solid and liquid material moving at different rates). Transit in the colon is also nonuniform; dosage forms are often at rest spending up to 30 min periods with no or minimal propagation [6]. In general, however, the transit time is relatively long compared to the upper gastrointestinal tract and this could confer significant advantages for drug delivery to this site.

Colonic drug delivery has its origins in the 1950s when the prodrug sulfasalazine was introduced as a treatment for rheumatoid arthritis and later IBD. However, the mechanism of action of sulfasalazine was not appreciated until studies in the 1970s. This prodrug consists of sulfapyridine azo-bonded to a molecule of 5-aminosalicylic acid (5-ASA, mesalazine, mesalamine). This can pass through the stomach and small intestine intact, being cleaved by colonic bacteria to release the active moiety (mesalamine). This was the standard of care for IBD (besides steroid treatments) until the 1990s when several new drugs were developed and approved: olsalazine (1990) and new mesalamine formulations (Asacol [1992], Pentasa [1993]). Olsalazine is another prodrug in which two linked mesalamine molecules are cleaved by colonic bacteria to the two active moieties. Asacol and Pentasa are modified release formulations; Pentasa controls release via an ethylcellulose coat whereas Asacol has a pH-triggered release via a coating which dissolves at >pH 7. In 2000, balsalazide (Colazol) was approved. This is cleaved to release mesalamine and an inactive molecule (4-aminobenzoly-beta-alanaine). After this, focus turned to developing new formulations rather than new drug molecules and further mesalamine formulations were released in 2007 (Lialda) and 2008 (Asacol HD and Apriso). Lialda has a pH-triggered mechanism of release combined with a slow release mechanism.

Apriso is another delayed/extended release formulation based on pH-triggered release. The recent shift in clinical prescribing towards higher doses of anti-inflammatory medications has fuelled the development of high dose products, such as Lialda and Asacol HD (Fig. 18.2). Beyond the mesalamine-based drugs, steroid treatments have also been targeted to the colon. An example is Entocort which uses pH and water-insoluble polymers to control the release of budesonide in the gastrointestinal tract [7]. A further budesonide product is Budenofalk which consists of pellets with pH responsive polymer coatings. These latter two products are multiunit systems and illustrate the shift away from single unit dosage forms due to the latter's inconsistency and susceptibility to failure as modified release preparations [8, 9]. Many other colonic drug delivery systems have been developed, or are in clinical trials, and some of these are discussed in this chapter.

Fig. 18.2 Historical development of mesalamine formulations in the US

18.2 Specific Benefits of Colonic Delivery

The colon has several physiological advantages, beyond the requirement for topical therapy, which provide opportunities for systemic drug delivery. One example that could be exploited is active drug absorption and epithelial drug metabolism. Efflux transporters present in the gut have different distributions in the colon than in the small intestine. Two such transporters are P-glycoprotein (P-gP) and breast cancer resistance protein (BCRP), which are membrane-bound ATP-binding cassette (ABC) transporters. P-gp is the transporter protein encoded for by the multidrug resistant gene 1 (MDR-1 or ABCB1), whereas BCRP is encoded by ABCG2 (ATP-binding cassette subfamily G2). Efflux transporters function to prevent absorbed substances reaching the blood stream; if an unwanted molecule breaches the intestinal epithelium it can be expelled back into the gut lumen by these transporters. They have the effect of decreasing the bioavailability of drugs which are substrates for them.

Levels of some efflux transporters are lower in the colon than in the upper small intestine which could potentially lead to improved bioavailability of selected drugs. The relative levels of such transporters have been summarized in a review by McConnell et al. [10]. Drug bioavailability from the gut is not just dependent on passive uptake and active influx and efflux mechanisms. The intestinal epithelium is also home to a family of membrane-bound metabolic enzymes known as cytochromes. Cytochromes, most notably cytochrome P450, are capable of metabolizing a host of drug molecules, potentially reducing their bioavailability. With a few exceptions, the levels of cytochrome enzymes are generally lower in the lower intestine than in the small intestine, and targeting drugs there could lead to improved bioavailability. This concept was demonstrated in a recent study using simvastatin. The bioavailability of simvastatin was three times higher when it was delivered to the distal gastrointestinal tract (delayed release) relative to the upper gut (immediate release) [11].

As well as the perceived physiological advantages of systemic drug delivery via the colon (lower efflux transporter levels, lower metabolic and proteolytic enzyme levels) there are other reasons to consider this mode of delivery.

- *Dose reduction*. Improved bioavailability from delivery to the distal gut (as occurred with the simvastatin example) could allow a lower initial dose to be administered. This has positive medical and economic implications.
- *Chronotherapy*. Diseases which are worse in the morning, for example, asthma, angina pectoris, and rheumatoid arthritis could benefit from colonic drug delivery in which the dose is taken the night before and reaches the colon to release drug before awakening. A clinical trial (CAPRA-1; 12 weeks, 288 patients) was carried out on the efficacy of a modified-release preparation of prednisolone vs. immediate release prednisolone for the treatment of rheumatoid arthritis [12]. The modified-release product was designed to be taken around 10 p.m., disintegrate after 4 h at which point it should be in the distal small intestine/colon. The change in duration of morning stiffness (stiffness in joints which is worst upon waking) was significantly higher with the modified release preparation and there were no safety issues.

Despite the apparent difficult physiology, some drugs have good bioavailability from colonic delivery and even, as illustrated earlier, better bioavailability than from the small intestine. Table 18.1 lists drugs which have been reported to have good colonic absorption. This list cannot be considered exhaustive: many drugs have not been investigated for colonic delivery and others may have been tested but information is not in the public domain.

Good colonic absorption is often necessary for drugs developed as sustained release or modified release dosage forms. Furthermore, BCS Class II, III, and IV molecules, having low solubility, low permeability, or both, which may not be well absorbed in the small intestine, may be absorbed to some extent in the colon.

Table 18.1 Drugs which are well absorbed in the colon (for full references see review [10])

5-Flurouracil	Nefazodone
Budesonide	Nifedipine
Diclofenac	Nisoldipine
Glibenclamide	Nitrendipine
Hydroxy-fasudil	Ondansetron
Ibuprofen	Oseltamivir
Ipsapirone	Oxprenolol
Isosorbide-5-mononitrate	Rivastigmine
Lefradafiban	Rofecoxib
Lumiracoxib	Theophylline
Metoprolol	Ursodeoxycholic acid

18.3 System Design: Reaching the Colon

The main challenge to deliver active ingredients to the colon by orally administered dosage forms is maintaining site specificity. This requires that little or no disintegration or dissolution occurs in the upper gut, but that drug is released in the colon. Despite the challenge, there are several products commercially available. Strategies for colonic targeting are concerned with identifying and exploiting the unique properties of the colon and aligning these with the relevant physicochemical and biological properties of the drug. The primary strategies for colonic targeting are illustrated in Fig. 18.3. Figure 18.4 illustrates the historical progression of key technologies developed for colonic delivery.

18.3.1 pH-Responsive Delivery

pH dependent systems targeting the lower bowel utilize polymers sensitive to the pH gradient along the intestinal tract. This approach represents the majority of commercial products formulated for colonic delivery (other examples being prodrugs, i.e., drug specific approaches). Targeting the colon with such polymers is conceptually sound, but can be difficult. Polymers that avoid dissolution at the low pH of the stomach and the pH of the small intestine (6.6 ± 0.5 in the proximal) are used. These then need to dissolve at the higher, near neutral pH of the distal gut (7.5 ± 0.4 in the distal small intestine). pH is maximal around the ileocecal junction [13] but can fall on entry to the colon due to the production of short-chain fatty acids by resident microbiota. Intersubject pH at a specific GI site can also span a range of up to 2 pH units [14]: Intraindividual variability is also considerable [15]. However, pH responsive approaches to colonic delivery have provided successful commercial products.

Fig. 18.3 Strategies for colon targeting via the oral route

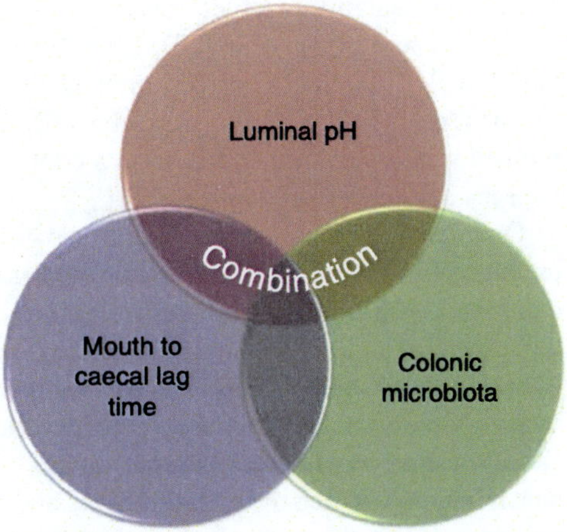

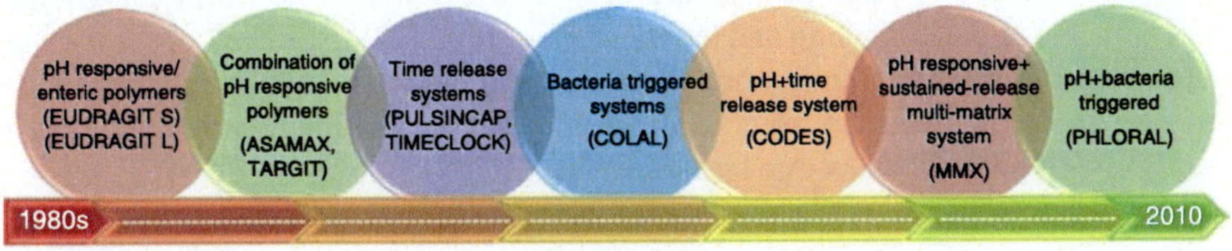

Fig. 18.4 Evolution of colonic delivery system design

A methacrylic acid/methyl methacrylate copolymer (EUDRAGIT S), with a dissolution threshold of pH 7 was first used to target the colon in the 1980s [16]. A number of commercially available tablet dosage forms were then developed based on this principle. These were employed to deliver mesalamine to the distal gut for the oral treatment of IBD. However, failure to disintegrate in vivo has been noted with some tablets and similar preparations [8, 9] possibly due to intraindividual variability and lower colonic pH in IBD patients [17, 18]. Thus, precise site specificity remains elusive with such coated formulations.

Other approaches to pH-responsive drug delivery to the distal gut include using polymers which have a lower pH threshold. EUDRAGIT L dissolves at lower pH (>6) and has also been applied as coatings for colonic targeting. Salofalk is a EUDRAGIT L-coated mesalamine tablet for the treatment of ulcerative colitis, although more likely to deliver to the mid- to distal small intestine. Combinations of two methacrylic acid copolymers EUDRAGIT S and EUDRAGIT L or EUDRAGIT S and EUDRAGIT L 100–55 (pH >5.5) can be used to manipulate drug release profiles within the pH range of 5.5–7.0 by changing the polymer ratios [19]. Such systems can theoretically deliver drug to any desired region of the GI tract. A gastro resistant mesalamine tablet formulation (Asamax) utilizes a EUDRAGIT L/S mixture as coating material [20]. An in vivo gamma scintigraphy study showed that initial tablet disintegration occurred at the ileocecal junction and the ascending colon. TARGIT™ technology uses EUDRAGIT L/S (3:1) mixture to coat injection-molded starch capsules which form a continuous unit between the lid and the body for easy application of coatings [21].

One recent development for improved colonic delivery has been the multimatrix or MMX™ system. The concept of MMX is based on pH-triggered drug release combined with a diffusion-based release mechanism to achieve sustained release; the technology has been designed to release drug throughout the colon. The system is based on a tablet formulation in which the active ingredient is dispersed in an inner lipophilic matrix and covered by an outer hydrophilic matrix generated by in situ hydration of selected polymer chains. A gastric resistant, pH-dependent film coat is applied. When the coat dissolves and fluid is imbibed into the core, a viscous gel mass forms through which drug diffuses and is released. The technology was applied to mesalamine and budesonide, and is now being applied to other drug molecules (e.g., heparin) albeit still in early development. The mesalamine formulation was first evaluated by pharmaco-scintigraphy study; initial erosion or disintegration was apparent in the ascending and the transverse colon [22]. However, absorption of mesalamine commenced in the small intestine and ileum and before relevant tablet erosion could be identified from scintigraphic images. This suggested that such a mode of delivery still lacks true site specificity [23].

18.3.2 *Time-Based Delivery*

Time-dependent systems utilize the time delay between dosage form ingestion and colonic arrival for colon-specific targeting. This is generally achieved by various coating mechanisms which erode over a predetermined period of time. Time-based approaches work on the concept that a dosage form will spend up to 2 h in the stomach in the fasted state, followed by 3–4 h in the small intestine. Using erodible polymers a lag time can be built in to allow drug release after this time. This approach has the disadvantage that it ignores the huge variability seen in gastrointestinal transit. It is often assumed that most of the variability in gastrointestinal transit lies with gastric emptying, and it is true that gastric emptying time is highly variable, being influenced by dosage form size and density, and by food intake [24, 25]. However, small intestinal transit time is also variable. After emptying into the small intestine, to achieve colonic delivery, a dosage form will have to traverse the length of the small intestine. Although this happens in a mean time of 3 ± 1 h, the actual transit time can range from only 30 min to over 9 h [26]. Furthermore, stagnation times at the ileocecal junction can be in the range of 2–10 h [27]. Feeding can shorten this ileocecal junction stasis time as the gastroenteric response ushers food into the large intestine [28]. These cumulative effects make the colonic arrival time of a dosage form difficult to predict and reproducibly control. Thus, site specificity for many time-dependent colonic delivery systems is poor, even though a well-designed system can accurately release its drug load after a preset lag time in vitro. One time-dependent system is the Pulsincap™ technology, in which a hydrogel plug, in an otherwise impermeable capsule, swells, and then ejects, allowing drug release [29]. Osmotic pump devices have also been developed which release drug under increasing osmotic pressure as water is imbibed through the GI tract; the oral osmotic system for colonic targeting (OROS-CT) has a 3–4 h delay before drug release occurs. However, no such systems have been commercialized, possibly due to the unpredictability of transit time.

18.3.3 Microbially Triggered Drug Delivery

Resident microbiota are a highly specific environmental feature of the colon. Consistently high levels of bacteria in the colon, relative to the upper regions of the GI tract, present more reliable opportunities for colonic delivery than the more variable pH. Prodrugs such as sulfasalazine rely on colonic bacteria to break down the inactive precursor and release the active drug moiety (mesalamine). *Polysaccharidase*-producing bacteria could facilitate use of polysaccharides as colonic delivery systems. Such materials are cheap, safe, and biodegradable. Polysaccharides can avoid degradation in the small intestine, but are a substrate of the colonic microbiota, e.g., amylose, chitosan, and pectin. Such materials could be used as coatings or matrix systems. Although polysaccharide-based systems showed promise in targeting to the colon, only few have been evaluated clinically and none to date have been commercialized. A system based on amylose mixed with the water insoluble polymer ethylcellulose (known as COLAL) has shown positive results in phase I and II clinical trials. Amylose is a starch polysaccharide; starch polysaccharides come in many forms, several of which are indigestible by human pancreatic enzymes but act as a food source for colonic bacteria [30]. This combination of colon-specific polysaccharide and insoluble polymer (to prevent swelling and premature drug release) has achieved consistent colonic targeting with a variety of drug molecules [30–33].

18.3.4 Combination Approaches

The variability in colonic physiology has led to a focus on multiple parallel approaches for colon-specific targeting, i.e., a combination of two or more physiological trigger mechanisms in one dosage form. Examples in this chapter have shown the fallibility of using pH or time-dependent triggers alone. Microbial triggers seem more reliable, but they too may be influenced by changes in microbial populations or numbers in disease. Combining approaches to ensure that a "back-up" release mechanism is included in the dosage form seems prudent.

A novel colon-targeted delivery system (CODES™) relies on colonic bacterial activity to affect drug release. CODES™ also incorporates a pH-sensitive element, although the concept differs from that described previously. The technology utilizes a lactulose-containing tablet core with a layered coating system [34]. The core is first coated with the acid-soluble polymer EUDRAGIT E and then subsequently overcoated with an enteric coating of EUDRAGIT L. The outer enteric coat protects the tablets in the stomach and then dissolves in the small intestine. At this point, the integrity of the tablets remains due to the insolubility of the acid-soluble EUDRAGIT E coating at the pH of the small intestine. Upon arriving in the colon the lactulose is dissolved, leaches out through the coating and is fermented into short chain fatty acids by colonic bacteria. This reduces the pH of the external environment causing the final layer (EUDRAGIT E) to dissolve. A gamma scintigraphy study showed that the majority of CODES™ tablets disintegrated in the proximal colon in both fasted and fed state [35].

A new concept in colonic drug targeting was introduced by combining pH-responsive and bacterially triggered mechanisms [36] in a single layer matrix film. The technology comprises a mixture of pH-responsive polymer (EUDRAGIT S) and biodegradable polysaccharide (resistant starch) and is known as PHLORAL™. The EUDRAGIT S component in the coating has two functions: it prevents the disintegration of the film in the upper gastrointestinal tract and controls the swelling of starch. The resistant starch in the coating resists digestion by mammalian amylase enzymes secreted by the pancreas but is readily digestible by colonic bacterial enzymes. Once entering the colon, both trigger mechanisms contribute to the dosage form disintegration and act as back up or fail safe to ensure appropriate drug targeting. A gamma scintigraphy study showed that the system provides colon specificity. Consistent disintegration of tablets coated with the technology was seen at the ileocecal junction or large intestine [36].

18.4 Drug and Disease Candidates for Colonic Delivery

Drug delivery systems described so far have focused on delivery for IBD. The primary drug candidates in use or under investigation are the mesalamine derivatives and the steroids prednisolone, budesonide, and beclometasone. Many other drugs and diseases are postulated as candidates for colonic delivery, whether these be local treatment or are aiming to achieve systemic delivery. Cytotoxic agents are of great interest. Colon cancer is the third most common form of cancer and is prevalent in the western world. Current treatment is generally with cytotoxic agents such as 5-fluorouracil, radiotherapy or surgical intervention, or a combination of these. The feasibility of local targeting of conventional and nonconventional cytotoxics has been postulated [37] but it is likely that this would be more useful as an adjunct therapy than a stand-alone treatment, due to the serious and potentially aggressive nature of the disease.

Other anti-inflammatories may also prove useful candidates for treating IBD, e.g., thromboxanes, prostaglandins, and leukotrienes. Heparins and nicotine could also prove beneficial in the treatment of IBD [38–41]. Low molecular weight heparin has been suggested as a candidate for the treatment of IBD by local delivery to the colon. Promising results with parnaparin in rats (improved histology, less mucosal damage, and reduction in colonic weight increase) [42] led to a small clinical trial in man using the colon-specific MMX system [43]. Parnaparin-MMX decreased clinical disease activity indices at 8 weeks, decreased disease activity index at 8 weeks, and did not interfere with hemocoagulative parameters. They suggest that such preliminary data warranted a randomized, controlled clinical trial.

Disease states can affect the environment and cellular mechanisms in the colon. This can affect how drugs are metabolized or absorbed. Colonic permeability is increased in patients with IBD [44], celiac disease, and other conditions [45]. Enzyme levels are also affected in disease, for example, the CYP3A4 metabolism may be higher in inflamed gut tissue of colitis patients than in healthy volunteers [46]. Similarly, MDR1 expression can change in inflammatory conditions of the bowel [47] and P-gp expression and BCRP expression were shown to be reduced in individuals with active inflammation of ulcerative colitis [47, 48]. Enzyme levels may differ from normal levels in tumor tissue, although the evidence to date has been conflicting [49, 50]. Influx transporters are also susceptible to disease states; the solute carrier family is downregulated during *Vibrio cholerae* infection [51].

Physical environment can also differ in disease states. Fluid volume is affected by pathology. Constipation results from increased water resorption in the gut leading to more viscous colonic contents. Chronic diarrhea is common in the active phase of IBD and is linked to transit as well as fluid volumes. The makeup of the colonic contents can also be affected; the pH can be lower in the diseased (IBD) colon [18, 52, 53]. Recent work by [54] shows that there are variations in short chain fatty acids in ulcerative colitis. There are also differences in buffer capacity, osmolality, and protein levels between healthy patients and relapsed or remitting ulcerative colitis patients. Such differences need to be taken into account when designing drugs and delivery systems for colonic targeting.

18.5 Conclusion

The colon is a diverse and dynamic environment, which is not yet fully studied or understood. The clinical applications of oral colonic drug delivery reach back to the 1950s but have been limited to the local treatment of IBD. There are, however, other local diseases of the large intestine which could benefit from topical delivery to the colonic mucosa, e.g., colonic cancer. Systemic delivery can also be achieved by this route and some drugs may benefit from this delivery site. Applications of colonic delivery are greatly influenced by the physiological environment, in particular the low levels of free fluid, high bacterial levels, and long and often variable transit time. However, there are specific benefits of the colon which could be exploited to improve bioavailability, for example, the lower levels of efflux transporters or metabolic enzymes. This could provide dose reduction for some drugs, although it should be acknowledged that this is a very drug-specific effect and many drugs have low bioavailability in the colon. The physiological environment of the colon is exploited for drug delivery; the lag time to the colon combined with time-delayed delivery systems could be utilized for chronotherapy, and the pH or bacterial gradients can be used to achieve colon-specific drug targeting. Although only pH-responsive systems have achieved commercial success, a range of polymeric- and polysaccharide-based approaches to pH and microbiotal triggered drug delivery have shown promise. Some of the most promising systems employ combination approaches to colonic targeting. Given that diseases such as IBD and colonic cancer are significant problems for the medical world, these new strategies and technologies should be exploited to improve drug efficacy and patient care.

References

1. Rajilic-Stojanovic M, Smidt H, de Vos WM (2007) Diversity of the human gastrointestinal tract microbiota revisited. Environ Microbiol 9:2125–2136
2. Sousa T, Paterson R, Moore V, Carlsson A, Abrahamsson B, Basit AW (2008) The gastrointestinal microbiota as a site for the biotransformation of drugs. Int J Pharm 363:1–25
3. Wilding I (2001) The enterion capsule: a novel technology for understanding the biopharmaceutical complexity of new molecular entities (NMEs). Drug Deliv Technol 1:8–11
4. Tuleu C, Basit AW, Waddington WA, Ell PJ, Newton JM (2002) Colonic delivery of 4-aminosalicylic acid using amylose-ethylcellulose-coated hydroxypropyl methylcellulose capsules. Aliment Pharmacol Ther 16:1771–1779
5. Hardy JG, Wilson CG, Wood E (1985) Drug delivery to the proximal colon. J Pharm Pharmacol 37:874–877
6. Weitschies W, Kosch O, Monnikes H, Trahms L (2005) Magnetic marker monitoring: an application of biomagnetic measurement instrumentation and principles for the determination of the gastrointestinal behavior of magnetically marked solid dosage forms. Adv Drug Deliv Rev 57:1210–1222
7. Edsbacker S, Andersson T (2004) Pharmacokinetics of budesonide (Entocort EC) capsules for Crohn's disease. Clin Pharmacokinet 43:803–821
8. Schroeder KW, Tremaine WJ, Ilstrup DM (1987) Coated oral 5-aminosalicylic acid therapy for mildly to moderately active ulcerative colitis. A randomized study. N Engl J Med 317:1625–1629
9. Sinha A, Ball BJ, Connor AL, Nightingale J, Wilding IR (2003) Intestinal performance of two mesalamine formulations in patients with active ulcerative colitis as assessed by gamma scintigraphy. Pract Gastroenterol 27:56–69
10. McConnell EL, Liu F, Basit AW (2009) Colonic treatments and targets: issues and opportunities. J Drug Target 17:335–363
11. Tubic-Grozdanis M, Hilfinger J, Amidon G, Kim J, Kijek P, Staubach P, Langguth P (2008) Pharmacokinetics of the CYP 3A substrate simvastatin following administration of delayed versus immediate release oral dosage forms. Pharm Res 25(7):1591–1600
12. Buttgereit F, Doering G, Schaeffler A, Witte S, Sierakowski S, Gromnica-Ihle E, Jeka S, Krueger K, Szechinski J, Alten R (2004) Efficacy of modified-release versus standard prednisone to reduce duration of morning stiffness of the joints in rheumatoid arthritis (CAPRA-1): a double-blind, randomised controlled trial. Lancet 371:205–214

13. Evans DF, Pye G, Bramley R, Clark AG, Dyson TJ, Hardcastle JD (1988) Measurement of gastrointestinal pH profiles in normal ambulant human subjects. Gut 29:1035–1041
14. Fallingborg J, Christensen LA, Ingeman-Nielsen M, Jacobsen BA, Abildgaard K, Rasmussen HH (1989) pH-profile and regional transit times of the normal gut measured by a radiotelemetry device. Aliment Pharmacol Ther 3:605–613
15. McConnell EL, Fadda HM, Basit AW (2008) Gut instincts: explorations in intestinal physiology and drug delivery. Int J Pharm 364:213–226
16. Dew MJ, Hughes PJ, Lee MG, Evans BK, Rhodes J (1982) An oral preparation to release drugs in the human colon. Br J Clin Pharmacol 14:405–408
17. Nugent SG, Kumar D, Rampton DS, Yazaki E, Evans DF (2000) Gut pH and transit time in ulcerative colitis appear sufficient for complete dissolution of pH-dependent mesalazine-containing capsules. Gut 46:A9
18. Fallingborg J, Christensen LA, Jacobsen BA, Rasmussen SN (1993) Very low intraluminal colonic pH in patients with active ulcerative colitis. Dig Dis Sci 38:1989–1993
19. Khan MZI, Prebeg Z, Kurjakovic N (1999) A pH-dependent colon targeted oral drug delivery system using methacrylic acid copolymers – I. Manipulation of drug release using Eudragit (R) L100-55 and Eudragit (R) S100 combinations. J Control Release 58:215–222
20. Brunner M, Lackner E, Exler PS, Fluiter HC, Kletter K, Tschurlovits M, Dudczak R, Eichler HG, Muller M (2006) 5-aminosalicylic acid release from a new controlled-release mesalazine formulation during gastrointestinal transit in healthy volunteers. Aliment Pharmacol Ther 23:137–144
21. Watts P, Smith A (2005) TARGIT technology: coated starch capsules for site-specific drug delivery into the lower gastrointestinal tract. Expert Opin Drug Deliv 2:159–167
22. Brunner M, Assandri R, Kletter K, Tschurlovits M, Corrado ME, Villa R, Eichler HG, Muller M (2003) Gastrointestinal transit and 5-ASA release from a new mesalazine extended-release formulation. Aliment Pharmacol Ther 17:395–402
23. Brunner M, Ziegler S, di Stefano AF, Dehghanyar P, Kletter K, Tschurlovits M, Villa R, Bozzella R, Celasco G, Moro L, Rusca A, Dudczak R, Muller M (2006) Gastrointestinal transit, release and plasma pharmacokinetics of a new oral budesonide formulation. Br J Clin Pharmacol 61:31–38
24. Devereux JE, Newton JM, Short MB (1990) The influence of density on the gastrointestinal transit of pellets. J Pharm Pharmacol 42:500–501
25. Podczeck F, Course NC, Newton JM, Short MB (2007) The influence of non-disintegrating tablet dimensions and density on their gastric emptying in fasted volunteers. J Pharm Pharmacol 59:23–27
26. Davis SS, Hardy JG, Fara JW (1986) Transit of pharmaceutical dosage forms through the small intestine. Gut 27:886–892
27. Davis SS (1989) Small intestinal transit. In: Hardy JG, Davis SS, Wilson CG (eds) Drug delivery to the gastrointestinal tract. Ellis Horwood, Chichester
28. Spiller RC, Brown ML, Phillips SF (1987) Emptying of the terminal ileum in intact humans. Influence of meal residue and ileal motility. Gastroenterology 92:724–729
29. Hebden JM, Wilson CG, Spiller RC, Gilchrist PJ, Blackshaw E, Frier ME, Perkins AC (1999) Regional differences in quinine absorption from the undisturbed human colon assessed using a timed release delivery system. Pharm Res 16:1087–1092
30. Basit AW, Short MD, McConnell EL (2009) Microbiota-triggered colonic delivery: robustness of the polysaccharide approach in the fed state in man. J Drug Target 17(1):64–71
31. Cummings JH, Milojevic S, Harding M, Coward A, Gibson GR, Botham L, Ring SG, Wraight EP, Stockham MA, Allwood MC, Newton JM (1996) In vivo studies of amylose- and ethylcellulose-coated [13 C] glucose microspheres as a model for drug delivery to the colon. J Control Release 40:123–131
32. Bloor JR, Basit AW, Chatchawalsaisin JN, Fish W, Newton JM, Wood EC, Wilding IR (1999) Targeting of ATL-2502 (Prednisolone Sodium Metasulphobenzoate) using COLAL Coating. AAPS PharmSci 1(4):S438

33. Basit AW, Podczeck F, Newton JM, Waddington WA, Ell PJ (2004) The use of formulation technology to assess regional gastrointestinal drug absorption in humans. Eur J Pharm Sci 21:179–189
34. Yang L, Watanabe S, Li J, Chu JS, Katsuma M, Yokohama S, Fix JA (2003) Effect of colonic lactulose availability on the timing of drug release onset in vivo from a unique colon-specific drug delivery system (CODES). Pharm Res 20:429–434
35. Katsuma M, Watanabe S, Takemura S, Sako K, Sawada T, Masuda Y, Nakamura K, Fukui M, Connor AL, Wilding IR (2004) Scintigraphic evaluation of a novel colon-targeted delivery system (CODES) in healthy volunteers. J Pharm Sci 93:1287–1299
36. Ibekwe VC, Khela MK, Evans DF, Basit AW (2008) A new concept in colonic drug targeting: a combined pH-responsive and bacterially-triggered drug delivery technology. Aliment Pharmacol Ther 28:911–916
37. Krishnaiah YSR, Satyanarayana V, Kumar BD, Karthikeyan RS, Bhaskar P (2003) In vivo pharmacokinetics in human volunteers: oral administered guar gum-based colon-targeted 5-fluorouracil tablets. Eur J Pharm Sci 19:355–362
38. Miner P, Wedel M, Bane B, Bradley J (2004) An enema formulation of alicaforsen, an antisense inhibitor of intercellular adhesion molecule-1, in the treatment of chronic, unremitting pouchitis. Aliment Pharmacol Ther 19:281–286
39. Miner PB Jr, Geary RS, Matson J, Chuang E, Xia S, Baker BF, Wedel MK (2006) Bioavailability and therapeutic activity of alicaforsen (ISIS 2302) administered as a rectal retention enema to subjects with active ulcerative colitis. Aliment Pharmacol Ther 23:1427–1434
40. van Deventer SJ, Wedel MK, Baker BF, Xia S, Chuang E, Miner PB Jr (2006) A phase II dose ranging, double-blind, placebo-controlled study of alicaforsen enema in subjects with acute exacerbation of mild to moderate left-sided ulcerative colitis. Aliment Pharmacol Ther 23:1415–1425
41. Miner PB Jr, Wedel MK, Xia S, Baker BF (2006) Safety and efficacy of two dose formulations of alicaforsen enema compared with mesalazine enema for treatment of mild to moderate left-sided ulcerative colitis: a randomized, double-blind, active-controlled trial. Aliment Pharmacol Ther 23:1403–1413
42. Celasco G, Moro L, Bozzella R, Mangano K, Quattrocchi C, Aiello C, Donia M, Fagone P, di Marco R (2008) Efficacy of intracolonic administration of low-molecular-weight heparin CB-01-05, compared to other low-molecular-weight heparins and unfractionated heparin, in experimentally induced colitis in rat. Dig Dis Sci 53:3170–3175
43. Pastorelli L, Saibeni S, Spina L, Signorelli C, Celasco G, de Franchis R, Vecchi M (2008) Oral, colonic-release low-molecular-weight heparin: an initial open study of Parnaparin-MMX for the treatment of mild-to-moderate left-sided ulcerative colitis. Aliment Pharmacol Ther 28:581–588
44. Piche T, Barbara G, Aubert P, Bruley des Varannes S, Dainese R, Nano JL, Cremon C, Stanghellini V, De Georgio R, Galmiche JP, Neunlist M (2008) Impaired intestinal barrier integrity in the colon of irritable bowel syndrome patients: involvement of soluble mediators. Gut 58:196–201
45. Demeo MT, Mutlu EA, Keshavarzian A, Tobin MC (2002) Intestinal permeation and gastrointestinal disease. J Clin Gastroenterol 34:385–396
46. Pellequer Y, Weissenborn V, Lamprecht A (2007) Decreased drug penetration in inflamed tissue related to changed mucosal metabolism in experimental colitis. J Pharm Sci 96:2145–2153
47. Blokzijl H, van der Borght S, Bok LI, Libbrecht L, Geuken M, van den Heuvel FA, Dijkstra G, Roskams TA, Moshage H, Jansen PL, Faber KN (2007) Decreased P-glycoprotein (P-gp/MDR1) expression in inflamed human intestinal epithelium is independent of PXR protein levels. Inflamm Bowel Dis 13:710–720
48. Englund G, Jacobson A, Rorsman F, Artursson P, Kindmark A, Ronnblom A (2007) Efflux transporters in ulcerative colitis: decreased expression of BCRP (ABCG2) and Pgp (ABCB1). Inflamm Bowel Dis 13:291–297

49. Canaparo R, Nordmark A, Finnstrom N, Lundgren S, Seidegard J, Jeppsson B, Edwards RJ, Boobis AR, Rane A (2007) Expression of cytochromes P450 3A and P-glycoprotein in human large intestine in paired tumour and normal samples. Basic Clin Pharmacol Toxicol 100: 240–248
50. Bergheim I, Bode C, Parlesak A (2005) Decreased expression of cytochrome P450 protein in non-malignant colonic tissue of patients with colonic adenoma. BMC Gastroenterol 5:34
51. Flach CF, Qadri F, Bhuiyan TR, Alam NH, Jennische E, Holmgren J, Lonnroth I (2007) Differential expression of intestinal membrane transporters in cholera patients. FEBS Lett 581:3183–3188
52. Nugent SG, Kumar D, Rampton DS, Yazaki E, Evans DF (2000) Gut pH and transit time in ulcerative colitis appear sufficient for complete dissolution of pH-dependent mesalazine-containing capsules. Gut 46:A9
53. Sasaki Y, Hada R, Nakajima H, Fukuda S, Munakata A (1997) Improved localizing method of radiopill in measurement of entire gastrointestinal pH profiles: colonic luminal pH in normal subjects and patients with Crohn's disease. Am J Gastroenterol 92:114–118
54. Vertzoni M, Goumas K, Söderlind E, Abrahamsson B, Dressman JB, Poulou A, Reppas C (2010) Characterization of the ascending colon fluids in ulcerative colitis. Pharm Res 27(8):1620–1626.

About the Editors

Patrick Crowley (http://www.callumconsultancy.com)
Patrick Crowley has over 40 years of experience in dosage form development in the pharmaceutical industry, in both the UK and the USA. He has played key roles in the development of 16 commercial products, including fermentation products, semi-synthetics, synthetic structures and a biopharmaceutical product and holds several formulation-related patents. His interests and publications include drug and product stability, enhancing performance by dosage form design and the application of quality-by-design principles and practices in dosage form development, manufacture and control.

Clive G. Wilson, Ph.D., F.A.P.S. (Hons), F.C.R.S.
Clive Wilson is the J.P. Todd Professor of Pharmaceutics at Strathclyde University in Glasgow, Scotland, and the current president of the European Union Federation for Pharmaceutical Sciences. He is the chief scientist of Bio-Images Research Group, a phase II imaging specialist clinical unit in Glasgow Royal Infirmary.

His major areas of research have been the study of the behaviour of drug formulations in man. With colleagues at Nottingham, he pioneered applications of scintigraphy in the study of physiological and pathophysiological effects of transit on drug absorption and began the investigations into regional drug absorption in man. This was later applied to the design and optimisation of oral dosage forms, in silico modelling of formulations in man and studies on IVIVR. He has published more than 150 papers, 6 books and over 130 reviews and book chapters. He was made a fellow of the Controlled Release Society in June 2010 and a fellow of the Academy of Pharmaceutical Sciences in September 2010.

C.G. Wilson and P.J. Crowley (eds.), *Controlled Release in Oral Drug Delivery*, Advances in Delivery Science and Technology, DOI 10.1007/978-1-4614-1004-1,

Index

D

E

H

P